Vibrio cholerae

Genomics and Molecular Biology

Shah M. Faruque

Molecular Genetics Laboratory
International Centre for Diarrhoeal Disease Research, Bangladesh
Dhaka
Bangladesh

and

G. Balakrish Nair

National Institute of Cholera and Enteric Diseases
Kolkata
India

Caister Academic Press

Caister Academic Press
Norfolk, UK

www.caister.com

British Library Cataloguing-in-Publication Data
A catalogue record for this book is available from the British Library

ISBN: 978-1-904455-33-2

Printed and bound in Great Britain

Contents

Contributors

Afsar Ali
Department of Epidemiology & Preventive Medicine
School of Medicine
University of Maryland
Baltimore, MD
USA

aali@epi.umaryland.edu

Rupak K. Bhadra
Infectious Diseases & Immunology Division
Indian Institute of Chemical Biology
Kolkata
India

rupakbhadra@iicb.res.in

E. Fidelma Boyd
Department of Biological Sciences
University of Delaware
Newark, DE
USA

fboyd@udel.edu

Alejandro Cravioto
International Centre for Diarrhoeal Disease Research
Dhaka
Bangladesh

acravioto@icddrb.org

Bhabatosh Das
Infectious Diseases & Immunology Division
Indian Institute of Chemical Biology
Kolkata
India

dasiitkgp@yahoo.com

Gabriela Delgado
Department of Public Health
Faculty of Medicine
Universidad Nacional Autónoma de México
Mexico City
Mexico

delgados@servidor.unam.mx

Michelle Dziejman
Department of Microbiology and Immunology
University of Rochester School of Medicine and Dentistry
Rochester, NY
USA

michelle_dziejman@urmc.rochester.edu

Shah M. Faruque
Molecular Genetics Laboratory
International Centre for Diarrhoeal Disease Research, Bangladesh
Dhaka
Bangladesh

faruque@icddrb.org

Brooke A. Jude
Colby College
Waterville, ME
USA

Brooke.A.Jude@Dartmouth.edu

Roberto Kolter
Department of Microbiology & Molecular Genetics
Harvard Medical School
Boston, MA
USA

rkolter@hms.harvard.edu

John J. Mekalanos
Department of Microbiology and Molecular Genetics
Harvard Medical School
Boston, MA
USA

john_mekalanos@hms.harvard.edu

Rosario Morales
Department of Microbiology and Parasitology
Faculty of Medicine
Universidad Nacional Autónoma de México
Mexico City
Mexico

marosari@servidor.unam.mx

Asish K. Mukhopadhyay
National Institute of Cholera and Enteric Diseases
Kolkata
India

asish_mukhopadhyay@yahoo.com

G. Balakrish Nair
National Institute of Cholera and Enteric Diseases
Kolkata
India

nairgb@icmr.org.in

T. Ramamurthy
National Institute of Cholera and Enteric Diseases
Kolkata
India

tramu@vsnl.net

Elizabeth A. Shakhnovich
Department of Microbiology and Molecular Genetics
Harvard Medical School
Boston, MA
USA

eshakhno@gmail.com

O. Colin Stine
Department of Epidemiology and Preventive Medicine
University of Maryland School of Medicine
Baltimore, MD
USA

Ostin001@umaryland.edu

Ashrafus Safa
Department of Biology and Chemistry
City University of Hong Kong
Hong Kong

mrpsafa@yahoo.com

Yoshifumi Takeda
Cooperative Research Center of Okayama University for Infectious Diseases in India
National Institute of Cholera and Enteric Diseases
Kolkata
India

yoshitakeda@aol.com

Ronald K. Taylor
Dartmouth Medical School
Department of Microbiology/Immunology
Hanover, NH
USA

Ronald.K.Taylor@Dartmouth.edu

Fitnat H. Yildiz
Department of Environmental Toxicology
University of California
Santa Cruz, CA
USA

yildiz@etox.ucsc.edu

Preface

Vibrio cholerae, the causative agent of the devastating cholera epidemics, was first described by Filippo Pacini in 1854, and was the subject matter of the first classic in epidemiology and public health described by John Snow in the now famous 'Broad Street pump incident'. Following Robert Koch's authoritative description of the organism found in the stools of cholera patients in Egypt in 1883, the organism was also first cultured by him in Calcutta in 1884. About 125 years have elapsed since the first culture of *V. cholerae*, and much has been revealed of the pathogen. However, the past 10–15 years have witnessed the unravelling of remarkable new insights on the biology of this organism. These include the discovery of the filamentous phage that encodes cholera toxin, the existence of two chromosomes in *V. cholerae* and the whole genome sequencing of a *V. cholerae* O1 strain N16961. These pioneering works have led to even more rapid progress in understanding the genomics and molecular biology of *V. cholerae*, and this voluminous information flow has provided the impetus for putting this book together.

Two decades ago, when we began our research in *V. cholerae*, much less was known on this enteric pathogen. For example, it was not clear why most environmental strains of *V. cholerae* O1 were non-toxigenic while those that were excreted by cholera patients were toxigenic. The molecular basis for hypertoxigenicity of some strains was also not known, although certain strains clearly were more diarrhoeagenic than others. Research conducted during the last two decades has come up with firm explanations for such observations in *V. cholerae*. Likewise, there is a better understanding of how the El Tor strains, which were first described in 1905, are different from the El Tor strains that started the seventh pandemic in the island of Sulawasi and how the El Tor biotype differs from the classical biotype strains. At this time a whole set of new research has been directed towards understanding how newly emerged strains replace old strains, their survival strategy and competition in the environment, the molecular mechanism of cell to cell communication, and biofilm formation, as well as the genetic determinants of epidemic and pandemic properties. In this book, we have attempted to capture the essence of this mighty deluge of information regarding *V. cholerae*. The scope of this book, as emphasized by the publishers, is on the genomics and molecular biology of *V. cholerae*. Leading experts in the field have described various genetic facets of *V. cholerae* including its genomic organization, population genetics, molecular epidemiology, and synchronized regulation of gene expression, that allows the bacterium to survive under different environmental conditions. Other topics include the molecular basis for enhanced transmissibility of cholera during epidemics, survival of the pathogen in the environment, and above all the evolution of the species to attain increased fitness both as a pathogen and an environmental organism. We hope you find this book interesting and comprehensive.

Shah M. Faruque and G. Balakrish Nair

Introduction

1

Shah M. Faruque

Abstract

Vibrio cholerae, the causative agent of cholera, belongs to a group of organisms whose natural habitats are the aquatic ecosystems. The strains that cause cholera epidemics have evolved from non-pathogenic progenitor strains by acquisition of virulence genes, and *V. cholerae* represents a paradigm for this evolutionary process. A brief description to the virulence genes, their role in pathogenesis, and their possible mode of lateral transfer among different strains have been provided. The genomics and evolution of the species, population structure and multiple factors involved in the distinct epidemiological behaviour of *V. cholerae* have also been briefly introduced to make grounds for more detailed description of the topics presented in subsequent chapters.

Introduction

Toxigenic strains of the Gram-negative bacterium *Vibrio cholerae* is the causative agent of cholera, an acute life-threatening diarrhoeal disease which occurs in many developing countries, particularly in South Asia, Africa and Latin America (Faruque *et al.*, 1998; Kaper *et al.*, 1995). Cholera typically spreads in explosive epidemics affecting large number of people. The epidemic strains can also spread across countries and continents over time giving rise to cholera pandemics. Although *V. cholerae* is a human pathogen, these bacteria are part of the normal aquatic flora in estuarine and brackish waters (Colwell and Spira, 1992; Colwell, 2004). The genus *Vibrio* contains several bacterial species, of which *V. cholerae*, *V. parahaemolyticus* and *V. vulnificus* are the most important pathogens of humans. Of more than 200 known serogroups of *V. cholerae*, only two serogroups, O1 and O139, cause cholera (Faruque *et al.*, 1998), whereas *V. cholerae* belonging to serogroups other than O1 and O139 occasionally cause gastrointestinal or extra-intestinal illness utilizing similar or distinctly different virulence mechanisms (Dziejman *et al.*, 2005; Morris, 1994; Mukhopadhyay *et al.*, 1995). *V. cholerae* O1 belongs to two major serotypes, Ogawa and Inaba and a rarely reported Hikojima serotype. These serotypes have been further categorized into two biotypes, classical and El Tor, based on certain biochemical properties. Within the O1 and O139 serogroups, there are additional genetic variations that determine antibiotic resistance, phage resistance, and phenotypes that may be linked to environmental fitness (Guidolin and Manning, 1987; Takeya, 1974; Watnick *et al.*, 1999).

Pathogenesis

The pathogenesis of cholera is a complex process involving a number of virulence factors which aid the pathogen in its passage to reach the epithelium of the small intestine, colonize the epithelium, and produce a potent enterotoxin that disrupts ion transport by intestinal epithelial cells (Kaper *et al.*, 1995). Thus, the ability of *V. cholerae* strains to cause severe enteric infection depends on their virulence gene content, which varies between pathogenic and non-pathogenic strains. The two most important virulence factors in *V. cholerae* include the cholera toxin (CT) which is primarily responsible for the acute diarrhoea, and a colonization factor known as toxin coregulated pilus (TCP), so named because its expression is under the same genetic control as that of CT (Faruque *et al.*, 1998). Infection due to *V. cholerae* begins with the ingestion of contaminated food or water containing the organism. After passage through the acid barrier of the stomach, *V. cholerae* colonizes the small intestine using TCP, and produces CT, which is mainly responsible for the manifestation of the disease. CT acts as a classical A–B type toxin, leading to ADP-ribosylation of a small G protein, and constitutive activation of adenylate cyclase, thus giving rise to increased levels of cyclic AMP within the host cell. This results in rapid efflux of chloride ions and water from host intestinal cells, and leads to diarrhoea and vomiting. Massive outpouring of fluid and electrolytes leads to severe dehydration, electrolyte abnormalities and metabolic acidosis (Kaper *et al.*, 1995).

Since the natural environmental reservoir for *V. cholerae* is aquatic, ingestion by a human, passage through the stomach acid barrier and subsequent colonization of the intestine involve a highly regulated process, and the expression of many additional genes that are crucial for survival and adaptation of the bacterium in the gastrointestinal environment. However, the roles of these accessory virulence factors in cholera pathogenesis are not well established, and recent studies are beginning to reveal that at least some of these putative virulence factors also play a role in environmental fitness of the pathogen (Chiavelli *et al.*, 2001; Watnick *et al.*, 1999).

The *Vibrio cholerae* genome

Remarkably, the genome of *V. cholerae* is not a single chromosome but comprises two unique and separate circular mega-replicons (Trucksis *et al.*, 1998). The complete genome of a representative *V. cholerae* O1 El Tor biotype strain N16961 was sequenced (Heidelberg *et al.*, 2000), and was found to contain a total of 4,033,460 base pairs (bp). The two chromosomes designated as ChrI and ChrII consist of 2,961,146 bp and 1,072,314 bp respectively, and together encode 3885 open reading frames. The larger chromosome (ChrI) was found to contain most of the recognizable genes required for essential cell functions. These genes include those involved in DNA replication, transcription, translation and cell wall biosynthesis. Genes involved in pathogenesis, including those encoding toxins, surface antigens and adhesins, are also located on ChrI. In contrast, the small chromosome contains a larger fraction (59%) of hypothetical genes compared with the large chromosome (42%), and also contains many more genes that appear to have origins other than *Vibrio* spp. Various features of the small chromosome led to the proposal that it could originally have been a megaplasmid captured by an ancestral *Vibrio* sp. However, the GC content of the two chromosomes is almost the same, suggesting that the two chromosomes have been cohabitants for a significantly long time. Therefore, the origination of the small chromosome by excision from a single large ancestral chromosome also remains a possibility.

Virulence genes and chromosomal islands

The major virulence genes in *V. cholerae* are clustered in several regions of the bacterial chromosome. The *ctxAB* genes encoding the A and B subunits of CT are part of the genome of a lysogenic filamentous bacteriophage designated CTXϕ (Waldor and Mekalanos, 1996), whereas the genes for TCP are part of a larger cluster of genes known as the TCP pathogenicity island, alternatively referred to as the *Vibrio* pathogenicity island (VPI) (Faruque and Mekalanos, 2003). The CTXϕ genome has two regions, the 'core' and the RS2 (Waldor *et al.*, 1997). The 4.6 kb core region contains the *ctxAB* genes and genes (*psh*, *cep*, *pIIICTX*, *ace*, and *zot*) required for the morphogenesis of CTXϕ particles. The RS2 region consists of three genes *rstR*, *rstA*, and *rstB*, and two intergenic regions *ig1* and *ig2*. RstA is required for phage DNA replication, RstB is required for site-specific integration, and RstR is a repressor of *rstA* expression (Waldor *et al.*, 1997; Kimsey and Waldor, 1998). The *ctxAB* genes do not participate directly in the formation of phage particles, but these genes appear to be important for the phage to provide a survival advantage to its host bacteria in the human gastrointestinal environment (Faruque *et al.*, 1998).

Although the major subunit of TCP is encoded by the *tcpA* gene, the formation and function of the pilus assembly require the products of a number of other genes. At least 15 open reading frames constitute the *tcp* gene cluster which is located immediately adjacent and upstream to the *acf* gene cluster, encoding a putative accessory colonization factor (ACF). A genomic region of nearly 40 kb, including the TCP/ACF gene clusters flanked on both sides by a putative 20-bp attachment sequence, and carrying a putative integrase gene, and a transposase gene, constitutes the TCP island (Kovach *et al.*, 1996). The structure, GC content, and codon usage within the TCP island suggest that the acquisition of the pathogenicity island was a fairly recent event in the evolutionary history of pathogenic *V. cholerae*. Remarkably, the CTXϕ which encodes CT uses TCP as its receptor for infecting new strains, and thus these two horizontally moving elements are evolutionarily linked.

Genes or chromosomal islands that have putative additional roles in pathogenesis have also been reported. These include the RTX (repeat in toxin) gene cluster, the VPI-2 carrying *nanH* gene encoding neuraminidase, the *Vibrio* Seventh pandemic islands (VSP-1 and VSP-2), and genes for a new type IV pilus (Faruque and Mekalanos, 2003). The RTX gene cluster comprises four genes. Of them, *rtxA* encodes cytotoxic activity on Hep-2 cell, *rtxC* encodes toxin activator and *rtxB and rtxD* encode two proteins for toxin transportation. The RTX gene cluster is physically linked to the CTX prophage. The New type IV-A pilus gene cluster of *V. cholerae* constitutes a 5.4-kb *pilin* gene cluster that is similar to the *tap* gene cluster of *Aeromonas hydrophila*. A prepilin peptidase protein encoded by this gene cluster has been reported to be important for toxin secretion. Other genes encoding putative virulence factors that are presumed to play additional roles in pathogenesis include the *mshA* gene for mannose-sensitive haemagglutinin (MSHA), *hlyA* gene encoding a haemolysin, *ompU* gene for outer membrane protein, and the *stn* gene for a heat-stable enterotoxin of non-O1 *V. cholerae*.

Recently, comparative genomic analysis of diverse *V. cholerae* strains have been conducted using microarrays based on the completely sequenced genome of *V. cholerae* strain N16961 (Dziejman *et al.*, 2002; Heidelberg *et al.*, 2000). These analyses revealed that the El Tor biotype strain which is the causative agent of the ongoing seventh pandemic of cholera carries two chromosomal islands (VSP-1 and VSP-2) that were absent in classical *V. cholerae* strain of the sixth pandemic. It has been proposed that genes carried by the VSP islands might have been responsible for the epidemiological success of the seventh pandemic clone of *V. cholerae*.

Initial studies in animal models suggested that MSHA pili may have important role in the pathogenesis of El Tor biotype strains (Osek *et al.*, 1992, 1994). However, the initial report was later disputed by colonization potential studies with a *mshA* gene-deleted El Tor strain in the infant mouse model (Attridge *et al.*, 1996). A recent study claimed that MSHA pilus is used by environmental *V. cholerae* to adhere to plankton or to make biofilm on abiotic surfaces (Chiavelli *et al.*, 2001). In addition, the MSHA can act as a receptor for di*verse* filamentous bacteriophages (Jouravleva *et al.*, 1998; Faruque *et al.*, 2005), which presumably facilitates evolution. Studies with several outer membrane proteins including *IrgA*, whose expression is regulated by iron concentration, have also been shown to be important for virulence in infant mice model (Goldberg *et al.*, 1990; Sengupta *et al.*, 1992). Other genes that have been studied for their possible roles in pathogenesis include several putative adhesins such as lipopolysaccharides (Chitnis *et al.*, 1982) and flagellar proteins (Attridge and Rowley, 1983). However, the roles of these accessory virulence factors in cholera pathogenesis remain to be clearly established. Studies to date have confirmed that the major virulence genes that are required for pathogenesis in human as well as in animal models are the genes involved in the production of CT and TCP.

Regulation of virulence genes

Multiple systems are involved in the regulation of virulence-associated genes in *V. cholerae*. Several critical virulence genes are expressed by coordinate regulation so that multiple genes respond in a similar manner to environmental conditions (DiRita *et al.*, 1991; Skorpuski and Taylor, 1997). ToxR, a 32-kDa transmembrane protein, acts as the master regulator and is itself regulated by environmental signals. The activity of ToxR has been reported to be enhanced by another 19-kDa transmembrane protein, ToxS, which serves to assemble ToxR monomers into the dimeric form (DiRita and Mekalanos, 1991). Besides *ctxAB*, ToxR also regulates at least 17 distinct genes, which constitute the ToxR regulon. These include the TCP colonization factor (Taylor *et al.*, 1987), the accessory colonization factor (Peterson and Mekalanos, 1988), the outer membrane proteins OmpT and OmpU (Miller and Mekalanos, 1988), as well as three other lipoproteins (Parsot *et al.*, 1991). Among the genes of the ToxR regulon, only *ctxAB* is directly regulated by ToxR protein while the other genes are controlled by another regulatory factor called ToxT, a 32-kDa protein. ToxR controls the transcription of the *toxT* gene, which encodes ToxT a transcription activator protein (Higgins and DiRita, 1996). Consequently the enhanced expression of ToxT leads to activation of other genes in the ToxR regulon. Regulatory systems in *V. cholerae* apparently allow the bacterium to bring diversity in the expression of its genes to survive different environmental conditions, e.g. the human intestine and the estuarine environment.

Epidemiology

Cholera is a water-borne disease, and populations consuming surface water contaminated with pathogenic *V. cholerae* for drinking or preparing foods may be infected at a high rate. The rate of infection is highest among children 1 to 5 years of age in endemic areas. Other epidemiological features of cholera include a high degree of clustering of cases by location and season, frequent changes in antibiotic resistance patterns from year to year, and protection against the disease being afforded by improved sanitation/hygiene and pre-existing immunity (Faruque *et al.*, 1998).

Seven distinct pandemics of cholera have been recorded since 1817 (Pollitzer, 1959), and six of these pandemics began in the Indian subcontinent, usually the Ganges Delta region, and spread to other continents, extending over many years (Snow, 1855; Chambers, 1938; Pollitzer, 1959; Lavel, 1989). The sixth pandemic and presumably the earlier pandemics were caused by the classical biotype of *V. cholerae* O1, whereas the causative agent of the ongoing seventh pandemic is *V. cholerae* O1 of the El Tor biotype. The seventh pandemic started in 1961 on the island of Sulawesi in Indonesia, and spread to the entire Southeast Asian archipelago by the end of 1962 (Kamal, 1974). During 1963 to 1969, the pandemic spread to the Asian mainland and affected Malaysia, Thailand, Burma, Cambodia, Vietnam, India, Bangladesh, Pakistan. Thereafter outbreaks were also reported in Afghanistan, Iran, Iraq and nearby republics within the Soviet Union (Kamal, 1974). By 1970, El Tor cholera had invaded the Arabian Peninsula and reached sub-Saharan West Africa causing explosive outbreaks. These epidemics resulted in more than 400,000 cases with a high case fatality rate, due mainly to a lack of background immunity in the population and inadequacies in the health care infrastructure (Goodgame and Greenough, 1975).

The seventh pandemic reached South America in the form of an explosive epidemic that began in Peru in January 1991. Subsequently, epidemic cholera was reported in neighbouring Ecuador and then in Colombia and Chile (Levine, 1991; Pan American Health Organization, 1991; Ries *et al.*, 1992). In these countries populations with a low socioeconomic conditions who lacked proper drinking water and sanitation facilities were the worst victim (Pan American health organization, 1991). Cholera cases continued along the Pacific coast of South America and progressively entered more countries in South and Central America resulting 750,000 cases of cholera and 6500 deaths in the Americas (Pan American Health Organization, 1991). Cholera outbreak that occurred in July 1994, in the poverty-stricken Rwandan refugee camps in Goma, Eastern Zaire, was one of the worst in recent times. The outbreak led to the death of an estimated 12,000 Rwandan refugees during a 3-week period (Siddique *et al.*, 1995). The seventh pandemic continues to cause seasonal outbreaks in many developing countries, especially in Bangladesh and India.

Cholera epidemic with the clinical syndrome typical of cholera, was reported in different places in India and Bangladesh in the late 1992 and in 1993 (Cholera Woking Group, 1993; Ramamurthy *et al.*, 1993). The causative agent was a *V. cholerae* nonO1 strain which was later serogrouped as O139 (Shimada *et al.*, 1994). The

O139 serogroup of *V. cholerae* was recognized as the second causative agent of cholera after the O1 serogroup. Since then *V. cholerae* O139 has spread throughout India, and outbreaks or cases have been reported in Pakistan, Nepal, China, Thailand, Kazakhastan, Afghanistan, and Malaysia (Center for Disease Control, 1993; Chongsa-nguan, 1993; Public Health Laboratory Service, 1993; Swerdlow and Ries, 1993). Imported cases were also reported from the United Kingdom and the United States (Center for Disease Control, 1993; Public Health Laboratory Service, 1993). Outbreak of cholera due to *V. cholerae* O139 serogroup was presumed to be the initiation of an eighth pandemic (Swerdlow and Ries, 1993).

Genetic diversity and evolution

Molecular epidemiological surveillance of cholera has revealed temporal changes in the properties of toxigenic *V. cholerae*, and a continual emergence of new epidemic clones which often replace existing clones (Faruque *et al.*, 1998). The continual emergence and disappearance of different pathogenic *V. cholerae* clones, and their relative prevalence, has prompted numerous studies of the molecular basis of these changes. Multilocus enzyme electrophoresis and restriction fragment length polymorphism (RFLP) analysis of different genes have enabled the study of *V. cholerae* epidemiology on a global perspective (Wachsmuth *et al.*, 1994). RFLPs of conserved rRNA genes have been used to distinguish *V. cholerae* strains into different ribotypes. For example, ribotyping indicates that toxigenic US Gulf coast isolates are clonal and that they differ from other seventh pandemic isolates (Kaper *et al.*, 1982). Analysis of isolates from the 1991 Latin American epidemic showed that these were clearly related to seventh pandemic isolates from other parts of the world (Faruque and Albert, 1993).

The origination of different clones of pathogenic *V. cholerae* is assumed to be mostly associated with the acquisition of critical virulence genes and other gene clusters by diverse non-pathogenic progenitor strains. The mechanism by which horizontal transfer of CT genes occur has been reasonably defined with the discovery of CTXϕ, but the mechanism involved in the transfer of TCP pathogenicity island is still not clearly known. The O139 strains were derived from a *V. cholerae* O1 El Tor strain by genetic changes in serotype-specific genes (Bik *et al.*, 1995; Faruque *et al.*, 2003), presumably involving gene transfer and recombination events. Besides, frequent changes in the drug resistance pattern of clinical *V. cholerae* strains isolated in different countries have been noticed. The rapidly changing pattern of antibiotic resistance observed among toxigenic *V. cholerae* isolated from different outbreaks suggests that there is also substantial mobility of genetic elements encoding antibiotic resistance in *V. cholerae*. The *V. cholerae* genome has been found to contain a distinctive class of integrons, which are gene expression elements that acquire open reading frames and convert them to functional genes (Mazel *et al.*, 1998). This permits the bacteria to entrap genes from other microorganisms and thus constitute a mechanism for the clustering and spread of pathogenic genes as well as genes for other biochemical functions.

Microevolution among toxigenic *V. cholerae* may also contribute significantly to the emergence of strains with altered antigenic and pathobiological properties, and seasonal cholera epidemics may contribute to selective enrichment of genetic variants with unique properties that promote transmission or environmental persistence (Faruque and Mekalanos, 2003). For example, several researchers have documented the existence of different *tcpA* alleles in vibrios of various serogroups, including toxigenic and non-toxigenic strains from clinical and environmental sources (Boyd and Waldor, 2002). It has been postulated that the different *tcpA* alleles may have evolved in response to selective pressures that vary between the environment and the host. Similarly, a recent study of non-O1, non-O139 environmental isolates detected considerable genetic mosaicism in both the CTXϕ and the *tcp* gene cluster (Mukhopadhyay *et al.*, 2001). Some of these strains were shown to possess new alleles of the *tcpA*, *toxT*, and *tcpF* genes carried by the TCP pathogenicity island. Different alleles of the CTXϕ prophage repressor *rstR* were also found. Interestingly, some of the non-O1, non-O139 environmental strains possessing novel *tcpA*, *toxT*, and *tcpF* alleles were able to colonize indicating that non-canonical virulence alleles may still have virulence function. The fitness of strains in the environment may be favoured by different alleles of virulence genes other than those that increase fitness in the host. Moreover, some of these gene products including mannose-sensitive haemagglutinin (MSHA) and TCP homologues may have a function in attachment of environmental *V. cholerae* onto biotic and abiotic surfaces forming biofilms (Reguera and Kolter, 2005). Therefore, genes which have been described mainly as virulence genes in the context of human disease may also have important function in the environment. The functional virulence genes may have evolved from ancestral genes which exist as virulence gene homologues in some environmental *V. cholerae* strains. Virulence factor themselves might also be involved in environmental persistence given that it was recently shown that TCP pili are involved in biofilm formation on chitin surfaces (Reguera and Kolter, 2005). It seems logical that the epidemiological success of a pandemic strain may depend on its environmental fitness in addition to the ability to produce major virulence factors.

In summary, *V. cholerae* offers a genetic system to study the interactions of bacteria and accessory genetic elements encoding virulence-associated genes, leading to the origination of pathogenic bacteria with increased evolutionary fitness. Toxigenic *V. cholerae*, represents a paradigm for this process in that this organism evolved from environmental non-pathogenic *V. cholerae* by acquisition of virulence genes. Moreover, the pathogenic strains are selected and enriched through human infections, in the form of frequent large epidemics. The purpose of this book is to present available information and compile scientific data obtained from different aspects of *V. cholerae* studies particularly on the genomics and molecular biology of the pathogen. The 13 chapters of this book cover the latest developments, and provide directions for future investigations. Special emphasis has been paid to provide critical insights into the evolutionary biology of *V. cholerae* and virulence factors being encoded by accessory genetic elements, as well as the relationship between pathogenesis and the natural selection of a species so as to ensure its continued existence.

References

Attridge, S.R., Manning, P.A., Holmgren, J. and Jonson, G. (1996). Relative significance of mannose-sensitive hemagglutinin and toxin-coregulated pili in colonization of infant mice by *Vibrio cholerae* El Tor. Infect. Immun. *64*, 3369–3373.

Attridge, S.R. and Rowley, D. (1983). The role of the flagellum in the adherence of *Vibrio cholerae*. J. Infect. Dis. *147*, 864–872.

Bik, E.M., Bunschoten, A.E., Gouw, R.D. and Mooi, F. (1995). Genesis of the novel epidemic *Vibrio cholerae* O139 strain: evidence for horizontal transfer of genes involved in polysaccharide synthesis. EMBO J. *14*, 209–216.

Boyd, E.F. and Waldor, M.K. (2002). Evolutionary and functional analyses of variants of the toxin-coregulated pilus protein TcpA from toxigenic *Vibrio cholerae* non-O1/non-O139 serogroup isolates. Microbiology *148*, 1655–1666.

Centers for Disease Control. (1993). Imported cholera associated with a newly described toxigenic *Vibrio cholerae* O139 strain California, 1993. Morbid. Mortal. Weekly Rep. 42, 501–503.

Chambers, J.S. (1938). Cholera's first invasion of the new World. In The conquest of cholera. (New York, USA, Macmillan). pp. 24–44.

Chiavelli, D.A., Marsh, J.W. and Taylor, R.K. (2001). The mannose-sensitive hemagglutinin of *Vibrio cholerae* promotes adherence to zooplankton. Appl. Environ. Microbiol. *67*, 3220–3225.

Chitnis, D.S., Sharma, K.D. and Kamat, R.S. (1982). Role of somatic antigen of *Vibrio cholerae* in adhesion to intestinal mucosa. J. Med. Microbiol. *5*, 53–61.

Cholera Working group, International center for diarrhoeal disease research, Bangladesh. (1993). Large epidemic of cholera like disease in Bangladesh caused by *Vibrio cholerae* O139 synonym Bengal. Lancet *342*, 387–390.

Chongsanguan, M., Chaicumpa, W., Moolasart, P., Kandhasingha, P., Shimada, T., Kurazono, H. and Takeda, Y. (1993). *Vibrio cholerae* O139 Bengal in Bangkok. Lancet *342*, 430–431.

Colwell, R.R. and Spira, W.M. (1992). The ecology of *Vibrio cholerae*. In Cholera, D. Barua, and W.B. Greenough III ed. (New York, USA; Plenum Medical Book Co.) pp. 107–127.

Colwell, R.R. (2004). Infectious disease and environment: cholera as a paradigm for water-borne disease. Int. Microbiol. *7*, 285–289.

DiRita, V.J., Parsot, C., Jander, G. and Mekalanos, J.J. (1991). Regulatory cascade controls virulence in *Vibrio cholerae*. Proc. Natl. Acad. Sci. USA *88*, 5403–5407.

DiRita, V.J. and Mekalanos, J.J. (1991). Periplasmic interaction between two membrane regulatory proteins, ToxR and ToxS, results in signal transduction and transcriptional activation. Cell *64*, 29–37.

Dziejman, M., Serruto, D., Tam, V.C., Sturtevant, D., Diraphat, P., Faruque, S.M., Rahman, M.H., Heidelberg, J.F., Decker, J., Li, L., Montgomery, K.T., Grills, G., Kucherlapati, R. and Mekalanos, J.J. (2005). Genomic characterization of non-O1, non-O139 *Vibrio cholerae* reveals genes for a type III secretion system. Proc. Natl. Acad. Sci. USA *102*, 3465–3470.

Dziejman, M., Balon, E., Boyd, D., Fraser, C.M., Heidelberg, J.F. and Mekalanos, J.J. (2002). Comparative genomic analysis of *Vibrio cholerae*: genes that correlate with cholera endemic and pandemic disease. Proc. Natl. Acad. Sci. USA *99*, 1556–1561.

Faruque, S.M., Albert, M.J. and Mekalanos, J.J. (1998). Epidemiology, genetics and ecology of toxigenic *Vibrio cholerae*. Microbiol. Mol. Biol. Rev. *62*, 1301–1314.

Faruque, S.M. and Mekalanos, J.J. (2003). Pathogenicity islands and phages in *Vibrio cholerae* evolution. Trends Microbiol. *11*, 505–510.

Faruque, S.M., Iftekar, B.N., Kazukata, F., Diraphat, P., Chowdhury, N., Kamruzzaman, M., Qadri, F., Yamasaki S. and Mekalanos, J.J. (2005). Genomic sequence and receptor for the *Vibrio cholerae* phage KSF-1ϕ: Evolutionary divergence among filamentous vibriophages mediating lateral gene transfer. J. Bacteriol. *187*, 4095–4103.

Faruque, S.M. and Albert, M.J. (1992). Genetic relation between *Vibrio cholerae* O1 strains in Ecuador and Bangladesh. Lancet *339*, 740–741.

Faruque, S.M., Sack, D.A., Sack, R.B., Colwell, R.R., Takeda, Y. and Nair, G.B. (2003). Emergence and evolution of *Vibrio cholerae* O139. Proc. Natl. Acad. Sci. USA *100*, 1301–1309.

Goldberg, M.B., Boyko, S.A. and Calderwood, S.B. (1990). Transcripional regulation by iron of a *Vibrio cholerae* virulence gene and homology of the gene to the *Escherichia coli* Fur system. J. Bacteriol. *172*, 6863–6870.

Goodgame, R.W. and Greenough III, W.B. (1975). Cholera in Africa: a message for the West. Ann. Intern. Med. *82*, 101–106.

Guidolin, A. and Manning, P.A. (1987). Genetics of *Vibrio cholerae* and its bacteriophages. Micriobiol. Rev. *51*, 285–298.

Heidelberg, J.F., Eisen, J.A., Nelson, W.C., Clayton, R.A., Gwinn, M.L., Dodson, R.J., Haft, D.H., Hickey, E.K., Peterson, J.D., Umayam, L., Gill, S.R., Nelson, K.E., Read, T.D., Tettelin, H., Richardson, D., Ermolaeva, M.D., Vamathevan, J., Bass, S., Qin, H., Dragoi, I., Sellers, P., McDonald, L., Utterback, T., Fleishman, R.D., Nierman, W.C., White, O., Salzberg, S.L., Smith, H.O., Colwell, R.R., Mekalanos, J.J., Ventor, J.C. and Frasier, C.M. (2000). DNA sequence of both chromosomes of the cholera pathogen *Vibrio cholerae*. Nature, *406*, 477–483.

Higgins, D.E. and DiRita, V.J. (1996). Genetic analysis of the interaction between *Vibrio cholerae* transcription activator ToxR and ToxT promoter DNA. J. Bacteriol. *178*, 1080–1087.

Jermyn, W.S. and Boyd, E.F. (2002). Characterization of a novel *Vibrio* pathogenicity island (VPI-2) encoding neuraminidase (nanH) among toxigenic *Vibrio cholerae* isolates. Microbiology. *148*, 3681–3693.

Jouravleva, E.A., McDonald, G.A., Marsh, J.W., Taylor, R.K., Finkelstein, M.B. and Finkelstein, R.A. (1998). The *Vibrio cholerae* mannose-sensitive hemagglutinin is the receptor for a filamentous bacteriophage from *V. cholerae* O139. Infect. Immun. *66*, 2535–2539.

Kamal, A.M. (1974). The seventh pandemic of cholera, In Cholera. D. Barua, and W. Burrows ed. (Philadelphia. Saunders), pp. 1–14.

Kaper, J.B., Bradford, H.B., Roberts, N.C. and Falkow, S. (1982). Molecular epidemiology of *Vibrio cholerae* in the US Gulf Coast. J. Clin. Microbiol. *16*, 129–134.

Kaper, J.B., Morris, J.G., Jr and Levine, M.M. (1995). Cholera. Clin. Microbiol. Rev. *8*, 48–86.

Kimsey, H.H. and Waldor, M.K. (1998). CTXϕ immunity: Application in the development of cholera vaccines. Proc. Natl. Acad. Sci. USA *95*, 7035- 7039.

Kovach, M.E., Shaffer, M.D. and Peterson, K.M. (1996). A putative integrase gene defines the distal end of a large cluster of ToxR-regulated colonization genes in *Vibrio cholerae*. Microbiology *142*, 2165–2174.

Laval, E. (1989). El colera en Chile (1886–1888). Rev. Chil. Infect. *6*, 96–99.

Levine, M.M. (1991). South America: the return of cholera. Lancet *338*, 45–46.

Mazel, D., Dychinco, B., Webb, V.A. and Davies, J. (1998). A distinctive class of integron in the *Vibrio cholerae* genome. Science *280*, 605–608.

Miller, V.L. and Mekalanos, J.J. (1988). A novel suicide vector and its use in construction of insertion mutations: osmoregulation of outer membrane proteins and virulence determinants in *Vibrio cholerae* requires *ToxR*. J. Bacteriol. *170*, 2575–2583.

Morris, J.G. (1994). Non-O1 group 1 *Vibrio cholerae* strains not associated with epidemic disease, In *Vibrio cholerae* and cholera: molecular to global perspectives. I.K. Wachsmuth, P.A. Blake, and O. Olsvik ed, (Washington DC, USA, ASM Press), pp. 103–116.

Mukhopadhyay, A.K., Saha, P.K., Garg, S., Bhattacharya, S.K., Shimada, T., Takeda, T., Takeda, Y. and Nair, G.B. (1995). Distribution and virulence of *Vibrio cholerae* belonging to serogroups other than O1 and O139: a nation survey. Epidemiol. Infect. *114*, 65–70.

Mukhopadhyay, A.K., Chakraborty, S., Takeda, Y., Nair, G.B. and Berg, D.E. (2001). Characterization of VPI pathogenicity island and CTXphi prophage in environmental strains of *Vibrio cholerae*. J. Bacteriol. *183*, 4737–4746.

Osek, J., Svennerholm, A.M. and Holmgren, J. (1992). Protection against *Vibrio cholerae* El Tor infection by specific antibodies against mannose-binding hemagglutinin pili. Infect. Immun. *60*, 4961–4964.

Osek, J., Jonson, G., Svennerholm, A.M. and Holmgren, J. (1994). Role of antibodies against biotype-specific *Vibrio cholerae* pili in protection against experimental classical and El Tor cholera. Infect. Immun. *62*, 2901–2907.

Pan American Health Organization. (1991). Cholera situation in the Americas. Epidemiol. Bull. *12*, 1–4.

Parsot, C., Taxman, E. and Mekalanos, J.J. (1991). ToxR regulates the production of lipoproteins and the expression of serum resistance in *Vibrio cholerae*. Proc. Natl. Acad Sci. USA *88*, 1641–1645.

Peterson, K.M. and Mekalanos, J.J. (1988). Characterization of the *Vibrio cholerae* ToxR regulon: identification of novel genes involved in intestinal colonization. Infect. Immun. *56*, 2822–2829.

Pollitzer, R. (1959). Cholera. World Health Organization Monograph *42*, Geneva, Switzerland.

Public Health Laboratory Service. (1993). *Vibrio cholerae* O139 and epidemic cholera. Commun. Dis. Rep. Weekly *3*, 173.

Ramamurthy, T., Garg, S., Sharma, R., Bhattacharya, S.K., Nair, G.B., Shimada, T., Takeda, T., Karasawa, T., Kurazano, H., Pal, A. and Takeda, Y. (1993). Emergence of a novel strain of *Vibrio cholerae* with epidemic potential in Southern and Eastern India. Lancet *341*, 703–704.

Reguera, G. and Kolter, R. (2005). Virulence and the environment: a novel role for *Vibrio cholerae* toxin-coregulated pili in biofilm formation on chitin. J. Bacteriol. *187*, 3551–5.

Ries, A.A., Vugia, D.J., Beingolea, L., Palacios, A.M., Vasquez, E., Wells, J.G., Baca, N.G., Swerdlow, D.L., Pollack, M., Bean, N.H., Seminario, L. and Tauxe, R.V. (1992). Cholera in Piura, Peru: a modern urban epidemic. J. Infect. Dis. *166*, 1429–1433.

Sengupta, D.K., Sengupta, T.K. and Ghose, A.C. (1992). Major outer membrane proteins of *Vibrio cholerae* and their role in induction of protective immunity through inhibition of intestinal colonization. Infect. Immun. *60*, 4848–4855.

Shimada, T., Arakawa, E., Itoh, K., Okitsu, T., Matsushima, A., Asai, Y., Yamai, S., Nakajato, T., Nair, G.B., Albert, M.J. and Takeda, Y. (1994). Extended serotyping scheme for *Vibrio cholerae*. Curr. Micribiol. *28*, 175–178.

Siddique, A.K., Salam, A., Islam, M.S., Akram, K., Majumdar, R.N., Zaman, K., Fronczak, N. and Laston, S. (1995). Why treatment centres failed to prevent cholera deaths among Rwandan refugees in Goma, Zaire. Lancet *345*, 359–361.

Skorupski, K. and Taylor, R.K. (1997). Control of the ToxR virulence regulon in *Vibrio cholerae* by environmental stimuli. Mol. Microbiol. *25*, 1003–1009.

Snow, J. (1855). On the mode of communication of cholera, 2nd edn (London, UK, J. Churchill).

Swerdlow, D.L. and Ries, A.A. (1993). *Vibrio cholerae* nonO1 the eighth pandemic? Lancet *342*, 382–383.

Taylor, R.K., Miller, V.L., Furlong, D.B. and Mekalanos, J.J. (1987). Use of *phoA* gene fusions to identify a pilus colonization factor coordinately regulated with cholera toxin. Proc. Nat. Acad. Sci. USA *84*, 2833–2837.

Takeya, K. (1974). Lysogeny in El Tor vibrios and prophage typing, In Cholera. D. Barua, and W. Burrows ed. (Philadelphia, W.B. Saunders Co.), pp. 74–83.

Trucksis, M., Michalski, J., Deng, Y.K. and Kaper, J.B. (1998). The *Vibrio cholerae* genome contains two unique circular chromosomes. Proc. Natl. Acad. Sci. USA *95*, 14464–14469.

Wachsmuth, I.K., Olsvik, O., Evins, G.M. and Popovic, T. (1994). Molecular epidemiology of cholera, In K. Wachsmuth, P.A. Blake, and O. Olsvik ed., *Vibrio cholerae* and cholera: Molecular to Global Perspectives. (Washington DC, American Society of Microbiology), pp. 357–370.

Waldor M.K. and Mekalanos, J.J. (1996). Lysogenic conversion by a filamentous bacteriophage encoding cholera toxin. Science *272*, 1910–1914.

Waldor, M.K., Rubin, E.J., Pearson, G.D.N., Kimsey, H. and Mekalanos, J.J. (1997). Regulation, replication, and integration functions of the *Vibrio cholerae* CTXϕ are encoded by regions RS2. Mol. Microbiol. *24*, 917–926.

Watnick, P.I., Fullner, K.J. and Kolter, R. (1999). A role for the mannose sensitive hemagglutinin in biofilm formation by *Vibrio cholerae* El Tor. J. Bacteriol. *181*, 3606–3609.

Genomics of *Vibrio cholerae* and its Evolution

2

Elizabeth A. Shakhnovich and Michelle Dziejman

Abstract

The 4.0-Mbp genome of N16961, an O1 serogroup, El Tor biotype, seventh pandemic strain of *V. cholerae*, comprises two circular chromosomes of unequal size that are predicted to encode a total of 3885 genes. The genomic sequence of this representative strain has facilitated global experimental approaches that have furthered our understanding of the genetic and phenotypic diversity found within the species *V. cholerae*. Sequence data have been used to identify horizontally acquired sequences, dissect complex regulatory and signalling pathways, and develop computational approaches to predict patterns of gene expression and the presence of metabolic pathway components. In addition, these data have served as a basis for the construction of microarrays to study the evolution of the organism through comparative genomic analyses. Genomic sequencing of additional strains, subtractive hybridization studies and the introduction of new model systems have also contributed to the identification of novel sequences and pathogenic mechanisms associated with other strains. The sequence of strain N16961 has therefore resulted in an expanded view of the genetic repertoire of *V. cholerae* and focused our attention on the progressive evolution of this marine bacterium that can also be a human pathogen.

Introduction

It is well established that multiple mechanisms of DNA acquisition, rearrangement, loss, and mutation operate in pathogenic bacterial species and perpetuate the evolutionary fluidity associated with prokaryotic genomes (Lawrence, 2005). Such fluidity generates a pool of genetic variants, and over 100 years ago, Darwin first suggested a relationship between stress and the selection of organisms possessing advantageous traits (Darwin, 1872). Similarly, it has been proposed that the ability of a microorganism to survive in different hosts, environmental reservoirs, and abiotic niches relies on selection for adaptive changes (Faruque *et al.*, 2004; Sokurenko *et al.*, 2006). Although the causalities between adaptation, evolution and genetic change remain controversial even today (Kugelberg *et al.*, 2006; Roth *et al.*, 2006), it is certain that for *V. cholerae*, genomic plasticity has resulted in extremely diverse strains that inhabit environmental waters worldwide and in some cases, colonize the human gastrointestinal tract and cause disease.

Classification of *V. cholerae* strains presents several challenges. Multiple systems exist, but each typically describes only a subset of traits that may or may not overlap among the different classification schemes. As an example, serogroup differentiation is based on the structure of the somatic O-antigen, yet strains of all serogroups can be non-pathogenic or pathogenic for humans (Chatterjee and Chaudhuri, 2003). However, of the more than 200 different serogroups identified to date, only O1 and O139 serogroup strains have caused epidemic and pandemic disease (Chatterjee and Chaudhuri, 2003; Wachsmuth *et al.*, 1994). Isolates of other serogroups, collectively referred to as non-O1/non-O139 strains, can cause sporadic cholera cases but rarely cause epidemic disease (Faruque *et al.*, 2004; Barua, 1992; Dalsgaard *et al.*, 1995). Further differentiation of O1 strains according to biotype (classical vs. El Tor) is based on a number of immunological and biochemical properties (Chatterjee and Chaudhuri, 2003; Wachsmuth *et al*, 1994). Although the molecular basis for biotype differences is not well understood, this differentiation is important since current, seventh pandemic O1 El Tor strains have displaced O1 classical, sixth pandemic strains both as the cause of disease and in the environmental reservoir (Barua, 1992; Wachsmuth *et al.*, 1994).

Pathogenic *V. cholerae* can also be classified by the presence or absence of two canonical virulence factors, the toxin co-regulated pilus (TCP) and cholera toxin (CT). Epidemic O1 and O139 serogroup strains share common pathogenic mechanisms mediated by TCP and CT, whereas non-pathogenic isolates of all serogroups typically lack these virulence factors (Faruque *et al.*, 1998). Some non-O1/non-O139 strains cause sporadic disease in a TCP/CT-dependent manner, but the majority of pathogenic non-O1/non-O139 strains do not carry known virulence factors and challenge the paradigm that pathogenic *V. cholerae* causes disease solely by TCP and CT mediated mechanisms (Faruque *et al.*, 2004).

Historically, cholera has been the subject of epidemiological studies for over 100 years and molecular investigations have focused on aspects of O1 and O139 strain pathogenesis, including virulence factor identification, function and regulation (Faruque and Nair, 2002; Krukonis and DiRita, 2003; Reidl and Klose, 2002). Recently, interest has expanded to include studies of pathogen evolution influenced by horizontal gene transfer, the effects of global environmental changes on vibrios as integral components of the marine ecosystem, and analyses of isolated phenotypes such as biofilm formation, motility, chitin degradation, aquatic survival or distinct signalling networks (Faruque and Mekalanos, 2003; Colwell, 1996; Watnick and Kolter, 2000; Butler and Camilli, 2005). These studies have contributed significantly to our understanding of how *V. cholerae* adapts and survives when presented with specific conditions. However, a comprehensive understanding of *V. cholerae* ecology, epidemiology, pathogenesis and survival in the marine environment will ultimately require the integration of studies to construct a network modelling the coordinated functions operating seamlessly across space and time.

In reductionist terms, this network is represented by the genomic sequence of an organism; the DNA sequence encodes all possible phenotypes and regulatory functions to orchestrate gene expression. Elucidating the genetic content of a pathogenic organism has therefore been considered a watershed for scientists in terms of designing, executing and interpreting experiments consistent with a systems biology approach to research (Raskin *et al.*, 2006). The genomic sequence of *V. cholerae* O1 serogroup El Tor biotype strain N16961, published in 2000, was no exception (Heidelberg *et al.*, 2000). Recognized for its enormous potential, the N16961 genomic sequence also challenges us to extrapolate data for a single strain to the very diverse species encompassed by *V. cholerae.*

Genome analysis

Physical maps of the *V. cholerae* chromosome lagged behind those of *E. coli* and *Salmonella* in construction and refinement, remaining largely elusive until the late 1990s (Majumder *et al.*, 1996; Chatterjee *et al.*, 1998; Nandi *et al.*, 1997; Parker *et al.*, 1979). Although critical virulence factors had been identified, the global mapping of genes relative to one another was, in part, limited by the nature of P factor mediated conjugation. Relatively efficient DNA transfer was possible, but the lack of stable P factor integration into the chromosome precluded the use of high frequency (Hfr) donors that had proved essential for gene mapping in other organisms (Datta *et al.*, 1973; Parker and Romig, 1972). The difficulty of mapping studies was more fully appreciated in retrospect when it was discovered that *V. cholerae* contains two circular chromosomes (Trucksis *et al.*, 1998).

Although not unique, multiple chromosomes were initially reported only for members of the α-proteobacteria family such as *Brucella* spp., *Agrobacterium* spp. and *Rhodobacter* spp. (Halling *et al.*, 2005; Wood *et al.*, 2001; Choudhary *et al.*, 2007) *Vibrionacae* are classified as γ-proteobacteria, and all *Vibrio* spp. analysed to date contain two circular chromosomes (Tagomori *et al.*, 2002). Several other γ-proteobacteria carry two chromosomes, including *Photobacterium* (a member of the *Vibrionacae*) and *Pseudoalteromonas*, and recent analysis of the β-proteobacterial *Burkholderia* spp. has revealed three circular chromosomes (Vezzi *et al.*, 2005; Medigue *et al.*, 2005; Chain *et al.*, 2006). Clearly, our view of bacterial genome diversity continues to expand as sequencing efforts identify additional species that carry multiple chromosomes.

Genomic sequence analysis of strain N16961, a seventh pandemic clinical isolate from Bangladesh, confirmed the presence of two circular chromosomes comprising a ~4.0 Mb genome that is predicted to encode 3887 open reading frames (ORFs) (Heidelberg *et al.*, 2000). The basic characteristics of the N16961 genome are summarized in Table 2.1. Chromosome I (ChrI), also referred to as the large chromosome, is 2.96 Mb in size, whereas chromosome II (ChrII, the small chromosome) is 1.07 Mb in size. Genes on each chromosome were assigned a four digit number, with VC designating those found on ChrI and VCA preceding those on ChrII. Each chromosome has a G+C content of approximately 47%, similar to that of other enteric pathogens such as *Yersinia*, *E. coli*, *Shigella* and *Salmonella* spp. (~48–52%). ChrII contains a number of interesting features, including a disproportionately high number of ORFs predicted to encode proteins having no known homology in current databases. The genomic sequence also confirmed the presence of a 127-kb integron island on ChrII that functions as a 'gene-capture' system for the integration of exogenous DNA (Mazel *et al.*, 1998).

Functional categories

Genes encoding known functions or having predicted functions based on the homology of their encoded products can be grouped into various categories. These functional classifications can be found at the genome page for *V. cholerae* strain N16961 maintained as part of the Comprehensive Microbial Resource at http://cmr.jcvi.org/tigr-scripts/CMR/CmrHomePage.cgi. A discussion of several categories and subcategories follows.

Toxins and other virulence factors

The *ctxAB* genes encode cholera toxin, an ADP-ribosylating enzyme whose activity results in the severe dehydrating diarrhoea characteristic of the disease (Mekalanos, 1985). These genes are carried by a filamentous phage known as CTXϕ, which uses the toxin co-regulated pilus (TCP) as its receptor to enter cells and establish lysogeny (Waldor and Mekalanos, 1996). The genomic sequence of strain N16961 confirmed that a single copy of the toxin genes and associated phage sequences are located on ChrI flanked by 17-bp *attRS1* integration sites. Previous studies identi-

Table 2.1 Genome characteristics of *V. cholerae* O1 strain N16961

Characteristics	Total genome	Chromosome I	Chromosome II
Sequence length		2,961,149 bp	1,072,315 bp
Coding sequences		2,632,214 bp (88.89%)[a]	928,840 bp (86.62%)
Intergenic sequences		328,935 bp (11.10%)	143,475 bp (13.37%)
G+C		1,412,315 bp (47.69%)	503,061 bp (46.9%)
Genes	4009		
Protein-coding genes	3887	2773	1114
Conserved hypothetical genes	628	460 (16.58%)	168 (15.08%)
Hypothetical genes	944	525 (18.93%)	419 (37.61%)
Paralogous gene families	465		
tRNA genes	98	94	4
rRNA genes	24	24	0

[a]Per cent of total for each chromosome.

fied duplicate copies of the *ctxAB* genes within the genome of *V. cholerae* O1 classical biotype isolates, consistent with data showing that the classical strains produce more toxin than El Tor strains (Mekalanos, 1983).

Although the pathogenesis of *V. cholerae* O1 and O139 strains is largely mediated by TCP and CT, other gene products have also been assumed to contribute to colonization and disease. These include the previously identified genes *rtxA* (VC1451), *acfA-D* (VC0840, VC0841, VC0844 and VC0845), *hlyA* (VCA0219) and *hapA* (VCA0865). Genome annotation confirmed that the *rtxA* gene, encoding the actin binding RTX toxin responsible for eukaryotic cell rounding *in vitro*, is the largest ORF in the N16961 genome (13,677 bp) (Fullner and Mekalanos, 1999). The accessory colonization factors encoded by *acf* genes were identified as co-regulated genes within the TCP pathogenicity island, and their deletion results in a colonization defect in the infant mouse model (Peterson and Mekalanos, 1988). The HlyA haemolysin has enterotoxic activity, whereas the *hap* gene product is a metalloprotease that disrupts intestinal epithelial cell tight junctions (Mitra *et al.*, 2000; Coelho *et al.*, 2000; Wu *et al.*, 1996; Alm *et al.*, 1988). Both *hlyA* and *hapA* are found on ChrII. Annotation also identified multiple putative proteases and at least seven other genes whose products are predicted to encode haemolysins (Heidelberg *et al.*, 2000). Whether these predicted haemolysins function to disrupt cellular membranes in the human host or have alternative functions important for the aquatic lifestyle or association with marine organisms remains to be determined. N16961 does not encode ST, a heat-stable toxin that is produced by some pathogenic TCP$^-$ CT$^-$ non-O1/non-O139 strains (Heidelberg *et al.*, 2000).

Type IV pili

The N16961 genome sequence is predicted to encode four proteins having homology to type IV pilins: TCP, MSHA, PilA and a currently uncharacterized predicted pilin (VC0502). All the genes necessary for TCP, MSHA and PilA pilus biogenesis and assembly are found on the large chromosome (Heidelberg *et al.*, 2000). In general, these pili have each been associated with multiple functions, described below. The uncharacterized putative pilin gene is located within the Vibrio Seventh Pandemic Island-II (VSP-II), and is therefore most likely associated only with seventh pandemic strains as described below. However, pilin production and function have not yet been demonstrated (Dziejman *et al.*, 2002; O'Shea *et al.*, 2004).

The TCP is composed of repeating subunits of *tcpA* and is encoded within the 45.3-kb TCP pathogenicity island (Taylor *et al.*, 1987; Manning, 1997). Several functions have been assigned to TCP and suggest that it has both environmental and host-associated roles. TCP is an essential colonization factor in the infant mouse model and in humans. It is the receptor for the CTX phage, and it appears to mediate advantageous bacterial interactions during biofilm formation on chitinaceous surfaces (Thelin and Taylor, 1996; Waldor and Mekalanos, 1996; Taylor *et al.*, 2004; Asaduzzaman *et al.*, 2004; Reguera and Kolter, 2005). The genes necessary for TCP assembly lie within the TCP island, along with the *toxT* gene that encodes a transcriptional regulator essential for TCP and CT expression (Brown and Taylor, 1995). The *tcpA* gene can vary in sequence by as much as ~18% among different strains, having O1 biotype specific sequences and unique polymorphisms in non-O1/non-O139 strains (Rhine and Taylor, 1994; Boyd and Waldor, 2002). Although these sequence polymorphisms have been useful for epidemiological and typing studies, we have only recently begun to appreciate the subtle differences in function conferred by resulting amino acid changes. These differences have been elegantly illustrated using EM studies revealing that pili associate to form relatively ordered, twisted bundle-like arrangements whose properties can be altered by single amino acid substitutions (Kirn *et al.*, 2000; Craig *et al.*, 2003).

The mannose-sensitive haemagglutinin (MSHA) plays an important role in biofilm formation, but not in colonization (Watnick *et al.*, 1999; Thelin and Taylor, 1996; Tacket *et al.*, 1998). In fact, recent studies indicate that repression of the MSHA is essential for infection in the infant mouse model (Hsiao *et al.*, 2006). MSHA also serves as a phage receptor (Faruque *et al.*, 2005a; Jouravleva *et al.*, 1998; Jonson *et al.*,

1991; Jonson *et al.*, 1994). Although the genes necessary for MSHA biosynthesis are present in nearly all strains of *V. cholerae*, only El Tor strains appear to form intact pili (Jonson *et al.*, 1991; Jonson *et al.*, 1994). However, it has been reported that related pilins are produced by non-O1/non-O139 strains, and classical strains are known to possess a fucose-sensitive haemagglutinin activity (Kuroki *et al.*, 2001; Yamashiro and Iwanaga, 1996; Franzon *et al.*, 1993).

The *pilABCD* gene cluster encodes PilA, which in other organisms functions in twitching motility (Nudleman and Kaiser, 2004). This phenomenon has not yet been demonstrated for *V. cholerae*, although some genes predicted to encode proteins involved in twitching associated pilus retraction are present. Rather, PilA has been identified as a protein expressed during human infection and also as a chitin-induced pilus essential for mediating the recently described natural competence of *V. cholerae* (Hang *et al.*, 2003; Meibom *et al.*, 2005).

Motility and chemotaxis

V. cholerae is motile in both the host and the aquatic environment, and more than 40 proteins are essential for producing a single polar flagella. All the flagellar structural and motor genes are found on ChrI. Current annotation has identified approximately 48 methyl-accepting chemotaxis proteins, encoded on both chromosomes. The large number of chemotaxis genes is consistent with the hypothesis that throughout its evolution, *V. cholerae* has acquired and/or duplicated genes whose functions have expanded the organism's ability to literally navigate and survive in diverse environments. Several investigators have discussed the role of chemotaxis in infection, and the multitude of signals received and processed by this large family of proteins remains an active area of investigation (Butler and Camilli, 2005).

DNA repair

Based on sequence similarities with *E. coli*, strain N16961 is predicted to encode several DNA repair and DNA damage response pathways, including the *uvr*, *mut* and SOS systems (Heidelberg *et al.*, 2000). In most cases, verification of these pathways and the specific scenarios requiring DNA repair await further investigation. However, it is reasonable to speculate that *V. cholerae* is exposed to damaging ultraviolet light in its environmental reservoir or during its transit into the human host. Consistent with the need to repair UV-induced damage, genetic and biochemical analyses have identified one of three putative photolyase homologues as a cyclobutane pyrimidine dimer photolyase (Worthington *et al.*, 2003). The major components of recombination associated pathways are also present, including RecA. Their role in the integration of horizontally acquired elements (e.g. CTXϕ) has been of significant interest, and *recA* has been an especially important consideration in the construction of non-toxigenic and non-recombinogenic candidate vaccine strains (McLeod and Waldor, 2004; Taylor *et al.*, 1994).

Strain N16961 also encodes one RadC and two 'RadC-like' homologues, whose precise function(s) in any organism is unclear since previous associations with a RecG helicase-like activity were challenged (Lombardo and Rosenberg, 2000). However, it is of interest to note that in *V. cholerae*, the RadC-like proteins are encoded within regions of the chromosome that appear to have been horizontally acquired. It is therefore tempting to speculate that their original role may have involved the chromosomal integration of the horizontally acquired sequences that encode them. In support of this hypothesis, a fourth RadC-like homologue is found within the SXT element. This 100-kb conjugative transposon-like mobile element encodes multiple antibiotic resistance genes, and is chromosomally integrated in most epidemic strains (Waldor *et al.*, 1996). A large deletion of the region containing the *radC*-like gene reduced the frequency of SXT transfer 1000-fold, although the precise contribution of individual genes within this region is unknown (Beaber *et al.*, 2002).

Metabolism

Initial observations suggested that ChrI carried most of the genes necessary for growth and survival of the organism, including those for the TCA cycle, electron transport and the PTS system. However, genes encoding ribosomal proteins L20 and L35 are found only on ChrII, as are genes necessary for some metabolic pathways

(Heidelberg *et al.*, 2000). VCA0572, encoding an enzyme necessary for peptidoglycan precursor synthesis, was also defined as an essential gene located on ChrII (Judson and Mekalanos, 2000). More recent analysis using a library of random transposon insertions grown using standard *in vitro* conditions concluded that the majority of essential genes (defined by genes from which insertions could not be recovered) lie on the large chromosome (Cameron and Mekalanos, manuscript in preparation).

The N16961 genome encodes a large number of transport proteins having both broad substrate specificities and the necessary pathway components for degradation and utilization, consistent with the ability of *V. cholerae* to survive in diverse environments (Heidelberg *et al.*, 2000). ABC transporters for galactoside, oligopeptides and zinc are encoded on ChrI, whereas those for maltose, ribose, lactate, haemin, arginine and molybdenum are found on ChrII. Enzymes for trehalose and mannose degradative functions are found on ChrI, whereas those for ribose, lactate and sucrose utilization reside on ChrII. Other catabolic pathways, including glycolysis and gluconeogenesis, require genes located on both chromosomes. Some components of metabolic pathways appear to be redundant and are found on both chromosomes (e.g. genes encoding the components for ABC peptide, iron, phosphate, sulfate, magnesium and amino acid transporters). In many cases, it appears that one copy has arisen from a duplication event, but for others (e.g. genes encoding phosphate transporters), the genes are different enough to suspect that independent acquisition events led to apparently redundant functions (Heidelberg *et al.*, 2000).

Chitin utilization requires genes found on both chromosomes, and is perhaps one example of a link between niche specific bacterial metabolism and the induction of phenotypic traits. Chitin is recognized as a potentially important source of energy for the bacterium; it can be used as a carbon and nitrogen source, and is abundant in the natural environmental reservoir. Recent studies have also shown that chitin polymer can induce the expression of genes necessary for natural competence and for biofilm formation, including *pilA* (Meibom *et al.*, 2005; Meibom *et al.*, 2004).

LPS as a marker for strain diversity

The enzymes involved in O1-antigen synthesis for N16961 are encoded in the *wav* (VC0223-VC0240) and *wbe* (VC0241-VC0263) gene clusters on ChrI. The genetic organization and content of these regions can vary widely among strains, producing more than 200 different serogroups and illustrating the potential for diversity (Chatterjee and Chaudhuri, 2004; Faruque *et al.*, 2005b and c). It is not clear whether serogroup specificity by itself is associated with functions conferring advantageous traits, but evidence suggests that serogroup specificity of phages acting on *V. cholerae* contributes to the collapse of epidemics (Faruque *et al.*, 2005a,b). How such O-antigen diversity arose from a progenitor *V. cholerae* is unclear. Studies suggest that classical and El Tor O1 biotype strains differ significantly enough to propose that O1 antigen acquisition occurred independently (Karaolis *et al.*, 1994; 1995). In contrast, comparison of pathogenic O1 El Tor and O139 serogroup strains suggests a close relationship, and multiple studies conclude that the genes encoding the O139 O-antigen replaced those required for O1 antigen synthesis (Bik *et al.*, 1995; Comstock *et al.*, 1996; Faruque *et al.*, 1997).

Several reports have documented limited epidemics caused by TCP+CT+ O141 serogroup strains (Boyd and Waldor, 2002; Dalsgaard *et al.*, 1995, 1997, 2001). Ribotyping indicates that these strains may be related clones, although they appear unrelated to O1 and O139 strains. However, it is not clear whether independent events have led to parallel strain evolution for this serogroup or whether the increasingly global nature of our society contributes to the spread of certain strains via their physical transport to hospitable environments.

It is also unknown whether the remaining, diverse group of non-O1/non-O139 strains acquired the genes for the production of alternate O-antigens by horizontal transfer or other mechanisms. Furthermore, TCP+CT+ strains of these serogroups could have independently acquired these virulence factors, or common strain progenitors could have undergone serogroup switching to generate new strains. Evidence suggests that widespread serogroup switching involving O1 gene transfer in the environment is uncommon, however, and parameters that influence these events remain to be defined (Faruque *et al.*, 2004). In most cases, pathogenic TCP⁻ CT⁻ non-O1/non-O139 strains do not share common serogroups, but future identification of novel virulence factors may reveal some serogroup specificity (Faruque *et al.*, 2004).

A dynamic genome

The plasticity of a bacterial genome contributes to the dynamic evolution of a microbe and its adaptation to different environments. As mentioned earlier, the development of new traits can be the result of genetic gains, losses and alterations that include nucleotide polymorphisms. In *V. cholerae*, multiple mechanisms of genetic transfer have been documented, including conjugation, transposition events, acquisition of genes via a super-integron, phage transduction, and the recently reported ability of the organism to take up DNA by natural competence. Many of these mechanisms can function in intra and interspecies gene acquisition, but the frequency of exchange, the parameters that govern these events, and their impact on environmental adaptation of *V. cholerae* remain largely unexplored.

Chromosomal evolution

Complete genome sequences across species have facilitated computational approaches to identifying gene synteny and the clustering of functionally related genes, both of which can be used to measure genome evolution and predict gene function (Binnewies *et al.*, 2006). To date, the complete genomes of at least four different *Vibrio* species have been reported: *V. cholerae*, *V. fischerii*, *V. parahaemolyticus*, and *V. vulnificus* (Heidelberg *et al.*, 2000; Ruby *et al.*, 2005; Makino *et al.*, 2003; Chen *et al.*, 2003). Other species and strains whose genome sequencing is in progress are listed in the Genomes OnLine Database, including *V. alginolyticus*, *V. splendidus*, *V. angustum*, *V. harveyi*, *V. salmoncida* and multiple strains of *V. cholerae* (Liolios *et al.*, 2006). Gene order is generally not conserved across most species, and the evolutionary fluidity of prokaryotic genomes complicates such analysis since synteny is influenced by the large number of deletions, insertions, translocations and inversions that have likely occurred over time. However, prokaryotic genes are often organized into operons, which are commonly transferred as intact transcriptional units between species (Rogozin *et al.*, 2004; Binnewies *et al.*, 2006). While serving as one mechanism for preserving gene co-regulation, the operon structure can also maintain gene order by retaining orthologues in a single cluster.

Comparison of the DNA sequence of *V. cholerae* ChrI with sequences in the *E. coli* chromosome revealed a statistically significant conservation of sequence and position over the entire length of the chromosomes when viewed by genome diagonal plots (Eisen *et al.*, 2000). Similar results were observed when the reverse complement of *V. cholerae* ChrI was used for comparison, whereas comparisons using more distantly related species, such as *M. tuberculosis* and *E. coli*, did not produce statistically significant results. The symmetry was centred at the origin of replication for each chromosome, suggesting that matching sequences were found equidistant from the origin on both sides. One possible explanation for this observation is the frequent occurrence of inversions about the origin during the evolution and speciation of closely related organisms. The tandem duplication of genes followed by inversion events having their break points between duplicated sequences could also produce similar results (Eisen *et al.*, 2000).

Further evidence supporting the occurrence of multiple intra- and interchromosomal rearrangements in *Vibrio* evolution is derived from the results of similar *in silico* analyses using the genomes of *V. cholerae*, *V. parahaemolyticus*, and *V. vulnificus* (Reen *et al.*, 2006; Rogozin *et al.*, 2004). Genome diagonal plots indicating orthologous genes for *Vibrio* spp. have identified regions of synteny across each chromosome as well as chromosomal segments that have retained gene order but that have been exchanged between ChrI and ChrII. However, it is important to note that true colinearity of genomes can only be addressed when genomic sequence data is available for multiple strains within a given species, since genetic rearrangements are observed both between species and within strains of the same species.

Paralogous gene identification provides another measure of genome evolution since these gene families typically arise by the duplication of an ancestral gene, either within the genome of that species or from the earlier duplication of sequences in an ancestral genome (Sonnhammer and Koonin, 2002; Pushker *et al.*, 2004). In *V. cholerae*, there are ~465 paralogous gene families. Many of these families include genes found on both chromosomes, including the 38 genes encoding a family of proteins containing GGDEF domains. Recent evidence has associated a subset of these proteins with c-di-GMP mediated regulation of various processes including biofilm formation, motility, and virulence (Tamayo *et al.*, 2007). Other large paralogous families include the chemotaxis and methyl accepting chemotaxis proteins, the ABC transporters, and the LysR family of transcriptional regulators. Ninety-one paralogous families are composed solely of (conserved) hypothetical ORFs, and future studies that uncover their functions are expected to fill existing gaps in our understanding of how this large group of proteins contributes to *V. cholerae* phenotypes. Overall, the presence of many large paralogous families in *V. cholerae* suggests that gene duplication has been favored as one mechanism for generating functional divergence.

The small chromosome: its acquisition and role as a driving force of evolution

The presence of two chromosomes in *Vibrio* species suggests that this bipartite structure predates speciation from a primordial *Vibrio*. Heidelberg *et al.* (2000) hypothesized that ChrII was originally a megaplasmid captured by the ancestral *Vibrio* species and retained throughout the macro-evolutionary events that produced more than 75 *Vibrio* species. Several lines of evidence support this hypothesis, and also indicate that the megaplasmid has indeed evolved into a *bona fide* chromosome (Egan *et al.*, 2005). First, each chromosome has distinct replication requirements (as described later) that are coordinated with the cell cycle. Second, the presence of several toxin–antitoxin loci found on the small chromosome is consistent with a plasmid origin, but perhaps also provides a mechanism ensuring its retention (Szekeres *et al.*, 2007). Third, the asymmetric distribution of sequences encoding common, essential metabolic pathways suggests that the large chromosome is the 'dominant' or 'original' chromosome (Heidelberg *et al.*, 2000).

One of the more compelling lines of evidence suggesting that ChrII was originally a megaplasmid is provided by the integron island. Ancestral elements are found on antibiotic resistant plasmids (Rowe-Magnus *et al.*, 2001), and all *Vibrio* spp. appear to contain an integron island associated with the small chromosome (Rowe-Magnus *et al.*, 2003; Boucher *et al.*, 2006). Open reading frames are captured and integrated into the chromosome within the integron island, where their expression is regulated by a strong promoter located upstream (Rowe-Magnus *et al.*, 2003; Rowe-Magnus *et al.*, 1999, 2002). Many of the *V. cholerae* paralogous gene families include at least one ORF that is located within the integron island. This is consistent with the proposal that, following integron island mediated gene acquisition, gene duplication and sequence divergence can result in related yet highly specific protein functions encoded by paralogous genes. The integron island thus provides the potential for ChrII to continuously direct the evolution of *Vibrio* strains by recruiting and integrating novel sequences (and therefore functions), which can then be transferred or duplicated by other mechanisms (Boucher *et al.*, 2006). In fact, comparative genome analysis has shown that the greatest diversity among strains often exists within the integron island (Dziejman *et al.*, 2005; Rowe-Magnus *et al.*, 2003).

One might expect that chromosomal fusion would have occurred at some point in *V. cholerae* evolution, perhaps promoted by duplicate genes present on each replicon. As discussed below, it has been speculated that the different replication requirements of ChrI and ChrII have sustained the bipartite genome (Egan *et al.*, 2005). An alternative explanation suggests that replication of two smaller chromosomes vs. one large chromosome provides an evolutionary advantage by facilitating faster replication (Egan and Waldor 2003; Reen *et al.*, 2006). This is consistent with the extremely quick generation times (<10 min) observed for some *Vibrio* species (Aiyar *et al.*, 2002; Daniels *et al.*, 2000). As mentioned earlier, the presence of toxin–antitoxin loci may also ensure retention of ChrII. These scenarios are not mutually exclusive. Although restriction of individual chromosome replication under certain circumstances or partitioning of cells with only a single chromosome remain formal possibilities that may also confer an evolutionary or niche-specific advantage, neither of these phenomena have yet been demonstrated.

Horizontal gene transfer and strain evolution

Typically, several features are used to identify chromosomal sequences that may constitute horizontally transferred genetic information (Karlin, 2001). These include trinucleotide analysis or a G+C% bias that does not fit the profile for the whole chromosome, the presence of genes predicted to encode functions associated with mobility (e.g. transposases), and putative insertion sites flanked by repeated elements or near a tRNA gene. When associated with the ability of the organism to cause disease, these atypical regions are referred to as pathogenicity islands.

Both trinucleotide analysis and G+C% composition suggest that the TCP and CTX regions have been acquired by horizontal gene transfer, and as mentioned earlier, it is known that the CTX region comprises the genome of a filamentous phage (Heidelberg *et al.*, 2000; Waldor and Mekalanos, 1996). The ~40 kb TCP gene cluster carries putative transposase and integrase genes, contributing to the likelihood that it was acquired from an exogenous source (Karaolis *et al.*, 1998). Lysogenic conversion of strains to CT+ by the CTXϕ is well documented; however, efforts to identify the mechanism of TCP island acquisition have not produced definitive results (Rajanna *et al.*, 2003; Faruque *et al.*, 2003).

Several other gene clusters may have also arisen by integration of foreign DNA. These include the RTX (VC1447–VC1451) and *pilA* (VC2423–VC2426) gene clusters, the integron island (VCA0291–VCA0508), VPI-2 (VC1758-VC1809), and the Vibrio Seventh Pandemic (VSP) islands I (VC0175–0185) and II (VC0490–VC0516). The RTX and *pil* clusters and the integron island have been previously described. VPI-2 is a ~57-kb region that carries the genes for neuraminidase and amino acid sugar metabolism, but it is still unclear how this island contributes to pathogenicity (Jermyn and Boyd, 2002; Jermyn and Boyd, 2005). The VSP-I and II islands are gene clusters found almost exclusively in seventh pandemic (O1 El Tor and O139) strains, and are hypothesized to contributed to the fitness of these isolates in a way that remains undefined (Dziejman *et al.*, 2002; O'Shea *et al.*, 2004). These clusters were identified by comparative genomic analysis as described below. Orthologues of some VPI-2 and VSP-II genes are found in other *Vibrio* species, including genes predicted to encode proteins that are related to the integrase and transcriptional regulator of a cryptic bacteriophage from *E. coli* (Jermyn and Boyd, 2002; O'Shea *et al.*, 2004). Smaller regions of atypical GC content can also be found throughout the chromosome. While many of the genes within these regions are predicted to encode hypothetical proteins, some genes of interest include a predicted RNA directed DNA polymerase encoded within VC2385–VC2387, and a site-specific DNA methyltransferase encoded within VCA0196-VCA0202, (Banerjee and Chowdhury, 2006). Although, in most cases, confirmation of these sequences as transferred elements and the functions that they encode await further study, it is clear that the ability of *V. cholerae* to acquire DNA by multiple mechanisms has played a prominent role in strain evolution and the acquisition of new traits. This is consistent with the hypothesis that species diversification can be driven by a high rate of horizontal gene transfer (Lawrence and Ochman, 1998)

Multiple chromosomes: practical issues

Replication

The possession of two chromosomes elicits several mechanistic questions about how this genomic arrangement is maintained. Conventional explanations for prokaryotic chromosome management do not consider the requirements of multiple origins of replication, the equal partitioning of daughter chromosomes, and the spatial and temporal coordination of these events. Functional identification of the genes and sequences required for *V. cholerae* chromosomal replication determined that replication of each chromosome occurs synchronously, and is governed by overlapping, but distinct sets of regulatory factors and proteins (Egan and Waldor, 2003; Egan *et al.*, 2004). Briefly, both chromosomal origins of replication require the *dam* and *dnaA* genes, although the origin for ChrI is more similar to the oriC of *E. coli* than the origin for ChrII. The small chromosome origin has a number of previously unlinked features consistent with its proposed derivation from a megaplasmid, including a 12-bp repeat related to plasmid iterons that is flanked by two genes essential for replication, *rctA* and *rctB*. Subsequent studies have shown that *rctB* encodes a protein that binds within the origin, whereas experiments suggest that the critical function of *rctA* is encoded by an RNA (Duigou *et al.*, 2006). The presence of multiple plasmid-like features on a single replicative element suggests that perhaps ChrII has expanded its genetic content by the successive acquisition of plasmid-borne sequences.

Partitioning

Multiple studies indicate that chromosomal partitioning is accomplished by the action of an organized mechanism coordinated with replication and cell division, rather than passive segregation. Experiments have shown that for ChrI, partitioning involves the interaction of ParAB proteins with *parSI* sites in an active process that mimics mitotic-like forces, resulting in the asymmetric transport of the daughter chromosome to the opposite cell pole (Fogel and Waldor, 2006; Saint-Dic *et al.*, 2006; Fiebig *et al.*, 2006). In contrast, the segregation pattern of ChrII appears to begin at the centre of the cell, followed by symmetric movement to locations that are destined to become the cell centres after division (Fogel and Waldor, 2005; Fiebig *et al.*, 2006). These data suggest that partitioning of ChrII likely occurs by a mechanism(s) distinct from that utilized by ChrI.

Interchromosomal gene regulation

Unlike regulatory control in eukaryotes, prokaryotic gene expression has historically been considered only within the context of a single chromosome. While there are no *a priori* barriers to transcriptional regulation across chromosomes in *V. cholerae*, this concept warrants further examination. As described earlier, one hypothesis for the origin of the small chromosome postulates its acquisition as an accessory genetic element followed by evolution into a second chromosome. This event presumably occurred very far back in evolutionary time in a progenitor *Vibrio* strain, and more recent DNA acquisition events have occurred within the framework of a bipartite chromosome. Several explanations can therefore be proposed to account for regulatory control across chromosomes.

In the simplest case, gene expression is regulated by existing, conventional mechanisms, such as housekeeping sigma factors and common transcriptional machinery. As an example, genes for some metabolic pathways reside exclusively on ChrII, and their acquisition as a combined unit controlled by housekeeping functions is easy to comprehend. Another layer of complexity is added, however, in cases where genes for pathways reside on both chromosomes. Within the context of evolution over a long period of time, we can speculate that genomic rearrangements and exchanges between chromosomes resulted in their split distribution. Alternatively, acquisition of pathway components or gene duplication and diversification may have occurred over time until a complete pathway was constructed.

Envisioning the events that result in target genes and their dedicated, specific transcriptional regulators residing on different chromosomes requires more complex explanations. For example, the *hlyA* gene is found on ChrII, but its transcription is governed by HlyU, which is encoded on ChrI (Williams *et al.*, 1993, 1996). Similarly, genes required for AI-2 quorum sensing mechanisms are split between the two chromosomes (Miller *et al.*, 2002; Zhu *et al.*, 2002; Heidelberg *et al.*, 2000). Another example is provided by the ~150 genes belonging to the ToxR regulon (Bina *et al.*, 2003). The ToxR protein is a transmembrane transcriptional activator that is present in nearly all strains of *V. cholerae* (Faruque *et al.*, 1998; Krukonis and DiRita, 2003). Approximately 70 genes, found on both chromosomes, appear to be regulated directly by ToxR. In most cases, ToxR binding sequences and sites have remained elusive, although some shared motifs have been identified (Crawford *et al.*, 1998; Higgins and DiRita, 1996; Nye *et al.*, 2000) ToxR-regulated genes encode proteins involved in various functions such as glycerol metabolism and maltose utilization, and include genes encoding several ABC transporters and porins (Bina *et al.*, 2003). In addition, ToxR controls the expression of the horizontally acquired virulence factors TCP and CT via its transcriptional activation of the *toxT* gene (Yu and DiRita, 2002). Encoded within the TCP island, ToxT is an AraC-family transcriptional regulator that directly activates the promoters of virulence genes and accessory regulatory factors as part of a complex network of interactive proteins (Krukonis and DiRita, 2003; Krukonis *et al.*, 2000). The integration of signals through multiple quorum sensing networks whose components lie on both chromosomes also influences virulence gene expression, through the action of the LuxO and HapR transcriptional regulatory proteins. These proteins also regulate other cellular processes such as biofilm formation and motility through their interaction with sequences on both chromosomes. Collectively, these examples illustrate the complexity of understanding gene regulation across chromosomes in terms of the evolutionary events that resulted in their co-regulation.

Genomic experimental approaches

Bioinformatics

Multiple computational approaches for genome analysis have been used to predict the capacity of encoded genomic information. These include the creation of pathway-genome databases that use genomic sequence data to construct computationally predicted pathways (Shi *et al.*, 2006; Karlin and Mrazek 2000; Karlin *et al.*, 2001). VchoCyc is a pathway-genome database that used the N16961 genomic sequence data to predict highly expressed genes with supporting evidence from clinical tests and array experiments. This effort has provided an important tool for organizing genomic information to evaluate the metabolic capabilities of the organism, but verification of many complete pathways awaits experimental confirmation (Shi *et al.*, 2006). Notable exceptions include the biosynthetic pathways for enterobactin and vibriobactin, and the pathway for chitin utilization (Meibom *et al.*, 2004; Berg *et al.*, 2007; Wyckoff *et al.*, 2006).

Many methods of evaluating gene expression, such as microarray analysis, do not usually consider an evolutionary aspect. The hypothesis that a given genome has evolved codon usage patterns to optimize expression levels that best suit the habitat, energy sources and lifestyle of the microorganism has led to the investigation of codon bias analysis as a valid method for determining gene expression levels (Karlin and Mrazek, 2000). Combined with two-dimensional gel electrophoresis, such analysis has led to the development of algorithms to predict genes that might be highly expressed in several organisms, including *V. cholerae* (Karlin *et al.*, 2001). Categories of highly expressed genes include those encoding products for amino acid and nucleotide biosynthesis, and essential energy metabolism pathways. Classes predicted to have low levels of expression include regulatory genes, genes expressed during stationary phase or starvation, and most genes encoding repair functions. Data obtained from these *in silico* approaches have complemented findings for individual genes and studies that used a library of transposon based fluorescent reporter fusions to identify highly expressed genes by differential fluorescence sorting techniques after growth under specific conditions (Xu *et al.*, 2003).

Microarray development and comparative genomics

Several spotted microarrays have been designed and constructed based on the N16961 genome sequence, including ORF-length arrays generated by PCR amplification of individual predicted genes, and arrays composed of unique 70-mer sequences representing each ORF. Numerous expression-based studies have been performed, such as the comparison of expression profiles for strains grown under various conditions and for regulatory gene mutants, the identification of stage specific gene expression during biofilm formation and growth, and trancsriptome analyses of bacteria recovered from *in vivo* infections (Dziejman *et al.*, 2002; Bina *et al.*, 2003; Larocque *et al.*, 2005; Moorthy and Watnick, 2005; Yildiz *et al.*, 2004; Beyhan *et al.*, 2006a,b; Merrell *et al.*, 2002; Xu *et al.*, 2002, 2003).

The *V. cholerae* ORF based spotted microarray was also used to perform comparative genomic analyses (Dziejman *et al.*, 2002) to identify the molecular elements unique to seventh pandemic strains (that may also contribute to the presumed superior 'fitness' of El Tor strains over their classical predecessors). Experiments compared the genetic content of nine O1 and O139 serogroup strains isolated from diverse geographic locations over the last 100 years. These included environmental, classical and El Tor biotype strains. The results revealed less than 2% difference in the genetic content between these strains and strain N16961, consistent with other studies demonstrating close relationships among the representative strains (Karaolis *et al.*, 1995; Faruque *et al.*, 2004). More recent comparative genomic studies have used an expanded set of 30 O1 and O139 serogroup strains, isolated over the last 45 years of the seventh pandemic (Pang *et al.*, 2007). These studies also included non-toxigenic *V. cholerae* strains of both serogroups, and reported that when compared with stain N16961, ~15% of genes were absent from some strains. Consistent with earlier results, most of the absent genes were localized to the integron island, and non-toxigenic O1 strains differed significantly from their toxigenic counterparts (Faruque *et al.*, 2004; Dziejman *et al.*, 2002). Collectively, these studies confirm that the population of *V. cholerae* strains in the aquatic environment is quite heterogeneous. Although clonal relationships between pathogenic O1 and O139 strains are well established, it appears that seventh pandemic strains may not have maintained a static genome but rather have continued to alter their genetic content.

Comparative genomic studies also resulted in the identification of two gene clusters, named Vibrio Seventh Pandemic (VSP) I (~11 kb) and II (~27 kb), that were associated with seventh pandemic strains and not found in the classical, pre-pandemic O1 El Tor or environmental strains originally used for comparison (Dziejman *et al.*, 2002; O'Shea *et al.*, 2004). Both VSP-I and VSP-II encode many hypothetical proteins, but sequence homology predicts that these regions also encode a type IV pilin, two methyl-accepting chemotaxis proteins, an AraC-like transcriptional regulator, a ribonuclease, and the RadC-like and prophage-related protein mentioned earlier (Dziejman *et al.*, 2002; O'Shea *et al.*, 2004). Although the presence of VSP-II sequences is strongly correlated with seventh pandemic strains, several non-O1/non-O139 strains that contain VSP-II genes have been identified (Diraphat, Dziejman, Faruque and Mekalanos, unpublished) (Dziejman *et al.*, 2005; Pang *et al.*, 2007). Interestingly, a strain isolated at the beginning of the seventh pandemic in China was identical to N16961 and other seventh pandemic isolates, except that this strain lacked VSP-II. Phylogenetic analysis of this strain based on the sequence of four housekeeping genes suggests that its clonal derivation is separate from other toxigenic El Tor strains, and this strain has been proposed to represent an early precursor to seventh pandemic strains whose subsequent dominance may be associated with VSP-II acquisition (Pang *et al.*, 2007). An alternative explanation is that strain specific sequence polymorphisms could result in subtle functional differences that have restricted the full expansion of this strain to a pandemic clone. In addition, two recently isolated epidemic strains from China lacked many of the VSP-II genes, and their sequence analysis revealed an insertion sequence related to *V. cholerae* transposases (Pang *et al.*, 2007). However, it is not clear whether these exceptions represent intermediates in the pathogenic evolution of *V. cholerae* or are simply representatives of genetic diversity within the VSP-II locus.

Non-O1/non-O139 strain genomics

As a group, non-O1/non-O139 strains are presumed to share a niche with epidemic strains in the aquatic environment and in proximity to other microbial species. Presumably, this niche provides the opportunity for association with marine eukaryotes and the potential for horizontal gene transfer events from multiple sources. Historically, emphasis has been placed on studying the clinical isolates of O1 serogroup strains, but more recently, an increasing number of studies have focused on strains of all serogroups isolated from the environment. A recent increase in reports of sporadic non-O1/non-O139 disease raises many questions about this group of strains, including how they cause disease and whether these mechanisms are conserved in O1 and O139 strains. The factors that restrict non-O1/non-O139-related disease to sporadic episodes (and, conversely, those that promote epidemic expansion of strains) are currently unknown.

Molecular methods have revealed that non-O1/non-O139 strains are quite heterogeneous at the level of genomic content, perhaps most notable in the presence or absence of CTXϕ and the TCP island (Faruque *et al.*, 2004; Dalsgaard *et al.*, 1995, 2001; Guhathakurta *et al.*, 1999; Jiang *et al.*, 2003; Singh *et al.*, 2001). Comparative genomic microarray analyses of non-O1/non-O139 serogroup strains has shown greater numbers of absent genes compared with O1 and O139 serogroup strains (Dziejman *et al.*, 2005). Approximately half the absent genes are found on the small chromosome within the integron island, and although exceptions exist, VSP-I and VSP-II were absent in more than 99% of ~250 non-O1/non-O139 strains tested (Faruque, Dziejman and Mekalanos, unpublished). Collectively, the array results have identified chromosomal regions that are highly variable and further support the hypothesis that non-O1/non-O139 strains are evolutionarily distinct from O1 and O139 strains. In addition, these results have defined a 'core' set of genes that appear to be present in most *V. cholerae* strains, which when combined with core gene subgroups from different *Vibrio* spp. may be used for diagnostic purposes for broad-based pathogen detection (Vora *et al.*, 2005).

Identification of novel genes

While comparative genomic studies have been useful for identifying gene content relative to strain N16961, these studies can not reveal genetic information that has been acquired by other strains. As mentioned earlier, virulence factors associated with pathogenic TCP negative, CT negative non-O1/non-O139 strains remain largely unidentified. In addition, several incidents of limited epidemics caused by non-O1/non-O139 strains have been reported, suggesting that alternative virulence factors may play a role in the evolution of the pathogen (Dalsgaard *et al.*, 1995, 1997). Understanding the genes unique to these strains may therefore prepare us for future changes in the epidemiology of cholera.

N16961-based subtractive hybridization of libraries made from two non-O1/non-O139 environmental strains isolated from the California coast identified unique, toxin-like genes (that were not present in N16961) similar to *toxA* of *P. aeruginosa* and *cnf* of *E. coli* (Purdy *et al.*, 2005). These sequences appeared to be widespread among a group of non-O1/non-O139 isolates from a similar geographic location, suggesting that they were not restricted to individual strains. These findings support the hypothesis that environmental strains may serve as a reservoir for virulence gene pools, and the identification of many novel sequences contributes to the 'supragenome' of genetic information carried by *V. cholerae* strains (Faruque and Nair, 2002).

Additional genomic sequencing efforts have also resulted in the identification of genes unique to non-O1/non-O139 strains. A type three secretion system (T3SS) was identified by sequencing AM-19226, an O39 serogroup strain clinically isolated in 2001 in Bangladesh (Dziejman *et al.*, 2005). The genes encoding structural proteins that are homologous to the T3SS island II of *V. parahaemolyticus* are present in a wide panel of non-O1/non-O139 strains isolated from both environmental pools and diarrheal patients (Dziejman *et al.*, 2005; Chen *et al.*, 2007). The AM-19226 T3SS is functional and necessary for colonization in the infant mouse model, and one effector protein (VopF) has thus far been identified (Tam *et al.*, 2007). VopF appears to promote the formation of filo-form like projections in cultured mammalian cells and demonstrates an *in vitro* actin polymerization activity. Further investigation is required to identify other T3SS effectors and to understand how their molecular interactions with host cell proteins results in cholera-like symptoms.

New model systems have also been used to identify novel disease mechanisms. While the reversible intestinal tie adult rabbit diarrhoea (RITARD) model and the infant mouse model have been used to assess the ability of strains to cause diarrhoea and to colonize the host, several genetically tractable systems have recently been explored for their use as *V. cholerae* model systems. Use of a *Dictyostelium* host model system and an O37 serogroup clinical isolate from Sudan identified a novel secretion system (named type six, T6SS) responsible for translocating proteins into the eukaryotic host (Pukatzki *et al.*, 2006). While the *vas* genes encoding the T6SS are present in strain N16961, their altered regulation in strain V52 uncovered a cytotoxic phenotype that led to their identification. The *vas* genes share sequence similarity and organization with the IcmF associated homologous protein (IAHP) gene clusters of *P. aeruginosa, Y. pestis, E. coli* O157, *S. enterica, A. tumefaciens, E. ictaluri,* and *R. leguminosarum* that have diverse functions in mediating cell–host contact and disease, but are distinct from other secretion systems (Pukatzki *et al.*, 2006). The *V. cholerae* T6SS is responsible for secreting haemolysin coregulated protein (HCP), and the locus encodes several VgrG proteins, one of which contains an actin-cross-linking domain that can promote eukaryotic cell rounding upon translocation of the bacterial protein into the eukaryotic cytosol (Pukatzki *et al.*, 2007). Other model systems recently explored include *Drosophila, C. elegans* and the aquatic organism *Acanthamoeba* (Hilbi *et al.*, 2007; Park *et al.*, 2005; Blow *et al.*, 2005; Vaitkevicius *et al.*, 2006; Abd *et al.*, 2005). Bacteria have been shown to survive and replicate symbiotically in *Acanthamoeba*, raising the possibility that this eukaryote may facilitate *V. cholerae* environmental survival under certain circumstances (Abd *et al.*, 2007). Presumably, the expanded use of these new model systems will continue to identify additional functions related to pathogenesis and/or interactions with eukaryotic hosts in the environment.

Conclusions

The sequence of O1 El Tor strain N16961 captures a genetic snapshot of *V. cholerae* in its progressive evolution. More than 45 years after the start of the seventh pandemic, N16961 is recognized as a marker for the emergence of new strains: for its ability to displace the sixth pandemic, O1 classical strains, and to serve as a precursor for the evolution of 8th pandemic, O139 serogroup strains. Although O1 El Tor strains were not recognized as an epidemic threat until 1961, the formal possibility exists that O1 El Tor strains have caused widespread cholera prior to their association with the start of the seventh pandemic and perhaps, before recorded history of the disease.

An interesting consideration in modelling the emergence of pathogenic and epidemic strains is the bi-directional capability of horizontal gene transfer events. While these events are typically viewed as providing new or potentially advantageous functions for the organism, lost genetic information has the potential to drive evolution as well (Lawrence, 2005). Similarly, sequence polymorphisms can refine or expand existing gene functions, but in some cases may diminish or ablate protein activity. Although largely beyond the scope of this discussion, we are just beginning to appreciate the role of polymorphic sequences in conferring subtle difference in protein function for *V. cholerae*.

Our research goals now include a more global understanding of the organism and the expanded genetic repertoire of the species. Identifying genes common to all *V. cholerae* and those unique to strains at specific points in recent evolutionary time provides clues to associate epidemiological and clinical behaviours with genetic content. These studies can perhaps help us to monitor evolutionary change as it occurs, and provide insight into the events that drive the emergence and establishment of new strains. Already accepted as a diverse species, the subset of pathogenic *V. cholerae* strains is now also recognized as quite variable, and it is apparent that multiple pathways can lead to the transformation of this marine bacterium into a human pathogen. Understanding the genetic composition of *V. cholerae* strain N16961 has not only provided a wealth of data but has also led to a further understanding of the complexity within the species at the level of gene content and sequence polymorphisms. The goals of the current genomic era now include embracing a multidisciplinary approach to translate these complexities into discernable phenotypes that contribute to the pathogenic capabilities of an environmental organism.

References

Abd, H., Saeed, A., Weintraub, A., Nair, G.B. and Sandstrom, G. (2007). *Vibrio cholerae* O1 strains are facultative intracellular bacteria, able to survive and multiply symbiotically inside the aquatic free-living amoeba *Acanthamoeba castellanii*. FEMS Microbiol. Ecol. *60*, 33.

Abd, H., Weintraub, A. and Sandstrom, G. (2005). Intracellular survival and replication of *Vibrio cholerae* O139 in aquatic free-living amoebae. Environ. Microbiol. *7*, 1003–1008.

Aiyar, S.E., Gaal, T. and Gourse, R.L. (2002). rRNA promoter activity in the fast-growing bacterium *Vibrio natriegens*. J. Bacteriol. *184*, 1349–1358.

Alm, R.A., Stroeher, U.H. and Manning, P.A. (1988). Extracellular proteins of *Vibrio cholerae*: Nucleotide sequence of the structural gene (hlya) for the haemolysin of the haemolytic El Tor strain O17 and characterization of the hlya mutation in the non-haemolytic classical strain 569B. Mol. Microbiol. *2*, 481–488.

Asaduzzaman, M., Ryan, E.T., John, M., Hang, L., Khan, A.I., Faruque, A.S., Taylor, R.K., Calderwood, S.B. and Qadri, F. (2004). The major subunit of the toxin-coregulated pilus TcpA induces mucosal and systemic immunoglobulin a immune responses in patients with cholera caused by *Vibrio cholerae* O1 and O139. Infect. Immun. *72*, 4448–4454.

Barua, D. (1992). History of cholera. Cholera. D. Barua and W.B. Greenough III. New York, Plenum Publishing Corporation: 1–36.

Beaber, J.W., Hochhut, B. and Waldor, M.K. (2002). Genomic and functional analyses of SXT, an integrating antibiotic resistance gene transfer element derived from *Vibrio cholerae*. J. Bacteriol. *184*, 4259.

Berg, T., Schild, S. and Reidl, J. (2007). Regulation of the chitobiose-phosphotransferase system in *Vibrio cholerae*. Arch. Microbiol. *187*, 433–439.

Beyhan, S., Tischler, A.D., Camilli, A. and Yildiz, F.H. (2006a). Differences in gene expression between the classical and El Tor biotypes of *Vibrio cholerae* O1. Infect. Immun. *74*, 3633.

Beyhan, S., Tischler, A.D., Camilli, A. and Yildiz, F.H. (2006b). Transcriptome and phenotypic responses of *Vibrio cholerae* to increased cyclic di-GMP level. J. Bacteriol. *188*, 3600–3613.

Bik, E.M., Bunschoten, A.E., Gouw, R.D. and Mooi, F. (1995). Genesis of the novel epidemic *Vibrio cholerae* O139 strain: Evidence for horizontal transfer of genes involved in polysaccharide synthesis. EMBO J. *14*, 209–216.

Bina, J., Zhu, J., Dziejman, M., Faruque, S., Calderwood, S. and Mekalanos, J. (2003). ToxR regulon of *Vibrio cholerae* and its expression in vibrios shed by cholera patients. Proc. Natl. Acad. Sci. USA *100*, 2801–2806.

Binnewies, T.T., Motro, Y., Hallin, P.F., Lund, O., Dunn, D., La, T., Hampson, D.J., Bellgard, M., Wassenaar, T.M. and Ussery, D.W. (2006). Ten years of bacterial genome sequencing: comparative-genomics-based discoveries. Funct. Integr. Genomics. *6*, 165–185.

Blow, N.S., Salomon, R.N., Garrity, K., Reveillaud, I., Kopin, A., Jackson, F.R. and Watnick, P.I. (2005). *Vibrio cholerae* infection of Drosophila melanogaster mimics the human disease cholera. PLoS Pathog. *1*, e8.

Boucher, Y., Nesbo, C.L., Joss, M.J., Robinson, A., Mabbutt, B.C., Gillings, M.R., Doolittle, W.F. and Stokes, H.W. (2006). Recovery and evolutionary analysis of complete integron gene cassette arrays from Vibrio. BMC Evol. Biol. *6*, 3.

Boyd, E.F. and Waldor, M.K. (2002). Evolutionary and functional analyses of variants of the toxin-coregulated pilus protein TcpA from toxigenic *Vibrio cholerae* non-O1/non-O139 serogroup isolates. Microbiology *148*, 1655.

Brown, R.C. and Taylor, R.K. (1995). Organization of *tcp*, *acf*, and *toxt* genes within a ToxT-dependent operon. Mol. Microbiol. *16*, 425–439.

Butler, S.M. and Camilli, A. (2005). Going against the grain: Chemotaxis and infection in *Vibrio cholerae*. Nat. Rev. Microbiol. *3*, 611–620.

Chain, P.S., Denef, V.J., Konstantinidis, K.T., Vergez, L.M., Agullo, L., Reyes, V.L., Hauser, L., Cordova, M., Gomez, L., Gonzalez, M., Land, M., Lao, V., Larimer, F., LiPuma, J.J., Mahenthiralingam, E., Malfatti, S.A., Marx, C.J., Parnell, J.J., Ramette, A., Richardson, P., Seeger, M., Smith, D., Spilker, T., Sul, W.J., Tsoi, T.V., Ulrich, L.E., Zhulin, I.B. and Tiedje, J.M. (2006). Burkholderia xenovorans lb400 harbors a multi-replicon, 9.73-mbp genome shaped for versatility. Proc. Natl. Acad. Sci. USA *103*. 15280–15287.

Chatterjee, S.N. and Chaudhuri, K. (2003). Lipopolysaccharides of *Vibrio cholerae*. I. Physical and chemical characterization. Biochim. Biophys. Acta *1639*, 65–79.

Chatterjee, S.N. and Chaudhuri, K. (2004). Lipopolysaccharides of *Vibrio cholerae* II. Genetics of biosynthesis. Biochim Biophys Acta. *1690*, 93–109.

Chatterjee, S., Mondal, A.K., Begum, N.A., Roychoudhury, S. and Das, J. (1998). Ordered cloned DNA map of the genome of *Vibrio cholerae* 569B and localization of genetic markers. J. Bacteriol. *180*, 901–908.

Chen, C.-Y., Wu, K.-M., Chang, Y.-C., Chang, C.-H., Tsai, H.-C., Liao, T.-L., Liu, Y.-M., Chen, H.-J., Shen, A.B.-T., Li, J.-C., Su, T.-L., Shao, C.-P., Lee, C.-T., Hor, L.-I. and Tsai, S.-F. (2003). Comparative genome analysis of *Vibrio vulnificus*, a marine pathogen. Genome Res. *13*, 2577.

Chen, Y., Johnson, J.A., Pusch, G.D., Morris Jr., J.G. and Stine, O.C. (2007). The genome of non-O1 *Vibrio cholerae* NRT36S demonstrates the presence of pathogenic mechanisms that are distinct from those of O1 *Vibrio cholerae*. Infect. Immun. *75*, 2645–2647.

Choudhary, M., Zanhua, X., Fu, Y.X. and Kaplan, S. (2007). Genome analyses of three strains of Rhodobacter sphaeroides: Evidence of rapid evolution of chromosome II. J. Bacteriol. *189*, 1914.

Coelho, A., Andrade, J.R.C., Vincente, C.P. and DiRita, V.J. (2000). Cytotoxic cell vacuolating activity from *Vibrio cholerae* hemolysin. Infect. Immun. *68*, 1700–1705.

Colwell, R.R. (1996). Global climate and infectious disease: The cholera paradigm. Science *274*, 2025–2031.

Comstock, L.E., Johnson, J.A., Machalski, J.M., Morris, J.G. and Kaper, J.B. (1996). Cloning and sequence of a region encoding a surface polysaccharide of *Vibrio cholerae* O139 and characterization of the insertion site in the chromosome of *Vibrio cholerae* O1. Mol. Microbiol. *19*, 815–826.

Craig, L., Taylor, R.K., Pique, M.E., Adair, B.D., Arvai, A.S., Singh, M., Lloyd, S.J., Shin, D.S., Getzoff, E.D., Yeager, M., Forest, K.T. and Tainer, J.A. (2003). Type IV pilin structure and assembly: X-ray and EM analyses of *Vibrio cholerae* toxin-coregulated pilus and *Pseudomonas aeruginosa* PAK pilin. Mol. Cell. *11*, 1139–1150.

Crawford, J.A., Kaper, J.B. and DiRita, V.J. (1998). Analysis of ToxR-dependent transcription activation of ompU, the gene encoding a major envelope protein in *Vibrio cholerae*. Mol. Microbiol. *29*, 235–246.

Dalsgaard, A., Albert, M.J., Taylor, D.N., Shimada, T., Meza, R., Serichantalergs, O. and Echeverria, P. (1995). Characterization of *Vibrio cholerae* non-O1 serogroups obtained from an outbreak of diarrhea in Lima, Peru. J. Clin. Microbiol. *33*, 2715–2722.

Dalsgaard, A., Skov, M.N., Serichantalergs, O., Echeverria, P., Meza, R. and Taylor, D.N. (1997). Molecular evolution of *Vibrio cholerae* O1 strains isolated in Lima, Peru, from 1991 to 1995. J. Clin. Microbiol. *35*, 1151–1156.

Dalsgaard, A., Serichantalergs, O., Forslund, A., Lin, W., Mekalanos, J., Mintz, E., Shimada, T. and Wells, J.G. (2001). Clinical and environmental isolates of *Vibrio cholerae* serogroup O141 carry the CTX phage and the genes encoding the toxin-coregulated pili. J. Clin. Microbiol. *39*, 4086–4092.

Daniels, N.A., Ray, B., Easton, A., Marano, N., Kahn, E., McShan, A.L., 2nd, Del Rosario, L., Baldwin, T., Kingsley, M.A., Puhr, N.D., Wells, J.G. and Angulo, F.J. (2000). Emergence of a new Vibrio parahaemolyticus serotype in raw oysters: A prevention quandary. JAMA. *284*, 1541–1545.

Darwin, C. (1872). On the origin of species by means of natural selection. London, John Murray.

Datta, A., Parker, C.D., Wohlhieter, J.A. and Baron, L.S. (1973). Isolation and characterization of the fertility factor P of *Vibrio cholerae*. J. Bacteriol. *113*, 763–771.

Duigou, S., Knudsen, K.G., Skovgaard, O., Egan, E.S., Lobner-Olesen, A. and Waldor, M.K. (2006). Independent control of replication initiation of the two *Vibrio cholerae* chromosomes by DnaA and RctB. J. Bacteriol. *188*, 6419.

Dziejman, M., Serruto, D., Tam, V.C., Sturtevant, D., Diraphat, P., Faruque, S.M., Rahman, M.H., Heidelberg, J.F., Decker, J., Li, L., Montgomery, K., Grills G., Kucherlapati, R. and Mekalanos, J.J. (2005). Genomic characterization of non-O1 non-O139 *Vibrio cholerae* reveals genes for a novel Type III secretion system. Proc. Natl. Acad. Sci. USA *102*, 3465–70.

Dziejman, M., Balon, E., Boyd, D., Fraser, C.M., Heidelberg, J.F. and Mekalanos, J.J. (2002). Comparative genomic analysis of *Vibrio cholerae*: Genes that correlate with cholera endemic and pandemic disease. Proc. Natl. Acad. Sci. USA *99*, 1556–1561.

Egan, E.S. and Waldor, M.K. (2003). Distinct replication requirements for the two *Vibrio cholerae* chromosomes. Cell *114*, 521–530.

Egan, E.S., Lobner-Olesen, A. and Waldor, M.K. (2004). Synchronous replication initiation of the two *Vibrio cholerae* chromosomes. Curr. Biol. *14*, R501–502.

Egan, E.S., Fogel, M.A. and Waldor, M.K. (2005). Divided genomes: Negotiating the cell cycle in prokaryotes with multiple chromosomes. Mol. Microbiol. *56*, 1129.

Eisen, J.A., Heidelberg, J.F., White, O. and Salzberg, S.L. (2000). Evidence for symmetric chromosomal inversions around the replication origin in bacteria. Genome Biol. *1*, research 0011.1–0011.9.

Faruque, S.M. and Nair, G.B. (2002). Molecular ecology of toxigenic *Vibrio cholerae*. Microbiol. Immunol. *46*, 59–66.

Faruque, S.M. and Mekalanos, J.J. (2003). Pathogenicity islands and phages in *Vibrio cholerae* evolution. Trends Microbiol. *11*, 505–510.

Faruque, S.M., Nair, G.B. and Mekalanos, J.J. (2004). Genetics of stress adaptation and virulence in toxigenic *Vibrio cholerae*. DNA Cell Biol. *23*, 723–741.

Faruque, S.M., Naser, I.B., Islam, M.J., Faruque, A.S., Ghosh, A.N., Nair, G.B., Sack, D.A. and Mekalanos, J.J. (2005b). Seasonal epidemics of cholera inversely correlate with the prevalence of environmental cholera phages. Proc. Natl. Acad. Sci. USA *102*, 1702–7.

Faruque, S.M., Naser, I. Bin, Fujihara, K., Diraphat, P., Chowdhury, N., Kamruzzaman, M., Qadri, F, Yamasaki, S., Ghosh, A.N. and Mekalanos, J.J. (2005a). Genomic sequence and receptor for the *Vibrio cholerae* phage KSF-1Φ: Evolutionary divergence among filamentous vibriophages mediating lateral gene transfer. J. Bacteriol. *187*, 4095–4103.

Faruque, S.M., Zhu, J., Asadulghani, Kamruzzaman, M. and Mekalanos, J.J. (2003). Examination of diverse toxin-coregulated pilus-positive *Vibrio cholerae* strains fails to demonstrate evidence for vibrio pathogenicity island phage. Infect. Immun. *71*, 2993–2999.

Faruque, S.M., Ahmed, K.M., Siddique, A.K., Zaman, K., Alim, A.R. and Albert, M.J. (1997). Molecular analysis of toxigenic *Vibrio cholerae* O139 Bengal strains isolated in Bangladesh between 1993 and 1996: Evidence for emergence of a new clone of Bengal vibrios. J. Clin. Microbiol. *35*, 2299–2306.

Faruque, S.M., Albert, M.J. and Mekalanos, J.J. (1998). Epidemiology, genetics, and ecology of toxigenic *Vibrio cholerae*. Microbiol. Mol. Biol. Rev. *62*, 1301–1314.

Faruque, S.M., Islam, M.J., Ahmad, Q.S., Faruque, A.S., Sack, D.A., Nair, G.B. and Mekalanos, J.J. (2005c). Self-limiting nature of seasonal cholera epidemics: Role of host-mediated amplification of phage. Proc. Natl. Acad. Sci. USA *102*, 6119–6124.

Faruque, S.M., Chowdhury, N., Kamruzzaman, M., Dziejman, M., Rahman, M.H., Sack, D.A., Nair, G.B. and Mekalanos, J.J. (2004). Genetic diversity and virulence potential of environmental *Vibrio cholerae* population in a cholera-endemic area. Proc. Natl. Acad. Sci. USA *101*, 2123–2128.

Fiebig, A., Keren, K. and Theriot, J.A. (2006). Fine-scale time-lapse analysis of the biphasic, dynamic behaviour of the two *Vibrio cholerae* chromosomes. Mol. Microbiol. *60*, 1164.

Fogel, M.A. and Waldor, M.K. (2005). Distinct segregation dynamics of the two *Vibrio cholerae* chromosomes. Mol. Microbiol. *55*, 125–136.

Fogel, M.A. and Waldor, M.K. (2006). A dynamic, mitotic-like mechanism for bacterial chromosome segregation. Genes Dev. *20*, 3269.

Franzon, V.L., Barker, A. and Manning, P.A. (1993). Nucleotide sequence encoding the mannose-fucose-resistant hemagglutinin of *Vibrio cholerae* O1 and construction of a mutant. Infect. Immun. *61*, 3032–3037.

Fullner, K.J. and Mekalanos, J.J. (1999). Genetic characterization of a new Type IV-A pilus gene cluster found in both classical and El Tor biotypes of *Vibrio cholerae*. Infect. Immun. *67*, 1393–1404.

Guhathakurta, B., Sasmal, D., Pal, S., Chakraborty, S., Nair, G.B. and Datta, A. (1999). Comparative analysis of cytotoxin, hemolysin, hemagglutinin and exocellular enzymes among clinical and environmental isolates of *Vibrio cholerae* O139 and non-O1, non-O139. FEMS Microbiol. Lett. *179*, 401–407.

Halling, S.M., Peterson-Burch, B.D., Bricker, B.J., Zuerner, R.L., Qing, Z., Li, L.-L., Kapur, V., Alt, D.P. and Olsen, S.C. (2005). Completion of the genome sequence of *Brucella abortus* and comparison to the highly similar genomes of *Brucella melitensis* and Brucella suis. J. Bacteriol. *187*, 2715.

Hang, L., John, M., Asaduzzaman, M., Bridges, E.A., Vanderspurt, C., Kirn, T.J., Taylor, R.K., Hillman, J.D., Progulske-Fox, A., Handfield, M., Ryan, E.T. and Calderwood, S.B. (2003). Use of *in vivo*-induced antigen technology (IVIAT) to identify genes uniquely expressed during human infection with *Vibrio cholerae*. Proc. Natl. Acad. Sci. USA *100*, 8508–8513.

Heidelberg, J.F., Elsen, J.A., Nelson, W.C., Clayton, R.J., Gwinn, M.L., Dodson, R.J., Haft, D.H., Hickey, E.K., Peterson, J.D., Umayam, L., Gill, S.R., Nelson, K., Read, T.D., Tettelin, H., Richardson, D., Ermolaeva, M.D., Vamathevan, J., Bass, S., Qin, H., Dragoi, I., Sellers, P., McDonald, L., Utterback, T., Fleishmann, R.D., Nierman, W.C., White, O., Salzberg, S.L., Smith, H.O., Colwell, R.R., Mekalanos, J.J., Venter, J.C. and Fraser, C. (2000). DNA sequence of both chromosomes of the cholera pathogen *Vibrio cholerae*. Nature *406*, 477–484.

Higgins, D. and DiRita, V.J. (1996). Genetic analysis of the interaction between *Vibrio cholerae* transcription activator ToxR and toxT promoter DNA. J. Bacteriol. *178*, 1070–1087.

Hilbi, H., Weber, S.S., Ragaz, C., Nyfeler, Y. and Urwyler, S. (2007). Environmental predators as models for bacterial pathogenesis. Environ. Microbiol. *9*, 563–575.

Hsiao, A., Liu, Z., Joelsson, A. and Zhu, J. (2006). *Vibrio cholerae* virulence regulator-coordinated evasion of host immunity. Proc. Natl. Acad. Sci. USA *103*, 14542–14547.

Jermyn, W. and Boyd, E.F. (2002). Characterization of a novel Vibrio pathogenicity island (vpi-2) encoding neuraminidase (nanh) among toxigenic *Vibrio cholerae* isolates. Microbiology. *148*, 3681–3693.

Jermyn, W.S. and Boyd, E.F. (2005). Molecular evolution of Vibrio pathogenicity island-2 (VPI-2): Mosaic structure among *Vibrio cholerae* and *Vibrio mimicus* natural isolates. Microbiology *151*, 311–322.

Jiang, S., Chu, W. and Fu, W. (2003). Prevalence of cholera toxin genes (ctxA and zot) among non-O1/O139 *Vibrio cholerae* strains from Newport Bay, California. Appl. Environ. Microbiol. *69*, 7541–7544.

Jonson, G., Holmgren, J. and Svennerholm, A.M. (1991). Identification of a mannose-binding pilus on *Vibrio cholerae* El Tor. Microb. Pathogen. *11*, 433–441.

Jonson, G., Lebens, M. and Holmgren, J. (1994). Cloning and sequencing of *Vibrio cholerae* mannose-sensitive haemagglutinin pilin gene: Localization of mshA within a cluster of type 4 pilin genes. Mol. Microbiol. *13*, 109–118.

Jouravleva, E.A., McDonald, G.A., Marsh, J.W., Taylor, R.K., Boesman-Finkelstein, M. and Finkelstein, R.A. (1998). The *Vibrio cholerae* mannose-sensitive hemagglutinin is the receptor for a filamentous bacteriophage from *V. cholerae* O139. Infect. Immun. *66*, 2535.

Judson, N. and Mekalanos, J.J. (2000). TnAraOut, a transposon-based approach to identify and characterize essential bacterial genes. Nat. Biotechnol. *18*, 740–745.

Karaolis, D.K., Johnson, J.A., Bailey, C.C., Boedeker, E.C., Kaper, J.B. and Reeves, P.R. (1998). A *Vibrio cholerae* pathogenicity island associated with epidemic and pandemic strains. Proc. Natl. Acad. Sci. USA *95*, 3134–3139.

Karaolis, D.K., Lan, R. and Reeves, P.R. (1994). Molecular evolution of the seventh-pandemic clone of *Vibrio cholerae* and its relationship to other pandemic and epidemic *V. cholerae* isolates. J. Bacteriol. *176*, 6199–6206.

Karaolis, D.K., Lan, R. and Reeves, P.R. (1995). The sixth and seventh cholera pandemics are due to independent clones separately derived from environmental, nontoxigenic, non-O1 *Vibrio cholerae*. J. Bacteriol. *177*, 3191–3198.

Karlin, S. (2001). Detecting anomalous gene clusters and pathogenicity islands in diverse bacterial genomes. Trends Microbiol. *9*, 335–343.

Karlin, S. and Mrazek, J. (2000). Predicted highly expressed genes of diverse prokaryotic genomes. J. Bacteriol. *182*, 5238–5250.

Karlin, S., Mrazek, J., Campbell, A. and Kaiser, D. (2001). Characterizations of highly expressed genes of four fast-growing bacteria. J. Bacteriol. *183*, 5025–5040.

Kirn, T.J., Lafferty, M.J., Sandoe, C.M.P. and Taylor, R.K. (2000). Delineation of pilin domains required for bacterial association into microcolonies and intestinal colonization by *Vibrio cholerae*. Mol. Microbiol. *35*, 896–910.

Krukonis, E.S. and DiRita, V.J. (2003). From motility to virulence: Sensing and responding to environmental signals in *Vibrio cholerae*. Curr. Opin. Microbiol. *6*, 186–190.

Krukonis, E.S., Yu, R.R. and DiRita, V.J. (2000). The *Vibrio cholerae* ToxR/TcpP/ToxT virulence cascade: Distinct roles for two membrane-localized transcriptional activators on a single promoter. Mol. Microbiol. *38*, 67–84.

Kugelberg, E., Kofoid, E., Reams, A.B., Andersson, D.I. and Roth, J.R. (2006). Multiple pathways of selected gene amplification during adaptive mutation. Proc. Natl. Acad. Sci. USA *103*, 17319–17324.

Kuroki, H., Toma, C., Nakasone, N., Yamashiro, T. and Iwanaga, M. (2001). Gene analysis of *Vibrio cholerae* NAGV14 pilus and its distribution. Microbiol. Immunol. *45*, 417–424.

Larocque, R.C., Harris, J.B., Dziejman, M., Li, X., Khan, A.I., Faruque, A.S., Faruque, S.M., Nair, G.B., Ryan, E.T., Qadri, F., Mekalanos, J.J. and Calderwood, S.B. (2005). Transcriptional profiling of *Vibrio cholerae* recovered directly from patient specimens during early and late stages of human infection. Infect. Immun. *73*, 4488–4493.

Lawrence, J.G. (2005). Common themes in the genome strategies of pathogens. Curr. Opin. Genet. Dev. *15*, 584–588.

Lawrence, J.G. and Ochman, H. (1998). Molecular archaeology of the *Escherichia coli* genome. Proc. Natl. Acad. Sci. USA *95*, 9413–9417.

Liolios, K., Tavernarakis, N., Hugenholtz, P. and Kyrpides, N.C. ((2006). The genomes on line database (GOLD) v.2: A monitor of genome projects worldwide. Nucleic Acids Res. *34*, 332–334.

Lombardo, M.J. and Rosenberg, S.M. (2000). RadC102 of *Escherichia coli* is an allele of RecG. J. Bacteriol. *182*, 6287–6291.

Majumder, R., Sengupta, S., Khetawat, G., Bhadra, R.K., Roychoudhury, S. and Das, J. (1996). Physical map of the genome of *Vibrio cholerae* 569B and localization of genetic markers. J. Bacteriol. *178*, 1105–1112.

Makino, K., Oshima, K., Kurokawa, K., Yokoyama, K., Uda, T., Tagomori, K., Iijima, Y., Najima, M., Nakano, M., Yamashita, A., Kubota, Y., Kimura, S., Yasunaga, T., Honda, T., Shinagawa, H., Hattori, M. and Iida, T. (2003). Genome sequence of Vibrio parahaemolyticus: A pathogenic mechanism distinct from that of *V. cholerae*. Lancet *361*, 743–749.

Manning, P. (1997). The *tcp* gene cluster of *Vibrio cholerae*. Gene *192*: 63–70.

Mazel, D., Dychinco, B., Webb, V.A. and Davies, J. (1998). A distinctive class of integron in the *Vibrio cholerae* genome. Science. *280*, 605–608.

McLeod, S.M. and Waldor, M.K. (2004). Characterization of XerC- and XerD-dependent CTX phage integration in *Vibrio cholerae*. Mol. Microbiol. *54*, 935–947.

Medigue, C., Krin, E., Pascal, G., Barbe, V., Bernsel, A., Bertin, P.N., Cheung, F., Cruveiller, S., D'Amico, S., Duilio, A., Fang, G., Feller, G., Ho, C., Mangenot, S., Marino, G., Nilsson, J., Parrilli, E., Rocha, E.P., Rouy, Z., Sekowska, A., Tutino, M.L., Vallenet, D., von Heijne, G. and Danchin, A. (2005). Coping with cold: The genome of the versatile marine antarctica bacterium Pseudoalteromonas haloplanktis tac125. Genome Res. *15*, 1325–1335.

Meibom, K.L., Blokesch, M., Dolganov, N.A., Wu, C.Y. and Schoolnik, G.K. (2005). Chitin induces natural competence in *Vibrio cholerae*. Science *310*, 1824–1827.

Meibom, K.L., Li, X.B., Nielsen, A.T., Wu, C.Y., Roseman, S. and Schoolnik, G.K. (2004). The *Vibrio cholerae* chitin utilization program. Proc. Natl. Acad. Sci. USA *101*, 2524–2529.

Mekalanos, J.J. (1983). Duplication and amplification of toxin genes in *Vibrio cholerae*. Cell *35*, 253–263.

Mekalanos, J.J. (1985). Cholera toxin: Genetic analysis, regulation, and role in pathogenesis. Curr. Topics Microbiol. Immunol. *118*, 97–118.

Merrell, D.S., Butler, S.M., Qadri F., Dolganov, N.A., Alam, A., Cohen, M.B., Calderwood, S.B., Schoolnik, G.K. and Camilli, A. (2002). Host-induced epidemic spread of the cholera bacterium. Nature *417*, 642–645.

Miller, M.B., Skorupski, K., Lenz, D.H., Taylor, R.K. and Bassler, B.L. (2002). Parallel quorum sensing systems converge to regulate virulence in *Vibrio cholerae*. Cell *110*, 303–314.

Mitra, R., Figueroa, P., Mukhopadhyay, A.K., Shimada, T., Takeda, Y., Berg, D. and Nair, G.B. (2000). Cell vacuolation, a manifestation of the El Tor hemolysin of *Vibrio cholerae*. Infect. Immun. *68*, 1928–1933.

Moorthy, S. and Watnick, P.I. (2005). Identification of novel stage-specific genetic requirements through whole genome transcription profiling of *Vibrio cholerae* biofilm development. Mol. Microbiol. *57*, 1623–1635.

Nandi, S., Khetawat, G., Sengupta, S., Majumder, R., Kar, S., Bhadra, R.K., Roychoudhury, S. and Das, J. (1997). Rearrangements in the genomes of *Vibrio cholerae* strains belonging to different serovars and biovars. Int. J. Syst. Bacteriol. *47*, 858–862.

Nudleman, E. and Kaiser, D. (2004). Pulling together with Type IV pili. J. Mol. Microbiol. Biotechnol. *7*, 52–62.

Nye, M.B., Pfau, J.D., Skorupski, K. and Taylor, R.K. (2000). *Vibrio cholerae* H-NS silences virulence gene expression at multiple steps in the ToxR regulatory cascade. J. Bacteriol. *182*, 4295–4303.

O'Shea, Y.A., Finnan, S., Reen, F.J., Morrissey, J.P., O'Gara, F. and Boyd, E.F. (2004). The Vibrio Seventh Pandemic Island-II is a 26.9 kb genomic island present in *Vibrio cholerae* El Tor and O139 serogroup isolates that shows homology to a 43.4 kb genomic island in *V. vulnificus*. Microbiology *150*, 4053–4063.

Pang, B., Yan, M., Cui, Z., Ye, X., Diao, B., Ren, Y., Gao, S., Zhang, L. and Kan, B. (2007). Genetic diversity of toxigenic and nontoxigenic *Vibrio cholerae* serogroups O1 and O139 revealed by array-based comparative genomic hybridization. J. Bacteriol. *189*, 4837–4849.

Park, S.Y., Heo, Y.J., Kim, K.S. and Cho, Y.H. (2005). *Drosophila melanogaster* is susceptible to *Vibrio cholerae* infection. Mol. Cells *20*, 409–415.

Parker, C. and Romig, W.R. (1972). Self-transfer and genetic recombination mediated by P, the sex factor of *Vibrio cholerae*. J. Bacteriol. *112*, 707–714.

Parker, C., Gauthier, D., Tate, A., Richardson, K. and Romig, W.R. (1979). Expanded linkage map of *Vibrio cholerae*. Genetics *91*, 191–214.

Peterson, K.M. and Mekalanos, J.J. (1988). Characterization of the *Vibrio cholerae* ToxR regulon: Identification of novel genes involved in intestinal colonization. Infec. Immun. *56*, 2822–2829.

Pukatzki, S., Ma, A.T., Revel, A.T., Sturtevant, D. and Mekalanos, J.J. (2007). Type VI secretion system translocates a phage tail spike-like protein into target cells where it crosslinks actin. Proc. Natl. Acad. Sci. USA *104*, 15508–15513.

Pukatzki, S., Ma, A.T., Sturtevant, D., Krastins, B., Sarracino, D., Nelson, W.C., Heidelberg, J.F. and Mekalanos, J.J. (2006). Identification of a conserved bacterial protein secretion system in *Vibrio cholerae* using the *Dictyostelium* host model system. Proc. Natl. Acad. Sci. USA *103*, 1528–1533.

Purdy, A., Rohwer, F., Edwards, R., Azam, F. and Bartlett, D.H. (2005). A glimpse into the expanded genome content of *Vibrio cholerae* through identification of genes present in environmental strains. J. Bacteriol. *187*, 2992–3001.

Pushker, R., Mira, A. and Rodriguez-Valera, F. (2004). Comparative genomics of gene-family size in closely related bacteria. Genome Biol. 2004 5(4): R27.

Rajanna, C., Wang, J., Zhang, D., Xu, Z., Ali, A., Hou, Y.M. and Karaolis, D.K. (2003). The *Vibrio* pathogenicity island of epidemic *Vibrio cholerae* forms precise extrachromosomal circular excision products. J. Bacteriol. *185*, 6893–6901.

Raskin, D.M., Seshadri, R., Pukatzki, S.U. and Mekalanos, J.J. (2006). Bacterial genomics and pathogen evolution. Cell *124*, 703–714.

Reen, F.J., Almagro-Moreno, S., Ussery, D. and Boyd, E.F. (2006). The genomic code: inferring Vibrionaceae niche specialization. Nature Reviews Microbiology *4*, 697.

Reguera, G. and Kolter, R. (2005). Virulence and the environment: A novel role for *Vibrio cholerae* toxin-coregulated pili in biofilm formation on chitin. J. Bacteriol. *187*, 3551–3555.

Reidl, J. and Klose, K.E. (2002). *Vibrio cholerae* and cholera: Out of the water and into the host. FEMS Microbiol. Rev. *26*, 125–139.

Rhine, J.A. and Taylor, R.K. (1994). TcpA pilin sequences and colonization requirements for O1 and O139 *Vibrio cholerae*. Mol. Microbiol. *13*, 1013–1020.

Rogozin, I.B., Makarova, K.S., Wolf, Y.I. and Koonin, E.V. (2004). Computational approaches for the analysis of gene neighbourhoods in prokaryotic genomes. Brief Bioinform. *5*, 131–149.

Roth, J.R., Kugelberg, E., Reams, A.B., Kofoid, E. and Andersson, D.I. (2006). Origin of mutations under selection: The adaptive mutation controversy. Annu. Rev. Microbiol. *60*, 477–501.

Rowe-Magnus, D.A., Guerout, A.M. and Mazel, D. (1999). Super-integrons. Res. Microbiol. *150*, 641–651.

Rowe-Magnus, D.A., Guerout, A.M. and Mazel, D. (2002). Bacterial resistance evolution by recruitment of super-integron gene cassettes. Mol. Microbiol. *43*, 1657–1669.

Rowe-Magnus, D.A., Guerout, A.M., Biskri, L., Bouige, P. and Mazel, D. (2003). Comparative analysis of superintegrons: Engineering extensive genetic diversity in the Vibrionaceae. Genome Res. *13*, 428–442.

Rowe-Magnus, D.A., Guerout, A.M., Ploncard, P., Dychinco, B., Davies, J. and Mazel, D. (2001). The evolutionary history of chromosomal super-integrons provides an ancestry for multiresistant integrons. Proc. Natl. Acad. Sci. USA *98*, 652–657.

Ruby, E.G. Urbanowski, M., Campbell, J., Dunn, A., Faini, M., Gunsalus, R., Lostroh, P., Lupp, C., McCann, J., Millikan, D., Schaefer, A., Stabb, E., Stevens, A., Visick, K., Whistler, C. and Greenberg, E.P. (2005). Complete genome sequence of Vibrio fischeri: A symbiotic bacterium with pathogenic congeners. Proc. Natl. Acad. Sci. USA *102*, 3004–3009.

Saint-Dic, D., Frushour, B.P., Kehrl, J.H. and Kahng, L.S. (2006). A ParA homolog selectively influences positioning of the large chromosome origin in *Vibrio cholerae*. J. Bacteriol. *188*, 5626.

Shi, J., Romero, P.R., Schoolnik, G.K., Spormann, A.M. and Karp, P.D. (2006). Evidence supporting predicted metabolic pathways for *Vibrio cholerae*: Gene expression data and clinical tests. Nucleic Acids Res. *34*, 2438–2444.

Singh, D.V., Matte, M.H., Matte, G.R., Jiang, S., Sabeena, F., Shukla, B.N., Sanyal, S.C., Huq, A. and Colwell, R.R. (2001). Molecular analysis of *Vibrio cholerae* O1, O139, non-O1, and non-O139 strains: Clonal relationships between clinical and environmental isolates. Appl. Environ. Microbiol. *67*, 910–921.

Sokurenko, E.V., Gomulkiewicz, R. and Dykhuizen, D.E. (2006). Source-sink dynamics of virulence evolution. Nat. Rev. Microbiol. *4*, 548–555.

Sonnhammer, E.L.L. and Koonin, E.V. (2002). Orthology, paralogy and proposed classification for paralog subtypes. Trends in Genetics *18*, 619.

Szekeres, S., Dauti, M., Wilde, C., Mazel, D. and Rowe-Magnus, D.A. (2007). Chromosomal toxin–antitoxin loci can diminish large-scale genome reductions in the absence of selection. Mol. Microbiol. *63*, 1588–1605.

Tacket, C.O., Taylor, R.K., Losonsky, G., Lim, Y., Nataro, J.P., Kaper, J.B. and Levine, M.M. (1998). Investigation of the roles of toxin-coregulated pili and mannose-sensitive hemagglutinin pili in the pathogenesis of *Vibrio cholerae* O139 infection. Infect. Immun. *66*, 692–695.

Tagomori, K., Iida, T. and Honda, T. (2002). Comparison of genome structures of vibrios, bacteria possessing two chromosomes. J. Bacteriol. *184*, 4351–4358.

Tam, V.C., Serruto, D., Dziejman, M., Brieher, W. and Mekalanos, J.J. (2007). A Type III secretion system in *Vibrio cholerae* translocates a formin/spire hybrid-like actin nucleator to promote intestinal colonization. Host Cell Microbe *1*, 95.

Tamayo, R., Pratt, J.T. and Camilli, A. (2007). Roles of cyclic diguanylate in the regulation of bacterial pathogenesis. Annu. Rev. Microbiol. Epub ahead of print.

Taylor, D.N., Killeen, K.P., Hack, D.C., Kenner, J.R., Coster, T.S., Beattie, D.T., Ezzell, J., Hyman, T., Trofa, A., Sjogren, M.H., Friedlander, A., Mekalanos, J.J. and Sadoff, J.C. (1994). Development of a live, oral, attenuated vaccine against El Tor cholera. J. Infect. Dis. *170*, 1518–1523.

Taylor, R.K., Kirn, T.J., Meeks, M.D., Wade, T.K. and Wade, W.F. (2004). A *Vibrio cholerae* classical TcpA amino acid sequence induces protective antibody that binds an area hypothesized to be important for toxin-coregulated pilus structure. Infect. Immun. *72*, 6050–6060.

Taylor, R.K., Miller, V.L., Furlong, D.B. and Mekalanos, J.J. (1987). Use of phoa gene fusions to identify a pilus colonization factor coordinately regulated with cholera toxin. Proc. Natl. Acad. Sci. USA *84*, 2833–2837.

Thelin, K.H. and Taylor, R.K. (1996). Toxin-coregulated pilus, but not mannose-sensitive hemagglutinin, is required for colonization by *Vibrio cholerae* O1 El Tor biotype and O139 strains. Infect. Immun. *64*, 2853–2856.

Trucksis, M., Michalski, J., Deng, Y.K. and Kaper, J.B. (1998). The *Vibrio cholerae* genome contains two unique circular chromosomes. Proc. Natl. Acad. Sci. USA *95*, 14464–14469.

Vaitkevicius, K., Lindmark, B., Ou, G., Song, T., Toma, C., Iwanaga, M., Zhu, J., Andersson, A., Hammarstrom, M.L., Tuck, S. and Wai, S.N. (2006). A *Vibrio cholerae* protease needed for killing of *Caenorhabditis elegans* has a role in protection from natural predator grazing. Proc. Natl. Acad. Sci. USA *103*, 9280–9285.

Vezzi, A., Campanaro, S., D'Angelo, M., Simonato, F., Vitulo, N., Lauro, F.M., Cestaro, A., Malacrida, G., Simionati, B., Cannata, N., Romualdi, C., Bartlett, D.H. and Valle, G. (2005). Life at depth: *Photobacterium profundum* genome sequence and expression analysis. Science. *307*, 1459–1461.

Vora, G.J., Meador, C.E., Bird, M.M., Bopp, C.A., Andreadis, J.D. and Stenger, D.A. (2005). Microarray-based detection of genetic heterogeneity, antimicrobial resistance, and the viable but nonculturable state in human pathogenic *Vibrio* spp. Proc. Natl. Acad. Sci. USA *102*, 19109–19114.

Wachsmuth, I.K., Olsvik, O., Evins, G.M. and Popovic, T. (1994). Molecular epidemiology of cholera. *Vibrio cholerae* and cholera: Molecular to global perspectives. I.K. Wachsmuth, P.A. Blake and O. Olsvik. Washington DC, American Society for Microbiology: 357–370.

Waldor, M.K. and Mekalanos, J.J. (1996). Lysogenic conversion by a filamentous phage encoding cholera toxin. Science *272*, 1910–1914.

Waldor, M.K., Tschäpe, H. and Mekalanos, J.J. (1996). A new type of conjugative transposon encodes resistance to sulfamethoxazole, trimethoprim, and streptomycin in *Vibrio cholerae* O139. J. Bacteriol. *178*, 4157–4165.

Watnick, P. and Kolter, R. (2000). Biofilm, city of microbes. J. Bacteriol. *182*, 2675–2679.

Watnick, P.I., Fullner, K.J. and Kolter, R. (1999). A role for the mannose-sensitive hemagglutinin in biofilm formation by *Vibrio cholerae* El Tor. J. Bacteriol. *181*, 3606–3609.

Williams, S.G., Varcoe, L.T., Attridge, S.R. and Manning, P.A. (1996). *Vibrio cholerae* HCP, a secreted protein coregulated with HlyA. Infect. Immun. *64*, 283–289.

Williams, S.G., Attridge, S.R. and Manning, P.A. (1993). The transcriptional activator HlyU of *Vibrio cholerae*: Nucleotide sequence and role in virulence gene expression. Mol. Microbiol. *9*, 751–760.

Wood, D.W., Setubal, J.C., Kaul, R., Monks, D.E., Kitajima, J.P., Okura, V.K., Zhou, Y., Chen, L., Wood, G.E., Almeida, N.F., Woo, Jr., L., Chen, Y., Paulsen, I.T., Eisen, J.A., Karp, P.D., Bovee, D., Chapman, Sr., P., Clendenning, J., Deatherage, G., Gillet, W., Grant, C., Kutyavin, T., Levy, R., Li, M.-J., McClelland, E., Palmieri, A., Raymond, C., Rouse, G., Saenphimmachak, C., Wu, Z., Romero, P., Gordon, D., Zhang, S., Yoo, H., Tao, Y., Biddle, P., Jung, M., Krespan, W., Perry, M., Gordon-Kamm, B., Liao, L., Kim, S., Hendrick, C., Zhao, Z.-Y., Dolan, M., Chumley, F., Tingey, S.V., Tomb, J.-F., Gordon, M.P., Olson, M.V. and Nester, E.W. (2001). The genome of the natural genetic engineer *Agrobacterium tumefaciens* c58. Science *294*, 2317–2323.

Worthington, E.N., Kavakli, I.H., Berrocal-Tito, G., Bondo, B.E. and Sancar, A. (2003). Purification and characterization of three members of the photolyase/cryptochrome family blue-light photoreceptors from *Vibrio cholerae*. J. Biol. Chem. *278*, 39143–39154.

Wu, Z., Milton, D., Nybom, P., Sjo, A. and Magnusson, K.E. (1996). *Vibrio cholerae* hemagglutinin/protease (ha/protease) causes morphological changes in cultures epithelial cells and perturbs their paracellular barrier function. Microbial Pathogen. *21*, 111–123.

Wyckoff, E.E., Mey, A.R., Leimbach, A., Fisher, C.F. and Payne, S.M. (2006). Characterization of ferric and ferrous iron transport systems in *Vibrio cholerae*. J. Bacteriol. *188*, 6515–6523.

Xu, Q., Dziejman, M. and Mekalanos, J.J. (2003). Determination of the transcriptome of *Vibrio cholerae* during intraintestinal growth and midexponential phase in vitro. Proc. Natl. Acad. Sci. USA *100*, 1286–1291.

Yamashiro, T. and Iwanaga, M. (1996). Purification and characterization of a pilus of a *Vibrio cholerae* strain: A possible colonization factor. Infect. Immun. *64*, 5233–5238.

Yildiz, F.H., Liu, X.S., Heydorn, A. and Schoolnik, G.K. (2004). Molecular analysis of rugosity in a *Vibrio cholerae* O1 El Tor phase variant. Mol. Microbiol. *53*, 497–515.

Yu, R.R. and DiRita, V.J. (2002). Regulation of gene expression in *Vibrio cholerae* by ToxT involves both antirepression and RNA polymerase stimulation. Mol. Microbiol. *43*, 119–134.

Zhu, J., Miller, M.B., Vance, R.E., Dziejman, M., Bassler, B.L. and Mekalanos, J.J. (2002). Quorum-sensing regulators control virulence gene expression in *Vibrio cholerae*. Proc. Natl. Acad. Sci. USA *99*, 3129–3134.

Population Genetics of *Vibrio cholerae* 3

Rosario Morales, Gabriela Delgado and Alejandro Cravioto

Abstract

The influence of evolutionary forces on the genetic diversity of natural populations of living organisms is the subject matter of population genetics. In the case of *Vibrio cholerae*, data obtained from detailed molecular studies of large populations of these bacteria have allowed for a better understanding of the epidemiology of diseases due to their presence in humans. The species has a high genetic diversity and a complex image of its population structure. There is also evidence of linkage disequilibrium and frequent intragenic and assortative recombination events in their housekeeping genes. Horizontal transfer of genes in *V. cholerae* is higher than those reported for *Escherichia coli* and *Salmonella enterica*. In spite of the frequent horizontal gene transfer, clonal lineages of *Vibrio cholerae* might persist for decades. The best example of this is the presence and survival of epidemic and pandemic clones over long periods of time. To date, there are four major genetic lines of toxigenic *V. cholerae* O1 biotype El Tor: an Australian clone (ET 1); the US Gulf Coast clone (ET 2); the seventh pandemic clone isolated in the South-East Asia together with the O139 'Bengal' clone (ET 3); and the clone that caused cholera in Latin America in the 1990s (ET 4). There are also isolated clones that have appeared over time under special conditions, e.g. serogroup O37 that was shown to have limited epidemic potential in the 1960s. Given the close evolutionary relationship between *V. cholerae* O1 and other non-O1 virulent serotypes and the fact that virulence genes can be transferred horizontally, new pathogenic strains of *V. cholerae* could arise in the future through the modification of existing clones that have the capacity to spread rapidly, and thus cause outbreaks of disease.

Introduction

The influence of evolutionary forces on the genetic diversity of natural populations of living organisms is the subject matter of population genetics. Through the application of different techniques that determine how genes change over time in large groups of individuals of the same species it is possible to determine how much variation exists in the population, the origin of these variations, and the factors that determine their persistence and their evolutionary importance (Maynard-Smith, 1996; Eguiarte, 1999). In the case of bacteria, data obtained in detailed molecular studies from large populations of organisms has allowed for a better understanding of the epidemiology of diseases related to their presence in humans (Eguiarte, 1999). Population genetics of pathogenic organisms involves the analysis of the ecological and evolutionary processes that take place in these populations within their natural environment.

Two different approaches have been proposed to study the population genetics of bacterial species. One approach focuses on determining the scale of evolutionary change in these organisms due to environmental factors that induce genetic variation, and the impact of these changes on their structure and in the organization of their genomes. The second approach focuses on detecting the appearance of new organisms over time as a result of evolutionary divergence in a particular population with little understanding of the ecological factors that have influenced these variations (Whittam, 1996). It is important to note that population genetics is a distinct and separate discipline to that of molecular epidemiology of infectious agents. Molecular epidemiology makes use of different conceptual tools and molecular methods to identify, define and delimit the relevant units, such as species, subspecies, strains, clones or genes in an effort to determine their molecular characteristics and genetic relationships in terms of their capacity to disseminate infectious diseases (Tibayrenc, 2005). In contrast to population genetics, molecular epidemiology focuses on the emergence and impact of new strains on the incidence of disease rather than on their capacity to survive as a species in a defined ecological niche.

Vibrio cholerae strains associated with severe diarrhoea in humans belong primarily to a single serotype designated as O1. Strains belonging to this serotype can be classified into two biotypes, classical or El Tor. Over the years, *V. cholerae* strains of classical or El Tor biotypes have been associated with widespread epidemics of cholera referred to as *pandemics* because of their simultaneous presence in several parts of the world and their persistence over prolonged periods of time.

Vibrio cholerae O1, classical and El Tor

Of the seven recorded cholera pandemics since 1817, the first six pandemics reportedly started in the state of Bengal, an area shared now by India and Bangladesh. In 1883, during the fifth pandemic, development in laboratory techniques allowed the microbiological identification of *V. cholerae* as the cause of the disease. Although strains with a classical biotype were responsible for the fifth and sixth pandemic (1898–1923), and presumably for the earlier pandemics of cholera, it was later replaced by a second biotype called El Tor, which to date is the most prevalent. However, strains of the classical biotype continued to be isolated in different parts of the world even after the appearance of the El Tor strains. In 1979, six classical biotype strains were isolated in Bangladesh, followed by three more in 1980 and two in 1981. In 1982, there was a significant rise in the number of strains with the classical biotype isolated in Bangladesh (Samadi *et al.*, 1983).

Due to the re-emergence of *Vibrio cholerae* of the classical biotype, Cook *et al.* (1984) compared the plasmid profiles, the chromosomal location of the *ctxA* gene encoding cholera toxin, and the presence and integration site of the VcA1 prophage into the chromosomes of several of these strains isolated in 1982 with those from the sixth pandemic, including some isolated between 1979 and 1981. The results indicated that there was very little genetic diversity within the classical biotype strains isolated at different times. These authors concluded that the cholera outbreak in Bangladesh in 1982 was due to the re-emergence of *V. cholerae* O1 of the classical biotype which had been the cause of the sixth pandemic. This strain had probably survived in the environment without alteration and remained undetected.

In 1990, Koblavi *et al.* used ribotyping to determine the clonal diversity of *V. cholerae* O1, and characterized a collection of classical and El Tor strains isolated from different geographical regions in Asia and Africa. The results showed a discreet diversity between strains of the classical biotype, which could be grouped into four ribotypes (B1 to B4). The strains isolated in Bangladesh in 1949, and later in the 1980s, belonged to the B1 ribotype, while those isolated in Asia between 1942 and 1962 belonged to the B2 ribotype, indicating the simultaneous presence of two different clones of classical biotype strains over several decades.

In a later study, Popovic *et al.* (1993) analysed a collection of *V. cholerae* strains isolated over 60 years from different geographical regions and found that the classical biotype strains could be grouped into seven ribotypes. The differences between these findings and those of Koblavi *et al.* (1980) can be explained by the large number of strains included in the Popovic study, as well as the diversity of regions and the timing of isolation of the strains. However, these results confirmed the simultaneous presence of several widely dispersed clones. Taken together, these studies also support the hypothesis that strains of the classical biotype have not been completely replaced by those of the El Tor biotype.

Vibrio cholerae O1 El Tor biotype and the seventh pandemic

The genetic relationships and molecular epidemiology of *V. cholerae* O1 El Tor biotype strains were studied by Kaper *et al.* (1982) using restriction fragment length polymorphism (RFLP). They analysed a collection of strains from human and environmental sources isolated both in the Gulf Coast of the United States, and in different countries in Asia during the early stages of the seventh pandemic. The strains isolated in the US Gulf Coast were found to belong to a single clone and were different from those isolated in Asia. Using more elaborate laboratory techniques, these findings were later confirmed in other studies (Almeida *et al.*, 1992; Goldberg *et al.*, 1983; Shandera *et al.*, 1983; Wachsmuth *et al.*, 1991, 1996).

Koblavi *et al.* (1990) used ribotyping to determine the genetic diversity of *V. cholerae* O1 El Tor strains isolated from different geographic locations of Asia and Africa. Their results showed that the El Tor biotype has a greater genetic diversity (13 ribotypes) as compared with the classical biotype. Some of these ribotypes had a wider geographic distribution, whilst others were limited to a precise geographic region and time. Based on these results the authors concluded that the seventh cholera pandemic was not associated with a single clone that had spread worldwide but that it was the result of a series of epidemics caused by different clones of El Tor strains occurring within the same time frame.

The re-emergence of cholera in Latin America in the 1990s, and the isolation of a non-O1 *V. cholerae* later serotyped as O139, which caused epidemics of a clinical syndrome identical to cholera in South-East Asia, offered new impetus to study the diversity of pathogenic *V. cholerae*. The isolation of non-toxigenic *V. cholerae* O1 strains from patients with diarrhoea and the environment added further interest to this effort. Research conducted in the 1980s using different laboratory techniques had already provided useful data on the structure and diversity of natural populations of *Vibrio cholerae*. The use of more advanced techniques, such as multilocus enzyme electrophoresis (MLEE), offered new opportunities for the advancement of this knowledge (García *et al.*, 1996; Lan *et al.*, 1998). Variations in electrophoretic mobility of several enzymes have been found for *V. cholerae* strains which can divide the species into multiple electrophoretic types (ETs). Using MLEE, Chen *et al.* (1991) determined the genetic relationship between endemic non-toxigenic O1 strains isolated in the US Gulf Coast and toxigenic O1 strains isolated from humans and the environment before 1991. Their results indicated that even though some of the non-O1 isolates belonged to the same clone as the toxigenic *V. cholerae*, the non-toxigenic strains formed a very heterogeneous group. The results also suggested that non-toxigenic strains could be precursors or even descendants of the toxigenic clones. Although difficult to prove at the time, the results of subsequent studies have given strong support to this hypothesis (Almeida *et al.*, 1993; Desmachelier *et al.*, 1988; Goldberg *et al.*, 1993; Popovic *et al.*, 1993).

Wachsmuth *et al.* (1993) analysed clonal relationships among *V. cholerae* O1 strains isolated from different parts of the world. Their results indicated that at least four clones of *V. cholerae* O1 biotype El Tor based on their electrophoretic types (ETs) were then in existence: an Australian clone (ET 1); the US Gulf Coast clone (ET 2); the seventh pandemic clone isolated in the South-East Asia (ET 3); and the clone causing cholera in Latin America in the 1990s (ET 4). This latter clone was presumed to be a variant of the one causing the seventh pandemic. Karaolis *et al.* (1994) arrived at the same conclusion by analysing the ribotypes of strains isolated throughout the seventh pandemic. Their study showed that this pandemic evolved in two waves; at the beginning, between 1961 and 1966, the original clone spread gradually throughout Asia, with a re-emergence in 1970. The causal agent of the second wave after 1970, which spread to Asia, Africa, Europe and Latin America, was a variant of the original clone probably resulting from changes in virulence or the adaptation of the microorganism to different environmental conditions. While all the other clones evolved from the same lineage, the Australian and US Gulf Coast clones appeared to have evolved separately (Desmarchelier *et al.*, 1995; Wachsmuth *et al.*, 1993). The toxigenic strains isolated in Australia (ET1) belonged to two dominant and exclusively Australian ribotypes (Desmarchelier *et al.*, 1995).

Ribotyping has been useful in determining the geographical distribution of the *V. cholerae* strains and to differentiate or relate among strains causing different epidemic outbreaks. For example, strains involved in the Latin American epidemic (ET 4) have been grouped into ribotype 5 (Popovic *et al.*, 1993). Interestingly, some strains of the seventh pandemic (ET 3) isolated over 15 years prior to the start of the Latin American epidemic from parts of the world other than Latin America have also been shown to belong to ribotype 5. Strains with a ribotype 8 have been isolated exclusively from central African countries, whilst strains of ribotype 2 have been isolated in the US Gulf Coast only. Ribotyping results have also shown that *V. cholerae* O1 strains isolated from the environment have a significantly higher genetic diversity than those isolated from humans (Desmarchelier *et al.*, 1995).

Pulsed field gel electrophoresis (PFGE) has also been used to study populations of *V. cholerae* (Cameron *et al.*, 1994). Studies using PFGE have shown a high degree of genetic diversity among clones ET 1, ET 2, and ET 3. These studies have also demonstrated that the strains isolated at the beginning of the Latin American epidemic (ET 4) had a unique PFGE profile, while those isolated in the later part of the epidemic belonged to a different *V. cholerae* O1 clone that was not related to ET 4. These results indicated the existence of a second source of *V. cholerae* of unknown origin causing disease simultaneously in the same population (Cameron *et al.*, 1994).

Researchers working at the CDC in Atlanta, Georgia (Evins *et al.*, 1995) analysed the genetic relationship between toxigenic and non-toxigenic *V. cholerae* O1 strains isolated from different parts of the Western hemisphere after 1991 and compared them with toxigenic El Tor strains of the Ogawa serotype isolated in Mexico in 1991. The Ogawa strains had a low prevalence at the beginning of the Latin American epidemic, but rose to a prevalence of 97% by 1992 (Giono *et al.*, 1994, 1995). These strains belonged to the seventh pandemic clone. A second *V. cholerae* O1 El Tor Ogawa strain with an uncommon antibiotic resistance pattern, and with a distinct ribotype (designated 6a) and a PFGE profile different from those of the original seventh pandemic clone, appeared in the latter part of the epidemic in Latin America and replaced the original clone almost completely (Evins *et al.*, 1995). The origin of this strain has never been determined although similar isolates have been found in Romania, India and South-East Asia. The results showed that the Latin American epidemic was caused by at least two different *Vibrio cholerae* O1 El Tor clones, one of which could be found on a global level.

It is possible that these isolates were analogous to those described by Cravioto *et al.* (1994) in a study using MLEE to determine the genetic diversity and clonal relationship between *V. cholerae* O1 strains isolated from different parts of Mexico after the start of a local epidemic in 1991. They reported the predominance of the Ogawa serotype, its clonal behaviour and its genetic relationship with the El Tor Inaba serotype strain that had been isolated at the beginning of the Latin American epidemic. The two strains differed in the migration of a single enzyme, leucine aminopeptidase (LAP). It is likely that the difference in this locus was due to one or more mutations that provoked a change in the net electrostatic charge of this enzyme, which was fixed in the population and probably spread over a number of years until its emergence during the seventh pandemic. Cravioto *et al.* (1994) also found that the *V. cholerae* O1 Ogawa El Tor strain causing the epidemic in Latin America was genetically identical to the *V. cholerae* O139 Bengal strain that had appeared in South-East Asia in 1992 (Fig. 3.1).

Despite all the information generated since the emergence of epidemic *V. cholerae* strains in Latin America, a number of issues concerning the population structure, genetic diversity of the different serotypes of *V. cholerae*, their capacity to cause disease and their ability to survive in the environment remained unanswered until more recently when various groups have addressed some of these issues (Beltran *et al.*, 1999; Farfan *et al.*, 2000; Freitas *et al.*, 2002; Salles *et al.*, 1991). In 1999, Beltran *et al.* used MLEE to determine the population structure and genetic diversity of 397

Figure 3.1 Geographical distribution of Vibrio cholerae O1 El Tor strains belonging to different electrophoretic types (ET) and major ribotypes.

V. cholerae strains, which had been isolated from humans and the environment between 1932 and 1995 and which belonged to different serotypes associated with diarrhoeal disease in humans, including cholera. This group of strains included the first 143 strains that formed the reference system at the time established by Shimada and collaborators for the serological characterization of *V. cholerae* (Shimada *et al.*, 1994, Yamai *et al.*, 1997). In general, the results showed a high degree of genetic diversity among these strains. A total of 279 ETs were identified among 397 strains, with a mean genetic diversity (H) per locus of 0.436. Of the 279 ETs identified, 275 were distributed into two principal divisions (I and II) that diverged from one another at a distance of 0.7 (a difference of 11 loci) (Fig. 3.2). Division I consisted of six subdivisions (Ia–If). Subdivision **Ia** included 214 ETs belonging to 316 strains, of which 204 ETs were divided into three groups, A, B and C, that diverged from each other at a genetic distance of 0.4 (Figs. 3.2 and 3.3).

Epidemic strains of *V. cholerae* O1 and O139 clustered in Group A. These strains formed a compact conglomerate that included: the toxigenic clone that was endemic in the US Gulf Coast (ET 2); the strains from the seventh pandemic and the clone causing cholera in Latin America that was first identified in Mexico in 1991 (ET 3); the clone causing the second wave of the Latin American epidemic, also isolated in Romania, and the O139 'Bengal' clone isolated in South-East Asia (ET 3); and the original or first-wave Latin American epidemic O1 El Tor Inaba clone isolated in Mexico, Guatemala and Peru (ET 4). The genetic distance between these ETs was 0.05 (a difference of one locus). The non-O1 *V. cholerae* strains analysed in this study clustered in groups B and C.

The results showed no association between the serotype of the strains and their biotype or their geographic origin since isolates belonging to the same serotype could be found in different ETs or in divergent lineages, while strains of the same ET were found to belong to different serogroups (Figs. 3.2 and 3.3). The association index (I_A), calculated for all 397 strains, was 1.248 ± 0.082, which suggested that the population studied had a clonal structure. Nevertheless, the I_A index for the individual groups (e.g. A, B, C) did not show significant levels of non-random

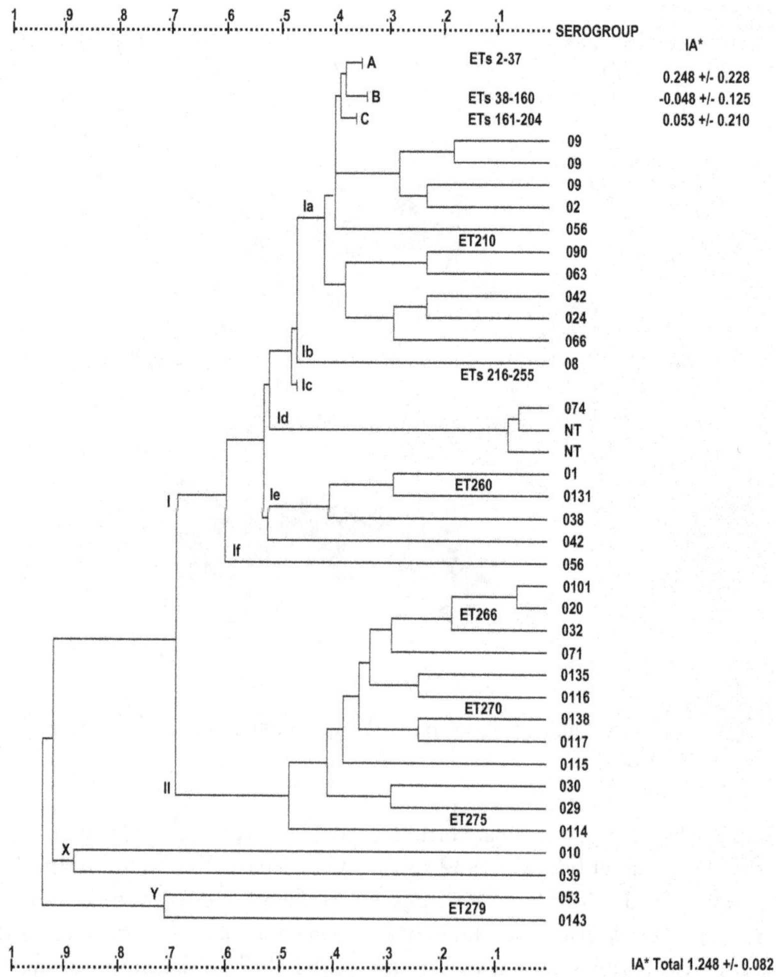

Figure 3.2 Dendrogram showing the genetic relationship among 279 electrophoretic types (ET) of V. cholerae. The majority of the strains are of ET in division I. The subdivisions Ia (groups A, B, C) and Ic are truncated. This figure also shows the association index (I_A) of the total population and of each group A, B and C (*$P \leq 0.05$).

association (Fig. 3.3). These findings indicated a high rate of horizontal transfer and recombination in the housekeeping genes of the strains that formed sub-division **Ia**, and that these events did not allow the population to develop and maintain distinctive alleles for long periods of time. It seemed likely that the frequency of intragenic and assortative recombination in the housekeeping genes in *V. cholerae* in general was higher than those found in other bacteria (Selander *et al.*, 1996, Whittam *et al.*, 1996). Through comparative analysis of the aspartate-semialdehyde dehydrogenase gene (*asd*) sequences of different *V. cholerae*, Karaolis *et al.* (1999) arrived at the same conclusion. Byun *et al.* (1999) used a comparative analysis of sequences of four basic metabolic genes to give further support to these findings.

The composition of subdivision **Ia** indicates that these clonal lineages can persist for decades. The more representative examples of these findings are those related to the similarities between pandemic strains belonging to the O1 and O139 serogroups. The results of Beltran *et al.* (1991) have also helped to identify the presence of different clones and clonal lineages of non-O1 *V. cholerae* strains with a wide geographic distribution. Some of them were found on a global scale and have been implicated as the cause of severe cholera-like disease in humans living in different parts of the world (Fig. 3.1).

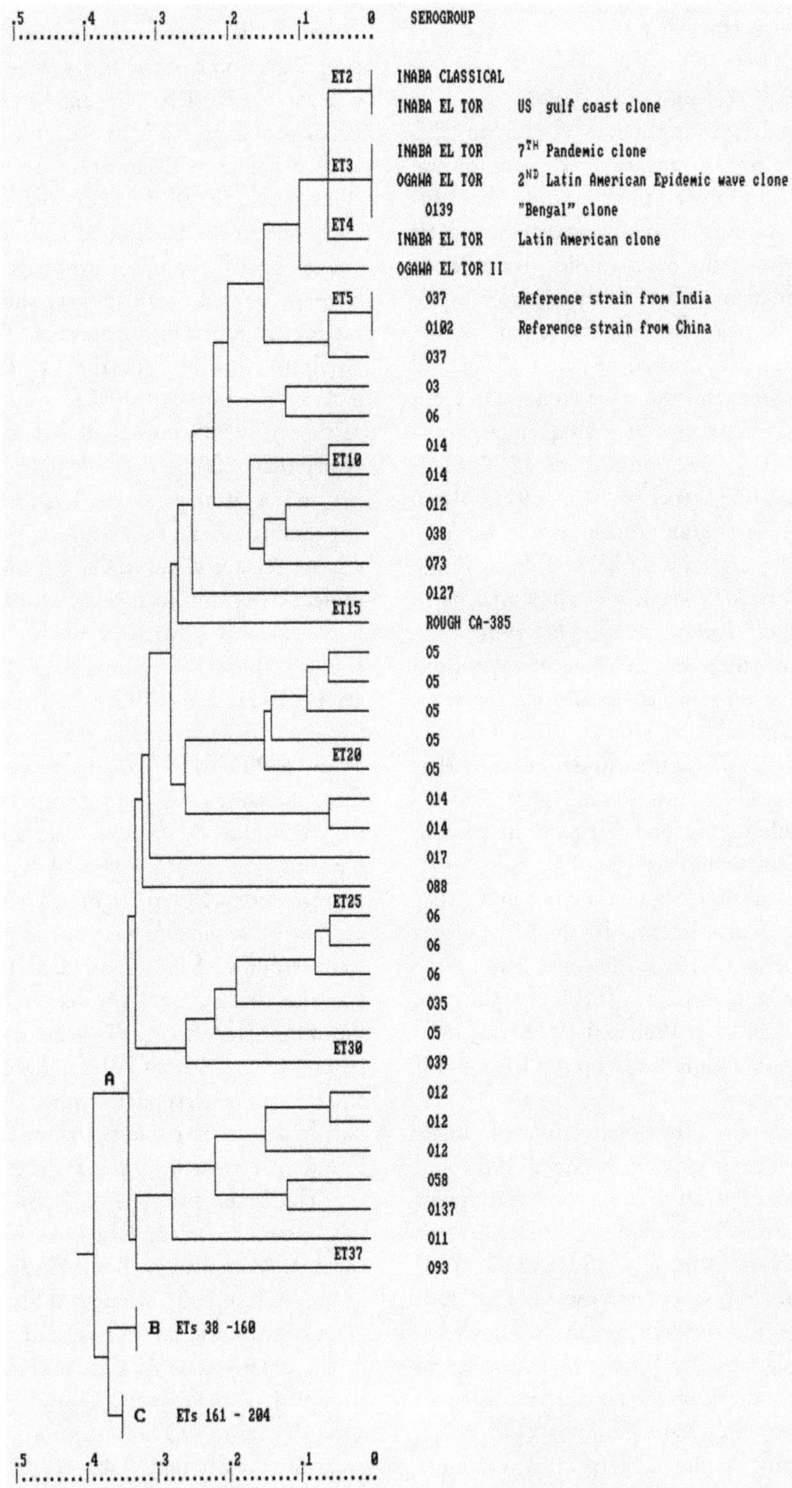

Figure 3.3 Dendrogram showing the genetic relationship among epidemic/pandemic clones, all being found in-group A. This is a tight group formed by 37 electrophoretic types. Groups B and C are truncated.

Lateral gene transfer

Conversion from Ogawa to Inaba

Chromosomal rearrangements found in different *V. cholerae* strains isolated from several geographical areas over time indicate that this microorganism has a highly variable genetic structure. One of the most studied examples of these arrangements relates to the changes in the serotype of *V. cholerae* O1 El Tor strains. While the conversion of *V. cholerae* from Ogawa to Inaba is not frequent, the inter-conversion from Inaba to Ogawa occurs at a frequency of approximately 10^{-5} (Bhaskaran *et al.*, 1957; Gangarosa *et al.*, 1967; Stroeher *et al.*, 1992; Colwell *et al.*, 1995). The immune response in humans against these antigens is probably a determining factor in the seroconversion of these strains *in vivo* (Sack *et al.*, 1969; Sheehy *et al.*, 1966). The *wbeT* gene which is required for the expression of the Ogawa serotype is responsible for serotype conversion in these strains and is normally associated with a single nucleotide mutation that changes the open reading frame (ORF) and inserts a pair of stop codons that truncate protein expression. The Inaba serotype of *V. cholerae* O1, predominant until 1989 in Calcutta and other parts of India, was displaced by the surge of the O139 serogroup in 1992. However, a toxigenic *V. cholerae* O1 biotype El Tor Inaba serotype reappeared in India in 1998 and 1999, almost 10 years after it had almost disappeared from this country.

The molecular characterization of Inaba and Ogawa strains isolated between 1989 and 1999 in different places in India showed different ribotype patterns. However, two Inaba strains isolated in New Delhi (DO182 and DO183) and two in Sewagram (SO86 and SO90) during 1998 and 1999 respectively, showed a ribotype designated RIII, which was similar to the ribotype pattern of the prevalent *V. cholerae* O1 Ogawa strain CO840 (Sharma *et al.*, 1997). The genotyping of the CTX prophage using a probe for *ctxA*, showed the presence of a single copy of the CTX prophage in all Inaba strains. However, hybridization with a probe for the RS1 element (repeated sequence, later shown to be the genome of a satellite phage), which is normally associated with the CTX prophage revealed different organizations of the CTX prophage. Some of the Inaba strains, had at least one copy of the RS1 element on each side of the CTX prophage, while in others it was organized in two copies with an arrangement in tandem (Garg *et al.*, 2000). Furthermore, strain DO183 and another strain isolated in Sewagram had a single copy of the CTX prophage with an RS1 element located upstream very similar to the arrangement found in the prevalent Ogawa strains. Analysis by PFGE of the O1 Ogawa strains isolated during 1998 showed an identical profile to that of Inaba strains isolated some years later (Garg *et al.*, 2000).

In an attempt to understand the sudden appearance of the Inaba serotype in India after a lapse of several years, Garg *et al.*, (2000) used molecular typing to investigate the relationship between local *V. cholerae* El Tor isolates. They showed that Inaba strain SO90, and Ogawa strains PG117 and PG81 had the same PFGE pattern. These strains as well as another Inaba strain SO84 were thought to have originated from the same clone. By contrast, strains CO840 (Ogawa) and V2 (Inaba) had different PFGE patterns. These data suggested that Inaba strains represented by SO90 originated from the prevailing Ogawa strain and were not a re-emergence of Inaba strain such as V2, that had been previously prevalent in that area. The occurrence of random mutations in the *wbeT* gene confirmed that strains SO84 and SO90 belonged to the same clone that was responsible for an epidemic that had lasted at least two years and that had been found in at least two separate locations.

The Inaba serotype of *V. cholerae* was also the cause of cholera outbreaks in 1996 in the town of Warangal in South India (Saha *et al.*, 1996), in July 2004 in north India (Taneja *et al.*, 2005), and between August and October 2005 in the coastal state of Orissa (Pal *et al.*, 2006). However, *V. cholerae* O1 Ogawa biotype El Tor and *V. cholerae* O139 continue to be mostly associated with outbreaks of cholera in India (Niyogi *et al.*, 1994; Pal *et al.*, 1995; Chhotray *et al.*, 2002).

Vibrio cholerae O139 serogroup

In late 1992, a strain of *V. cholerae* belonging to serogroup O139 (synonym Bengal) emerged as the causative agent of an epidemic outbreak of cholera in India and Bangladesh. This strain had a clonal relationship with *V. cholerae* O1 biotype El Tor (Cholera Working Group, 1993; Berche *et al.*, 1994; Bhadra *et al.*, 1994; Calia *et al.*, 1994; Bhadra *et al.*, 1995). The O139 Bengal strain was the result of an insertion of a ~35-kb DNA fragment encoding the O139-specific antigen into the *rfb* locus of an El Tor strain replacing the majority of the genes coding for the O1 antigen (Bik *et al.*, 1995; Comstock *et al.*, 1996). Population studies of *V. cholerae* based on MLEE have indicated that the O139 strains isolated during the first period of the epidemic (1992 to 1993) belonged to a single clone (Beltran *et al.*, 1990; Popovic *et al.*, 1995). However, the application of RFLP, PFGE, DNA sequencing, and AFLP techniques in later studies of the O139 strains have indicated that they possess a higher degree of clonal diversity, and belong to different ribotypes (Karaolis *et al.*, 1995; Faruque *et al.*, 1997; Byun *et al.*, 1999; Jiang *et al.*, 2000). More recent comparative studies of O139 strains isolated at different times during the epidemic have shown that although the re-emerging strains of *V. cholerae* O139 (from 1996 to 1997) had biochemical characteristics identical to those of strains isolated at the beginning of the epidemic, some of their molecular characteristics, e.g. the number of copies of CTX elements and ribotypes were different (Faruque *et al.*, 1997; Mukhopadhyay *et al.*, 1998, Basu *et al.*, 2000, Faruque *et al.*, 2000).

In 2002, Farfan *et al.* carried out a study to determine the clonal lineages of a set of 29 *V. cholerae* O139 strains isolated at the beginning of the epidemic (1992–1993). They sequenced an internal DNA fragment containing the information for six housekeeping genes (*asd, cadA, epd, idh-II* and *mdh* and *lap*) in these strains. A toxigenic O1 El Tor strain from the seventh pandemic and an environmental non-O1/non-O139 isolate were used for comparison. The genetic diversity of the *V. cholerae* strains used in this study was higher than that reported by Beltran *et al.* (1990) and Popovic *et al.* (1995) using MLEE. However, there was a lower genetic diversity among the O139 isolates, which clustered as a single clone in a single Et all the loci studied by Farfan *et al.* were polymorphic, demonstrating the existence of at least three distinct clones among the *V. cholerae* O139 isolates. The majority of the O139 strains analysed showed a close genetic relationship with the *V. cholerae* O1 El Tor strain used for comparison in the study. The observation that the two main groups of strains (ST1 and ST4) exhibited differences in the allelic profile of all six genes analysed suggested that *V. cholerae* O139 emerged independently from different progenitors that had acquired the ability to cause severe diarrhoea in humans. These data were consistent with the notion of a clonal population structure of the O139 strains. The strong linkage disequilibrium between these strains and the O1 strains could be explained by the fact that bacteria like *V. cholerae* constitute a metapopulation integrated by multiple ecological populations, which occupy different ecological niches, and have gone through different recombination events (Rius *et al.*, 2001). This process is likely to happen within the same population but rare or absent among different populations (Young, 1989; Maynard-Smith *et al.*, 1991; Reeves, 1992).

The allelic diversity in the O139 serogroup was further studied in 2003 by Garg *et al.*, who found putative recombinant alleles in all nine loci examined (*dnaE, lap, recA, pgm, gyrB, cat, chi, rstR*, and *gmd-*) in 96 *V. cholerae* O139 strains isolated between 1992 and 2000 from patients treated at the Infectious Diseases Hospital in Calcutta, India. On the basis of a simple computation of the accumulation of independent mutations, the number of base-pair differences among alleles was higher than expected (Garg *et al.*, 2003). These findings suggested that many of these events were due to lateral gene transfer (LGT). In *Vibrio cholerae,* as in *Neisseria* spp., *Streptococcus* spp., and other species, specific recombination of homologous genes appears to be common, and responsible for most of the alleles with multiple nucleotide differences (Maiden *et al.*, 1998; Enright *et al.*, 2000; Feil *et al.*, 2000; Dingle *et al.*, 2001). The genetic variability found in these studies helps to understand the evolution in these bacteria showing that most nucleotide changes occur by recombination that can alter any gene, and that *V. cholerae* O139 serogroup comprises strains with multiple epidemic lineages (Garg *et al.*, 2003).

The substantial genetic diversity among *V. cholerae* O139 isolates found in the first year of the epidemic may reflect a 'founder flush' phenomenon (Wallace *et al.*, 1981). During times of population expansion, i.e. a *flush*, any novel genotype with similar or even slightly deleterious fitness compared with the founder genotype will produce sufficient offspring to become established in the population. Although previous descriptions of this phenomenon were limited to insects, specifically butterflies (Ford *et al.*, 1930) and drosophilids (Carson *et al.*, 1968), the founder flush phenomenon may become the paradigm for epidemic bacterial expansion in individual patients and populations. This founder flush phenomenon, in turn, has implications for the interpretation of 'clonality' among epidemic isolates and for the understanding of factors that contribute to the emergence of new pathogenic strains.

Vibrio cholerae O37 serogroup

In 1968 a large outbreak of cholera occurred in Sudan caused by a *V. cholerae* strain belonging to the O37 serogroup. In their population studies, Beltran *et al.* (1999) included a strain of this serotype isolated from a patient in India in 1969, which probably belonged to the same clone of the O37 strain from Sudan. The O37 strain was closely related to *V. cholerae* O1 El Tor and to other epidemic strains (Fig. 3.3). This observation had been previously reported by other researchers (Bik *et al.*, 1996; Karaolis *et al.*, 1998) who analysed the O37 strain responsible for the cholera-like outbreak in Sudan using 'digital fingerprinting' with IS1004. Karaolis *et al.* determined that the O37 Sudan strain hosted a pathogenicity island in its chromosome, which was present in epidemic strains only. It is highly likely that this *V. cholerae* O37 strain represented a similar phenomenon to that of the *V. cholerae* O139 clone, where a new O serotype was established through the horizontal transfer and re-arrangements of the genes in the *rfb* region in a *V. cholerae* O1 clone. A strain used as reference for the *V. cholerae* O102 serotype isolated in 1988 from a patient with diarrhoea in China could be another example of these genetic events, since this O102 strain had an identical MLEE genotype to that of the O37 reference strain (Fig. 3.3).

Evolutionary genetic relationship between *Vibrio cholerae* strains

There has been a long-standing interest in the relationship between the *V. cholerae* strains isolated during the sixth and seventh pandemics. The biochemical characteristics of the classical and El Tor biotypes are different. While the El Tor strains are able to produce hemolysin, give a positive Voges–Proskauer reaction and produce haemagglutination, the classical biotype strains do not (Roy *et al.*, 1963; Finkelstein *et al.*, 1963; Feeley, 1965; Gallut, 1968, 1971; Barett *et al.*, 1981).

The study of the ancestry of the classical and El Tor biotypes of *V. cholerae* using different molecular methods has produced conflicting results (Chen *et al.*, 1991, Evins *et al.*, 1995; Karaolis *et al.*, 1995; Beltran *et al.*, 1999; Byun *et al.*, 1999; Boyd *et al.*, 2000; Farfan *et al.*, 2000; Stine *et al.*, 2000; Karaolis *et al.*, 2001; Li *et al.*, 2003; Kotetishvili *et al.*, 2003). Comparative nucleotide sequence analysis and multilocus virulence gene profile studies have been carried out to determine the evolutionary genetic relationship of these strains based on the analysis of two housekeeping genes, malate dehydrogenase (*mdh*) and a chaperonin (*groEL*) (O'Shea *et al.*, 2004). The phylogenetic distribution of 12 other regions associated with the virulence of these strains has also been determined. Taken together, the *mdh* sequence analysis and PCR-SSCP analyses at the *mdh* and *groEl* loci indicate that the *Vibrio cholerae* O1 classical and El Tor biotypes, as well as strains belonging to the O139 and O37 serogroups, form a highly homologous group representing a single clonal lineage that emerged during the sixth and seventh cholera pandemics by the successive acquisition of virulence regions. Interestingly, one study showed that several *Vibrio cholerae* non-O1/non-O139 strains could also be grouped in this epidemic cluster (O'Shea *et al.*, 2004). Genotypic and phenotypic analysis of two pre-seventh pandemic El Tor strains, RV79 and C5, isolated in Indonesia in 1937 and 1957 respectively, have shown that the RV79 strain is identical to other *V. cholerae* O1 strains at the *mdh* and *groEL* loci but that it lacks 3 of the 12 virulence regions, VSP-I, VSP-II, and VPI-2, present in subsequent isolates.

The C5 El Tor strain, similar to RV79, is identical to other *V. cholerae* O1 strains isolated 20 years later and lacks only one of the 12 virulence regions, RS1 (O'Shea *et al.*, 2004). Consistent with previous studies, O'Shea *et al.*, (2004) found that all classical strains examined lacked RS1, VSP-I and VSP-II and that these three regions were all present in the El Tor strains (Davis *et al.*, 2000, Dziejman *et al.*, 2002). In addition, these authors found that VSP-I was present in the O37 Sudan strain and that VSP-II was confined to strains isolated during the seventh pandemic. The authors concluded that a *V. cholerae* O1 strain initially acquired the TCP pathogenicity island, alternatively referred to as VPI, which encodes for TCP, the receptor for CTX phage (CTXϕ). This proposal was based on the close similarity between the sequence of the TCP island region in classical and El Tor strains (Karaolis *et al.*, 2001). The hypervariability found in the *tcpA* gene was likely to be the result of positive Darwinian selection in this region (Boyd *et al.*, 2002). A second pathogenicity island known as VPI-2, is found predominantly among O1 and in some O139 epidemic *V. cholerae* isolates. VPI-2, which contain genes involved in a restriction modification system and *N*-acetyl neuraminic acid utilization (Jermyn *et al.*, 2002), was most likely already present in an O1 strain that gave rise to the classical and El Tor biotypes.

Following acquisition of the TCP island and VPI-2 by an O1 progenitor strain, classical and El Tor isolates emerged and diverged from one another through the acquisition of VSP-I, VSP-II and RS1. Studies based on comparative nucleotide sequence analysis of the CTXϕ genes indicate that this region was acquired independently in classical and El Tor isolates (Boyd *et al.*, 2000). *V. cholerae* O1 El Tor acquired at least three genomic regions in addition to CTXϕ. These include RS1, which facilitates CTXϕ production, and VSP-I and VSP-II, whose participation in the virulence of *V. cholerae* remains unknown (Dziejman *et al.*, 2002). The *V. cholerae* O139 strains that emerged in 1992 were derived by O-antigen switching in an El Tor progenitor, likely facilitated by bacteriophages. The strain further evolved through the acquisition of a novel CTXϕ and the SXT constin encoding resistance to streptomycin, sulphamethoxazole and trimethoprim (Bik *et al.*, 1996; Davis *et al.*, 1999; Waldor *et al.*, 1997).

Genetic analyses of non-O1/non-O139 *Vibrio cholerae*

Genetic variability has also been found in the non-O1/non-O139 *V. cholerae* populations present in aquatic environments. *V. cholerae* is a natural inhabitant of these environments (Chan *et al.*, 1969; Colwell *et al.*, 1981; Feachem *et al.*, 1981; Ghosh *et al.*, 1994; Islam *et al.*, 1994; Huq *et al.*, 1996; Colwell *et al.*, 1997). The non-O1 strains not only survive in the water of rivers, estuaries and coastal areas around the world but also live in association with crustacean copepods and aquatic plants, either in a culturable state or in a viable but non-culturable state (Nalin, 1976; Sochard *et al.*, 1979; Grimes *et al.*, 1981; Miller *et al.*, 1985; Grimes *et al.*, 1986; Huq *et al.*, 1990, 1995; Islam *et al.*, 1990; Colwell *et al.*, 1994; Shukla *et al.*, 1995). These strains, however, continue to be considered of negligible epidemiological importance, since they have only been associated with illness in a low percentage of patients hospitalized with watery or bloody diarrhoea (Blake *et al.*, 1980; Sanyal, 1986; Mukhopadhyay *et al.*, 1996).

In the study carried out by Beltran *et al.* (1999), non-O1 *Vibrio cholerae* strains from the environment did not form a specific cluster and its evolutionary lineage was distributed along the length of the dendrograms. Some of the strains had the same electrophoretic types (ET) as those of *V. cholerae* non-O1 strains isolated from humans (Figs. 3.2 and 3.3). In 2000, Jiang *et al.*, characterized the temporal and spatial genetic diversity of 67 *V. cholerae* strains isolated between April and July 1998 from four different sampling sites in the Chesapeake Bay. The abundance of bacteria was strongly correlated with the temperature of the water at all four sites. The strains were fingerprinted by AFLP, which allowed the isolates to be distributed into three large clusters related to the date of isolation. Strains found in April and May were genetically similar, while those isolated in July were distantly related. The genetic diversity of the strains did not correlate with the sampling location. *V. cholerae* was shown to be a dynamic and genetically diverse species in the Chesapeake Bay. Environmental conditions, such as temperature, salinity, and/or interaction with other microorganisms in the water column affected the presence, abundance and genotypic changes of these strains.

The majority of strains isolated from the environment do not produce cholera toxin (CT) and lacks the *ctxAB* genes as well as those encoding zonula occludens toxin (Zot) and accessory cholera enterotoxin (Ace) (Kurazono *et al.*, 1995; Sharma *et al.*, 1998). In addition, most environmental *Vibrios* lack the toxin-co-regulated pilus (TCP), that is recognized as the major colonization factor in the pathogenicity of *V. cholerae.* Karaolis *et al.*, (1995) suggested that *V. cholerae* O1 isolates from the sixth and seventh pandemics and the US Gulf Coast were probably derived from non-toxigenic strains. In 2001, Singh *et al.*, compared the genomic organization of 26 *V. cholerae* strains, including O1, O139, and non-O1/non-O139 isolates, cultured from both clinical and environmental sources in different parts of India, to determine whether virulence and regulatory genes were present in all these isolates. Their results showed that, regardless of the serotype, all strains examined were positive by PCR for *hlyA*, the haemolysin gene, but that the majority of the non-O1/non-O139 strains had no secretory response in the initial assays. The environmental O1 and O139 isolates, like the clinical ones, were positive for *ctxA, zot, ace, tcpA*, and *ompU* indicating that the genes comprising the virulence cassette and those encoding for surface organelles required for intestinal adherence and colonization were retained by pathogenic *V. cholerae* strains even in the aquatic environment.

In a separate study four different genetic fingerprinting methods were used to measure the relationship between *V. cholerae* strains isolated from both clinical and environmental sources. El Tor and O139 strains yielded fingerprint patterns that were closely related but were different from those of the non-O1/non-O139 strains. The most interesting finding of this analysis was that two environmental isolates belonging to non-O1/non-O139 serogroups were also closely related to the toxigenic *V. cholerae* O1 and O139 strains.

Collectively, these data indicate that strains of *V. cholerae* that cause diarrhoea have a clonal origin in the aquatic environment. Using AFLP to uncover the genomic diversity of clinical and environmental isolates of *V. cholerae* O1, and non-O1/non-O139 obtained during an 11-year sampling period in Brazil, Thompson *et al.* (2003) found that the strains could be grouped into seven main clusters of genomes, that did not correlate with a particular date or place of isolation. The high genome plasticity of O1 and non-O1 *V. cholerae,* as revealed by this method, concurred with results found in previous studies (Farfan *et al.*, 2000, 2002). Furthermore, this study showed the genetic diversity of non-O1/non-O139 isolates found during the same period of the cholera epidemic in Brazil.

Comparative genomic analyses of sixth and seventh pandemic strains

The sequencing of two representative *V. cholerae* strains from the sixth and seventh pandemic O395 and N16961 respectively has identified differences in the numeric nucleotide position along the TCP pathogenicity island region of each strain (Karaolis *et al.*, 2001). In N16961, the TCP island was located at a position 41,272 bp, while in strain O395 the island was located at 41, 290 bp (excluding the *att* sites). The G+C composition of the TCP island in both these strains was found to be 35.5%, which is very different from that of the rest of the bacterial chromosome, and was around 47% (Karaolis *et al.*, 2001). In addition to various nucleotide and amino acid polymorphisms, there are several proteins whose predicted size differs significantly between the two strains, probably as a result of frame shift mutations. Data from these studies indicated that even if the two pandemic clones had emerged from a common ancestor, there had been important changes in the TCP island during the divergence of the two clones.

The complete genomic sequence of the seventh pandemic El Tor strain N16961 has provided an important tool to begin addressing questions about the evolution of *V. cholerae* as a human pathogen and as an environmental organism. To investigate the genetic similarity among strains of *V. cholerae* isolated from diverse geographical regions and periods of time, Dziejman *et al.*, (2002) constructed a *V. cholerae* genomic microarray based on the sequence of strain N16961. This array has been used to compare the gene content of classical, pre-pandemic El Tor, pandemic El Tor, and two non-toxigenic strains isolated over the past century. Although these comparative genomic analyses have revealed a high degree of genetic similarity between these strains, it has also helped to identify genes unique to each pandemic, as well as those present in El Tor and related O139 strains that are specific to the seventh pandemic. The presence or absence of specific genes has been identified and grouped into four main categories:

1. Genes that can differentiate classical from El Tor strains. These genes are absent only in the former, since the array was constructed using the N16961 strain (a seventh pandemic isolate), that would not have identified genes that were present only in the classical strains but absent in all other strains.
2. Genes that are present only in pandemic strains including classical isolates (group II);
3. Genes specific to the seventh pandemic, including epidemic and endemic El Tor isolates and the closely related O139 strain MO10 (group III).
4. Genes that are absent only in individual strains.

The close relatedness of the test strains to N16961 analysed by Dziejman *et al.* (2002) suggests that only a small group of *V. cholerae* strains are capable of evolving into human pathogens through the acquisition of the TCP pathogenicity island and the CTXϕ genome.

Conclusions

Various comparative studies that have been carried out may help us to understand the possible origin of new toxigenic clones of *V. cholerae*, and raise the possibility that all strains of *V. cholerae* have the potential to become agents of epidemic cholera (Faruque *et al.*, 1998). The frequent emergence of new strains of toxigenic *V. cholerae* and their selective enrichment during cholera outbreaks constitute an essential mechanism for the survival and evolution of this species and the genetic elements that mediate the transfer of virulence genes (Faruque *et al.*, 1998).

Given the genetic relationship between the pathogenic O1 strains and other non-O1 isolates, and the fact that genes that code for virulence factors can be transferred horizontally, it is possible that any *V. cholerae* can be 'transformed' into a potentially pathogenic strain, and even more into a potential epidemic strain (Mekalanos *et al.*, 1997; Beltran *et al.*, 1999). The O1 serotype is not indicative anymore of the capacity of the *V. cholerae* strains to cause disease since, antigenic conversion can take place by acquisition and/or loss of the serotype specific genes (Colwell *et al.*, 1995; Stroeher *et al.*, 1995).

Acknowledgements

The authors would like to thank José Luis Mendez for helping with the figures used in this paper, and Ian Shepherd for his critical review of the manuscript.

References

Aldova, E., Laznickova, K., Stepankova, E. and Lietava, J. (1968). Isolation of nonagglutinable vibrios from an enteritis outbreak in Czechoslovakia. J. Infect. Dis. *118*, 25–31.

Almeida, R.J., Cameron, D.N., Cook, W.L. and Wachsmuth, I.K. (1992). Vibriophage VcA-3 as an epidemic strain marker for the US Gulf coast *Vibrio cholerae* O1 clone. J. Clin. Microbiol. *30*, 300–304.

Barett, J. and Blake, P.A. (1981). Epidemiological usefulness of changes in hemolytic activity of *Vibrio cholerae* biotype El Tor during the seventh pandemic. J. Clin Microbiol. *13*, 126–128.

Basu, A., Garg, P., Datta, S., Chakraborty, S., Bhattacharya, T., Khan, A., Ramamurthy, T., Bhattacharya, S.K., Yamasaki, S., Takeda, Y. and Nair, G.B. (2000). *Vibrio cholerae* O139 in Calcuta, 1992–1998: evidence, antibiograms, and genotypes. Emerg. Infect. *6*, 139–147.

Beltran, P., Delgado, G., Navarro, A., Trujillo, F., Selander, R.K. and Cravioto, A. (1999). Genetic diversity and population structure of *Vibrio cholerae*. J. Clin. Microbiol. *37*, 581–590.

Berche, P., Poyart, C., Abachin, E., Lelievre, H., Vandepitte, J., Dodin, A. and Fournier, J.M. (1994). The novel epidemic strain O139 is closely related to the pandemic strain O1 of *Vibrio cholerae*. J. Infect. Dis. *170*, 701–704.

Bhadra, R.K., Roychoudhury, S., Banerjee, R.K., Kar, S., Majumdar, R., Sengupta, S., Chatterjee, S., Khetawat,

G. and Das, J. (1995). Cholera toxin CTX genetic element in *Vibrio cholerae* O139. Microbiology *141*, 1977–1983.

Bhadra, R.K., Roychowdhury, S. and Das, J. (1994). *Vibrio cholerae* O139 El Tor biotype. Lancet *343*, 728.

Bhaskaran, K. and Gorrill, R.H. (1957). A study of antigenic variaton in *Vibrio cholerae*. J. Gen. Microbiol. *16*, 721–729.

Bik E., Gouw, R.D. and Mooi, F.R. (1996). DNA fingerprinting of *Vibrio cholerae* strains with novel insertion sequence element: a tool to identify epidemic strains. J. Clin. Microbiol. *34*, 1453–1461.

Bik, E.M., Bunschoten, A.E., Grow, R.D. and Mooi, F.R. (1995). Genesis of the novel epidemic *Vibrio cholerae* O139 strain: evidence for horizontal transfer of genes involved in polysaccharide syntesis. EMBO J. *14*, 209–216.

Blake, P.A., Weaver, R.W. and Hollins D.G. (1980). Diseases of humans (other than cholera) caused by vibrios. Annu. Rev. Microbiol. *34*, 341–367.

Boyd, E.F. and Waldor, M.K. (2002). Evolutionary and functional analyses of variants of the toxin-corregulared pilus protein TcpA from toxigenic *Vibrio cholerae* non-O1/non-O139 serogroup isolates. Microbiology *148*, 1655–1666.

Boyd, E.F., Helipern, A.J. and Waldor, M.K. (2000). Molecular analyses of a putative CTXϕ precursor and evidence for independent acquisition of distinct CTXϕs by toxigenic *Vibrio cholerae*. J. Bacteriol. *182*, 5530–5538.

Boyd, E.F., Moyer, K.L., Shi, L. and Waldor, M.K. (2000). Infectious CTXϕ and the vibrio pathogenicity island prophage in *Vibrio mimicus*: evidence for recent horizontal transfer between *V. mimicus* and V. *cholerae*. Infect. Immun. *68*, 1507–1513.

Brazil, J.M., Alves, R.M., Rivera, I.N.G., Rodrigues, D.P., Karaolis, D.K.R. and Campos, L.C. (2002). Prevalence of virulence-associated genes in clinical and environmental *Vibrio cholerae* strains isolated in Brazil between 1991 and 1999. FEMS Microbiol. Lett. *215*, 15–21.

Byun, R., Elbourne, L.D.H., Lan, R. and Reeves, P.R. (1999). Evolutionary relationships of pathogenic clones of *Vibrio cholerae* by sequence analysis of four housekeeping genes. Infect. Immun. *67*, 1116–1124.

Calia, K.E., Muratagh, M., Ferrano, M.J. and Calderwood, S.B. (1994). Comparison of *Vibrio cholerae* O139 with *Vibrio cholerae* O1 classical and El Tor biotypes. Infect. Immun. *62*, 1504–1506.

Cameron, D., Khambaty, F.M., Wachsmuth, K., Tauxe, R.V. and Barret, T.J. (1994). Molecular characterization of *Vibrio cholerae* O1 strains by pulsed field gel electrophoresis. J. Clin. Microbiol. *32*, 1685–1690.

Carson, H.L. (1968). The population flush and its genetic consequences. In: Lewontin, R.C. ed. Population biology and evolution. Syracuse (NY): Syracuse University Press; p.123–137.

Center for Disease Control. (1991). Cholera outbreak-Peru, Ecuador and Colombia. MMWR. *40*, 225–227.

Center for Disease Control. (1992). Cholera-Western Hemisphere. MMWR *41*, 667–668.

Chakraborty, S., Mukhopadhyay, A.K., Bhadra, R.K., Ghosh, A.N., Mitra, R., Shimada, T., Yamasaki, S., Faruque, S.M., Takeda, Y., Colwell, R.R. and Nair, G.B. (2000). Virulence genes in environmental strains of *Vibrio cholerae*. Appl. Environ. Microbiol. *66*, 4022–4028.

Chan, E.G.S. and McManus, E.A. (1969). Distribution, characteritation, and nutrition of marine microorganisms from ten algae, *Polysiphonium ianosa*, and *Ascophylum nodosum*. Can. J. Microbiol. *15*, 409–420.

Chen, F., Evins, G.M., Cook, W.L., Almeida, R., Hargrett-Bean, N. and Wachsmuth, K. (1991). Genetic diversity among toxigenic and non toxigenic *Vibrio cholerae* O1 isolated from the Western hemisphere. Epidemiol. Infect. *107*, 225–33.

Cholera Working Group. (1993). Large epidemic of cholera-like disease in Bangladesh caused by *Vibrio cholera* O139 synonym Bengal. Lancet *342*, 387–390.

Chotray, G.P., Pal, B.B., Khuntia, H.K., Chowdhury, N.R., Chakraborty, S., Yamasaki, S., Ramamurthy, T., Takeda, Y., Bhattacharya, S.K. and Nair, G.B. (2002). Incidence and molecular analysis of *Vibrio cholera* associated with cholera outbreak subsequent to the supercyclone in Orissa, India. Epidemiol. Infect. *128*, 131–138.

Colwell, R.R. and Huq, A. (1994). Vibrios in the environment: viable but non-culturable *Vibrio* cholerae. In *Vibrio cholerae* and cholera-molecular to global perspectives, I.K. Wachsmuth, ø. Olsvik, P.A. Blake, and R. Olsvik ed. (Washinton, DC., USA, American Society for Microbiology) pp.117–133.

Colwell, R.R., Huq, A., Chowdhury, M.A.R., Brayton, P.R. and Xu, B. (1995). Serogroup conversion of *Vibrio cholerae*. Can J. Microbiol. *41*, 946–950.

Colwell, R.R., Kaper, J.B. and Joseph, S.W. (1997). *Vibrio cholerae, Vibrio parahaemolyticus*, and other vibrios: occurrence and distribution in Chesapeake Bay. Science. *198*, 394–396.

Colwell, R.R., Seidler, R.J., Kaper, J., Joseph, S.W., Garyes, S., Lockman, H., Maneval, D., Bradford, H., Roberts, N., Huq, I. and Huq, A. (1981). Occurrence of *Vibrio cholerae* serotype O1 in Maryland, and Louisiana estuaries. Appl. Environ. Microbiol. *41*, 555–558.

Comstock, L.E., Johnson, J.A., Michalski, J.M., Morris Jr., J.G. and Kaper, J.B. (1996). Cloning and sequence of a region encoding a surface polysaccharide of a *Vibrio cholerae* O139 and characterization of the insertion site in the chromosome of *Vibrio cholerae* O1. Mol. Microbiol. *19*, 815–826.

Cook, W.L., Wachsmuth, K., Johnson, S.R., Birkness, K.A. and Samadi, A.R. (1984). Persistence of plasmids, cholera toxin genes, and prophage DNA in classical *Vibrio cholerae* O1. Infect. Immun. *45*, 222–226.

Cravioto, A., Beltran, P., Delgado, G., Navarro, A., Eslava, C., León, S. and González, A.R. (1994). Non-O1 *Vibrio cholerae* O139 'Bengal' is genetically related to *Vibrio cholerae* O1 El Tor Ogawa isolated in Mexico. J. Infect. Dis. *169*, 1412–1413.

Davis, B.M., Kimsey, H.H., Chang, W. and Waldor, M.K. (1999). The *Vibrio cholerae* O139 Calcutta CTXϕ is infectious and encodes a novel repressor. J. Bacteriol. *181*, 6779–6787.

Davis, B.M., Kimsey, H.H., Kane, A.V. and Waldor, M.K. (2002). A satellite phage-encoded antirepressor induces repressor aggregation and cholera toxin gene transfer. EMBO J. *21*, 4240–4249.

Davis, B.M., Moyer, K.E., Boyd, E.F. and Waldor, M.K. (2000). CTX propages in classical biotype *Vibrio cholerae*: functional phage genes but dysfunctional phage genomes. J. Bacteriol. *182*, 6992–6998.

Desmarchelier, P.M., Momen, H. and Salles, C.A. (1988). A zymovar analysis of *Vibrio cholerae* isolated in Australia. Trans. R. Soc. Trop. Med. Hyg. *82*, 914–917.

Desmarchelier, P.M., Wong F.Y. and Mallard, K. (1995). An epidemiological study of *Vibrio cholerae* O1 in Australian environment based on rRNA gene polymorphisms. Epidemiol. Infect. *115*, 435–446.

Dingle, K.E., Colles, F.M., Wareing, D.R.A., Ure, R., Fox, A.J., Bolton, F.E. Bootsma, H.J., Willems, R.J.L., Urwin, R. and Maiden, M.C.J. (2001). Multilocus sequence typing system for *Campylobacter jejuni*. J. Clin. Microbiol. *39*, 14–23.

Dziejman, M., Balon, E., Boyd, D., Fraser, C.M., Heidelberg, J.F. and Mekalanos, J.J. (2002). Comparetive genomic analysis of *Vibrio cholerae*: genes that correlate with cholera endemic and pandemic disease. Proc. Natl. Acad. Sci. USA *99*, 1556–1561.

Eguiarte, L.E. (1999). Una guía para principiantes a la genética de poblaciones. In *La evolución biológica*, J. Nuñez Farfan and L.E. Eguiarte ed, (Mexico D. F., UNAM Press), pp.35–50.

Enright, M.C., Day, N.P., Davies, C.E., Peacock, S.J. and Spratt, B.G. (2000). Multilocus sequence typing for characterization of methicillin-resistant and methicillin-susceptible clones of *staphylococcus aureus*. J. Clin. Microbiol. *38*, 1008–1015.

Evins, G., Cameron, D., Wells, J., Green, K., Popovic, T., Giono, S., Wachsmuth, K. and Tauxe, R. (1995). The emerging diversity of electrophoretic types of *Vibrio cholerae* in the Western hemisphere. J. Infect. Dis. *172*, 173–179.

Farfan, M., Miñana-Galbis, D., Fusté, M.C. and Lorén, J.G. (2001). Allelic diversity and population structure in *Vibrio cholerae* O139 Bengal based on nucleotide sequence analysis. J. Bacteriol. *184*, 1304–1313.

Farfan, M., Miñana-Galbis, D., Fusté, M.C. and Lorén, J.G. (2000). Genetic relationships between clinical and environmental *Vibrio cholerae* isolates based on multilocus enzyme electrophoresis. Microbiology *146*, 2613–2626.

Faruque, S.M., Ahmed, K.M., Siddique, A.K., Zaman, K., Alim, A.R.M.A. and Albert, M.J. (1997). Molecular analysis of toxigenic *Vibrio cholerae* O139 Bengal strains isolated in Bangladesh between 1993 and 1996: evidence for emergence of a new clone of the Bengal vibrios. J. Clin. Microbiol. *35*, 2299–2306.

Faruque, S.M., Albert, M.J. and Mekalanos, J.J. (1998). Epidemiology, genetics, and ecology of toxigenic *Vibrio cholerae*. Microbiol. Mol. Biol. Rev. *62*, 1301–1314.

Faruque, S.M., Alim, A.R.M.A., Roy, S.K., Nair, G.B., Sack, R.B. and Albert, M.J. (1994). Molecular analysis of rRNA and cholera toxin genes carried by the new epidemic strain of toxigenic *Vibrio cholerae* O139 synonym Bengal. J. Clin. Microbiol. *32*, 1050–1053.

Faruque, S.M., Alim, A.R.M.A., Rahman, M.M., Siddique, A.K., Sack, R.B. and Albert, M.J. (1993). Clonal relationships among classical *Vibrio cholerae* O1 strains isolated between 1961 and 1992 in Bangladesh. J. Clin. Microbiol. *31*, 2513–2516.

Faruque, S.M., Asadulghani, Kamruzzaman, M., Nandi, R.K., Ghosh, A.N., Nair, G.B., Mekalanos, J.J. and Sack, D.A. (2002). RS1 element of *Vibrio cholerae* can propagate horizontally as a filamentous phage exploting the morphogenesis genes of CTXphi. Infect. Immun. *70*, 163–170.

Faruque, S.M., Saha, M.N., Asadulghani, Bag, P.K., Bhadra, R.K., Bhattacharya, S.K., Sack, R.B., Takeda, Y. and Nair, G.B. (2000). Genomic diversity among *Vibrio cholerae* O139 strains isolated in Bangladesh and India between 1992 and 1998. FEMS Microbiol. Lett. *184*, 279–284.

Faruque, S.M., Saha, M.N., Asadulghani, Sack, D.A., Sack, R.B., Takeda, Y. and Nair, G.B. (2000). The O139 serogroup of *Vibrio cholerae* comprises diverse clones of epidemic and non epidemic strains derived from multiple *Vibrio cholerae* O1 or non-O1 progenitors. J. Infect. Dis. *182*, 1161–1168.

Feachem, R., Miller, C. and Drasar, B. (1981). Enviroment aspect of cholera epidemiology. II. Ocurrence and survival of *Vibrio cholerae* in the environment. Trop. Dis. Bull. *78*, 865–880.

Feeley, J.C. (1965). Classification of *Vibrio cholerae* (*Vibrio comma*), including E1 Tor vibrios, by intrasubspecific characteristcs. J. Bacteriol. *89*, 665–678.

Feil, E.J., Smith, J.M., Enright, M.C. and Spratt, B.G. (2000). Estimating recombinational parameters in *Streptococcus pneumoniae* from multilocus sequence typing data. Genetics. *154*, 1439–1450.

Finkelstein, R.A. and Muckerjee, S. (1963). Hemagglutination: a rapid method for differentiating *Vibrio cholerae* and E1 Tor vibrios. Proc. Soc. Exp. Biol. Med. *112*, 335–339.

Ford, H.D. and Ford, E.B. (1930). Fluctuation in numbers, and its influence on variation, in *Mlitaea aurina*, Rott. (Lepidoptera). Trans. Roy. Entomol. Soc. Lond. *78*, 345–351.

Freitas, F., Momen, H. and Salles, C.A. (2002). The zymovars of *Vibrio cholerae*: multilocus enzyme electrophoresis of *Vibrio cholerae*. Mem. Inst. Oswaldo Cruz, Rio de Janeiro, *97*, 511–516.

Gallut, J. (1968). Actualite du cholera: evolution des problemes épidemiologiques et bacteriologiques. Bull. Inst. Pasteur *66*, 219–248.

Gallut, J. (1971). La septieme pandemic cholérique. Bull. Soc. Exot. (Paris). *64*, 551–560.

Gangarosa, E.J., Sonati, A., Saghari, H. and Feeley, J.C. (1967). Multiple serotypes of *Vibrio cholerae* from a case of cholera. Lancet *I*, 646–648.

García-Martínez, J., Martínez-Murcia, A.J., Rodríguez Varela, F. and Zorraquino, A. (1996). Molecular evidence supporting the existence of two major groups in uropathogenic *Escherichia coli*. FEMS Immunol. Med. Microbiol. *14*, 231–244.

Garg, P., Aydanian, A., Smith, D., Morris, J.G., Nair, G.B. and Stine, O.C. (2003). Molecular epidemiology of

O139 *Vibrio cholerae*: Mutation, lateral gene transfer, and founder flush. Emerg. Infect. Dis. *9*, 810–814.

Garg, P., Nandy, R.K., Chaudhury, P., Chowdhury, N.R, Ramamurthy, T., Yamasaki, S., Bhattacharya, S.K., Takeda, Y. and Nair, G.B. (2000). Emergence of *Vibrio cholerae* O1 biotype El Tor serotype Inaba from the prevailing O1 Ogawa serotype strains in India. J. Clin. Microbiol. *38*, 4249–4253.

Ghosh, A.R., Koley, H., De, D., Garg, S., Bhattacharya, M.K., Bhattacharya, S.K., Manna, B., Nair, G.B., Shimada, T., Takeda, T. and Takeda, Y. (1994). Incidence and toxigenity of *Vibrio cholerae* in fresh-water lake during the epidemic of cholerae caused by serogroup O139 Bengal in Calcutta, India. FEMS Microbiol. Ecol. *14*, 285–291.

Giono, S, Gutierrez, L., Rodriguez A.G., del Rio Zolezzi, A., Valdespino, J.L. and Sepúlveda, J. (1995). *Vibrio cholerae* O1 Fenotipos y Genotipos México. Gac. Med. Méx. *131*, 28–35.

Giono, S., Rodriguez, M.G., Gutierrez, L. and Valdespino, J.L. (1994). Phenotypic and genotypic characterization of *Vibrio cholerae* O1. Rev. Latinoam Microbiol. *36*, 243–51.

Goldberg, S. and Murphy, J.R. (1983). Molecular epidemiological studies of United States gulf coast *Vibrio cholerae* strains: Integration site of mutator vibriophage VCA-3. Infect. Immun. *42*, 224–230.

Grimes, D.J., Atwell, R.A., Brayton, P.R., Palmer, L.M., Rollins, D.M., Roszak, D.B., Singleton, F.L., Tamplin, M.L. and Colwell, R.R. (1986). The fate of enteric bacteria in estuarine and marine enviroment. Microb. Sci. *3*, 324–329.

Herrington, D.A., Hall, R.H., Losonsky, G.A., Mekalanos, J.J., Taylor, R.K. and Levine, M.M. (1988). Toxin, toxin-corregulated pili, and the *toxR* regulon are essential for *Vibrio cholerae* pathogenesis in humans. J. Exp. Med. *168*, 1487–1492.

Hobbs, M. and Mattick, J.S. (1993). Common components in the assembly of type 4 fimbriae, DNA transfer systems, filamentous phage and protein-secretion apparatus: a general systems, for the formation of surface-associated protein complexes. Mol. Microbiol. *10*, 233–243.

Hood, M.A., Ness, G.E. and Roderick, G.E. (1981). Isolation of *Vibrio cholerae* O1 from the eastern oyster *Crassostrea viginea*. Appl. Environ. Microbiol. *41*, 559–560.

Huq, A. and Colwell, R.R. (1996). Vibrios in the marine and estuarine environment: tracking of the *Vibrio cholerae*. J. Ecosyst. Health. *2*, 198–214.

Huq, A., Colwell, R.R., Chowdhury, M.A.R., Xu, B., Moniruzzaman, S.M., Islam, M.S., Yunus, M. and Albert, M.J. (1995). Co-existence of *Vibrio cholerae* O1 and O139 Bengal in plankton in Bangladesh. Lancet *345*, 1245.

Huq, A., Colwell, R.R., Rahman, R., Ali, A., Chowdhury, M.A.R., Parveen, S., Sack, D.A. and Russek-Cohen, E. (1990). Detection of *Vibrio cholerae* in the aquatic environmen measured by fluorecent antibody and culture method. Appl. Environ. Microbiol. *56*, 2370–2373.

Iredell, J.R. and Manning, P.A. (1994). The toxin-co-regulated pilus of *Vibrio cholerae* O1: a model for type 4 pylus biogenesis? Trends. Microbiol. *2*, 187–192.

Islam, M.S., Drasar, B.S. and Bradly, D.J. (1990). Survival of toxigenic *Vibrio cholerae* O1 with common duck weed, *Lemna minor*, in artificial aquatic ecosystem. Trans. R. Soc. Trop. Med. Hyg. *84*, 422–424.

Islam, M.S., Hasan, M.K., Miah, M.A., Yunus, M., Zaman, K. and Albert, M.J. (1994). Isolation of *Vibrio cholerae* O139 synonym Bengal from the aquatic environment in Bangladesh: implications for disease transmission. Appl. Enviroment. Microbiol. *60*, 1684–1686.

Jermyn, W.S. and Boyd, E.F. (2002). Characterization of a novel Vibrio pathogenicity island (VPI-2) enconding neuraminidase (nanH) among toxigenic *Vibrio cholerae* isolates. Microbiology *148*, 3681–3693.

Jiang, S.C., Matte, M., Matte, G., Huq, A. and Colwell, R.R. (2000). Genetic diversity of clinical and environmental isolates of *Vibrio cholerae* determined by amplified fragment length polymorphism fingerprinting. Appl. Environ. Microbiol. *66*, 148–153.

Jiang, S.C., Louis, V., Choopun, N., Sharma, A., Huq, A. and Colwell, R.R. (2000). Genetic diversity of *Vibrio cholerae* in Chesapeake Bay determined by amplified fragment length Polymorphism fingerprinting. Appl. Environ. Microbiol. *66*, 140–147.

Johnson, J., Salles, C.A., Panigrahi, P., Albert, M.J., Wright, A.C., Johnson, R.J. and Morris Jr., G. (1994). *Vibrio cholerae* O139 synonym Bengal is closely related to *Vibrio cholerae* El Tor but has important differences. Infect. Immun. *62*, 2108–2110.

Jonson, G., Holmgren, J. and Svennerholm, A.M. (1991). Identification of a mannose-binding pilus on *Vibrio cholerae* E1 Tor. Microb. Pathog. *11*, 433–441.

Kamal, A.M. (1971). Outbreak of gatro-enteritis by non-agglutinable (NAG) vibrios in the Republic of the Sudan. J. Egypt. Publ. Health Assoc. *46*, 125–159.

Kaper, J.B., Bradford, H.B., Roberts, N.C. and Falkow, S. (1982). Molecular epidemiology of *Vibrio cholerae* in the US gulf coast. J. Clin. Microbiol. *16*, 129–134.

Kaper, J.B., Morris Jr., J.G. and Levine, M.M. (1995). Cholera. Clin. Microbiol. Rev. *8*, 48–86.

Karaolis, D.K.R., Lan, R., Reeves, P.R. (1994). Molecular evolution of the seventh-pandemic clone of *Vibrio cholerae* and its relationship to other pandemic and epidemic *V. cholerae* isolates. J. Bacteriol. *176*, 6199–6206.

Karaolis, D.K.R., Johnson, J.A., Bailey, C.C., Boedeker, E.C., Kaper, J.B. and Reeves, P. (1998). A *Vibrio cholerae* pathogenicity island associated with epidemic and pandemic strains. Proc. Natl. Acad. Sci. USA *95*, 3134–3139.

Karaolis, D.K.R., Lan, R. and Reeves, P.R. (1996). The sixth and seventh cholera pandemics are due to independent clones separately derived from environmental non toxigenic, non-O1 *Vibrio cholerae*. J. Bacteriol. *177*, 3191–3198.

Karaolis, D.K.R., Somara, S., Maneval Jr., D.R., Johnson, J.A. and Kaper, J.B. (1999). A bacteriophage enconding a pathogenicity island, a type-IV pilus and a phage receptor in cholera bacteria. Nature *399*, 375–379.

Karaolis, K., Lan, R., Kaper, J.B. and Reeves, P.R. (2001). Comparation of *Vibrio cholerae* pathogenicity islands in sixth and seventh pandemic strains. Infect. Immun. *69*, 1947–1952.

Khetawat, G., Bhadra, R.K., Nandi, S. and Das, J. (1999). Resurgent *Vibrio cholerae* O139: Rearrangement of cholera toxin genetic elements and amplification of *rrn* operon. Infect. Immun. *67*, 148–154.

Koblavi, S., Grimont, F. and Grimont, P.A.D. (1990). Clonal diversity of *Vibrio cholerae* O1 evidenced by rRNA restriction patterns. Res. Microbiol. *141*, 645–657.

Kotetishvili, M., Stine, O.C., Chen, Y., Kreger, A., Sulakvelidze, A., Sozhamannan, S. and Morris Jr., J.G. (2003). Multilocus sequence typing has better discriminatory ability for typing *Vibrio cholerae* than does pulsed-field gel electrophoresis and provides a mesure of phylogenetic relatedness. J. Clin. Bacteriol. *41*, 2191–2196.

Kurazono, H., Pal, A., Bag, P.K., Nair, G.B., Karasawa, T., Mihara, T. and Takeda, Y. (1995). Distribution of genes encoding cholera toxin, zonula occludens toxin, accsesory cholera toxin, and E1 Tor hemolysis in *Vibrio cholerae* of diverse origins. Microb. Pathog. *18*, 231–235.

Lan, R. and Reeves, P. (1998). Recombination between rRNA operons created most of the ribotype variation observed in the seventh pandemic clone of *Vibrio cholerae*. Microbiology *144*, 1213–1221.

Li, M., Kotestishvili, M., Chen, Y. and Sozhamannan, S. (2003). Comparative genomic analyses of Vibrio pathogenicity island and cholera toxin prophage regions in nonepidemic serogroup strains of *Vibrio cholerae*. Appl. Environ. Microbiol. *69*, 1728–1738.

Lin, W., Fullner, K.J., Clayton, R., Sexton, J.A., Rogers, M.B., Calia, K.E., Calderwood, S.B., Fraser, C. and Mekalanos, J.J. (1999). Identification of a *Vibrio cholerae* RTX toxin gene cluster that is tightly linked to the cholera toxin prophage. Proc. Natl. Acad. Sci. USA *96*, 1071–1076.

Liu, S.L. and Sanderson, K.E. (1995). Rearrangements in the genome of the bacterium *Salmonella typhi*. J. Bacteriol. *177*, 3355–3357.

Maiden, M.C.J., Bygraves, J.A., Feil, E., Morelli, G., Russell, J.E., Urwin, R., Zhang, Q., Zhou, J., Zurth, K., Caugant, D.A., Feavers, I.M., Achtman, M. and Spratt, B.G. (1998). Multilocus sequence typing: a portable approach to the identification of clones within populations of pathogenic microorganisms. Proc. Natl. Acad. Sci. USA 95, 3140–3145.

Maynard Smith, J. (1996). Population Genetics: an Introduction, p. 2685–2690. *In* F. C. Neidhardt, R. Curtiss, J. L. Ingraham., E. C. C. Lin, K. B. Low, B. Magasanik, W. S. Reznikoff, M. Schaechter and H. E. Umbager (eds), *Escherichia coli and Salmonella: cellular and moleular biology*. ASM Press, Washington, DC.

Maynard Smith, J., Dowson, C.G. and Spratt, B.G. (1991). Localized sex in bacteria. Nature *349*, 29–31.

Mekalanos, J.J., Rubin. E.J. and Waldor. M.K. (1997). Cholera: molecular basis for emergence and pathogenesis. FEMS Immunol. Med. Microbiol. *18*, 241–248.

Mekalanos, J.J., Swatz, D.J. and Pearsons, G.D.N. (1983). Cholera toxin gene: nucleotide sequence, deletion analysis and vaccine development. Nature. *306*, 551–557.

Miller, C.J., Feachem, R.G. and Drasar, B.S. (1985). Cholera epidemiology in developed and developing countries: new troughts on transmission, seasonality, and control. Lancet *I*, 261–263.

Mitra, R., Basu, A., Dutta, D., Nair, G.B. and Takeda, Y. (1996). Resurgence of *Vibrio cholerae* O139 Bengal with altered antibiogram in Calcutaa, India. Lancet *348*, 1181.

Mukhopadhyay, A.K., Basu, A., Garg, P., Bag, P.K., Ghosh, A., Bhattacharya, S.K., Takeda, Y. and Nair, G.B. (1998). Molecular Epidemiology of reemergent *Vibrio cholerae* O139 Bengal in India. J. Clin. Microbiol. *36*, 2149–2152.

Mukhopadhyay, A.K., Chakraborty, S., Takeda, Y., Nair, G.B. and Berg, D.E. (2001). Characterization of VPI pathogenicity island and CTXϕ prophage in environmental strains of *Vibrio cholerae*. J. Bacteriol. *183*, 4737–4746.

Mukhopadhyay, A.K., Garg, S., Mitra, R., Basu, A., Dutta, D., Bhattacharya, S.K., Shimada, T., Takeda, T., Takeda, Y. and Nair, G.B. (1996). Temporal shift in traits of *Vibrio cholerae* strains isolated from hospitalized patients in Calcutta: 3 years (1993–1996) analysis. J. Microbiol. *34*, 2537–2543.

Nair, G.B., Faruque, S.M., Bhuiyan, N.A., Kamruzzaman, M., Siddique, A.K. and Sack, D.A. (2002). New variants of *Vibrio cholerae* O1 biotype El Tor with attributes of the classical biotype from hospitalized patients with acute diarrhea in Bangladesh. J. Clin. Microbiol. *40*, 3296–3299.

Nair, G.B., Qadri, F., Holmgren, J., Svennerholm, A.M., Safa, A., Bhuiyan, N.A., Shafi Ahmed, Q., Faruque, S.M., Takeda, Y. and Sack, D.A. (2006). Cholera due to altered El Tor strains of *Vibrio cholerae* O1 in Bangladesh. J. Clin. Microbiol. *44*, 4211–4213.

Nalin, D.R. (1976). Cholera, copepods and chitinase. Lancet 2, 958.

Nandi, S., Khetawat, G., Sengupta, S., Majumder, R., Kar, S., Bhadra, R.K., Roychoudhury, S. and Das, J. (1997). Rearrangements in the genome of *Vibrio cholerae* strains beloging to difference serovars and biovars. Int. J. Sys. Bacteriol. *47*, 858–862.

Niyogi, S.K., Mondal, S., Sarkar, B.L., Garg, S., Baneerjee, D. and Dey, G.N. (1994). Outbreak of cholera due to *Vibrio cholerae* O1 in Orissa state. Indian. J. Med. Res. *100*, 217–218.

O'shea, Y.A., Reen, F.J., Quirke, A.M. and Boyd, E.F. (2004). Evolutionary genetic analysis of the emergence of epidemic *Vibrio cholerae* isolates on the basic of comparative nucleotide sequence analysis and multilocus virulence gene profiles. J. Clin. Microbiol. *42*, 4657–4671.

Pal, B.B., Khuntia, H.K., Samal, S.K., Das, S.S. and Chhotray, P.C. (2006). Emergence of *Vibrio cholerae* O1 biotype E1 Tor serotype Inaba causing Outbreak of Cholera in Orissa, India. Jpn. J. Infect Dis. 59, 266–269.

Pal, B.B., Khuntia, K., Anuradha, M. and Chhotray, G.P. (2000). Emergence of *Vibrio cholerae* O139 during 1995 in Orissa, India- a retrospective study, Indian. J. Med. Microbiol. 18:195–196.

Popovic T, Bopp, C., Olsvik, O. and Wachsmuth, K. (1993). Epidemiologic application of a standardized ribotype scheme for *Vibrio cholerae* O1. J. Clin. Microbiol. *31*, 2474–2482.

Popovic, T., Fields, P.I., Olsvik, O., Wells, J.G., Evins, G.M., Cameron, D.N., Farmer III, J.J., Bopp, C.A., Wachsmuth, K., Sack, R.B., Albert, M.J., Nair, G.B., Shimada, T. and Feeley, J.C. (1995). Molecular subtyping of toxigenic *Vibrio cholerae* O139 causing epidemic cholera in India and Bangladesh, 1992–1993. J. Infect. Dis. *171*, 122–127.

Reeves, P.R. (1992). Variation in O-antigens, niche specific selection and bacterial populations. FEMS Microbiol. Lett. *100*, 509–516.

Rius, N., Fusté, M.C., Guasp, C., Lalucat, J. and Lorén, J.G. (2001). Clonal population structure of Pseudomonas stulzeri, a species with excepcional genetic diverssity. J. Bacteriol. *183*, 736–744.

Rivera, I.G., Chowdhury, M.A.R., Huq, A., Jacobs, D., Martins, M.T. and Colwell, R.R. (1995). Enterobacterial perspective intergenic consensus sequences and the PCR to generate fingerprints of genomic DNAs fom *Vibrio cholerae* O1, O139, and non-O1 strains. Appl. Environ. Microbiol. *61*, 2898–2904.

Rivera, I.N.G., Chun, J., Huq, A., Sack, R.B., and Colwell, R.R. (2001). Genotypes associated with virulence in environmental isolated of *Vibrio cholerae*. Appl. Environ. Microbiol. *67*, 2421–2429.

Rowe-Magnus, D.A., Guerout, A.M. and Maze, l.D. (2002). Bacterial resistence evolution by recrutiment of super-integron gene cassettes. Mol. Microbiol. *43*, 1657–1669.

Rowe-Magnus, D.A., Guerout, A.M., Ploncard, P., Dychinco, B., Davies, J. and Mazel, D. (2001). The evolutionary history of chromosomal super-integrons provides an ancestry for multiresistant integrons. Proc. Natl. Acad. Sci. USA *98*, 652–657.

Roy, C., Mukerjee, S. and Tanamal, S.J.W. (1963). Haemolytic and non-haemolytic E1 Tor vibrios. Ann. Biochem. Exp. Med. *23*, 553–558.

Rubin, E.J., Lin, W., Mekalanos, J.J. and Waldor, M.K. (1998). Replication and integration of a *Vibrio cholerae* cryptic plasmid linked to the CTX prophage. Mol. Microbiol. *28*, 1247–1254.

Sack, B. and Miller, C.E. (1969). Progresive chance of *Vibrio cholerae* serotypes in germ-free mice inoculated with *Vibrio cholerae*. J. Bacteriol *99*, 688–695.

Saha, P.K., Koley, H., Mukhopadhyay, A.K., Bhattacharya, S.K., Nair, G.B., Ramakrishnan, B.S., Krishnan, S., Takeda, T. and Takeda, Y. (1996). Nontoxigenic *Vibrio cholerae* O1 serotype Inaba biotype E1 Tor associated with a cluster of cases of cholera in southern India. J. Clin Microbiol. *34*, 1114–1117.

Salles, C.A. and Momen, H. (1991). Identification of *Vibrio cholerae* by enzyme electrophoresis. Trans R Soc Trop Med Hyg. *85*, 544–547.

Samadi, A.R., Hug, M.I., Shahid, N., Khan, M.U., Eusde, A., Rahman, A.S.M.M., Yunus, M. and Faruque, A.S.G. (1983). Classical *Vibrio cholerae* biotype displaces El Tor in Bangladesh. Lancet *1*, 805–807.

Sanyal, S.C. (1986). Vibrio toxin. p. 207–225 In F. Dorner and J. Drews (ed.), Pharmacology of bacterial toxins. International encyclopedia of pharmacology and therapeutics, vol. 119. Pergamon Press, Oxford, United Kingdom.

Selander, R.K., Li, J. and Nelson, K. (1996). Evolutionary genetics of Salmonella enterica. P. 2691–2707. In FC Neidhardt, R Curtis III, JL Ingraham, ECC Lin, KB Low, B Magasanik, WS Reznikov, M Riley, M Schaechter, HE Umbarger (ed). *Escherichia coli* and Salmonella typhimurium: cellular and molecular biology, 2nd edn, ASM Press. Washington DC.

Selander, R.K., Caugant, D.A., Ochman, H., Musser, J.M., Gilmour, M.N. and Whittam, T.S. (1986). Methods of multilocus enzyme electrophoresis for bacterial population genetics and systematics. Appl. Environ. Microbiol. *51*, 873–884.

Shandera, W.X., Hafkin, B., Martin, D.L., Taylor, J.P., Maserang, D.L., Wells, J.G., Kelly, M., Ghandi, K., Kaper, J.B., Lee, J.V. and Blake, P. (1983). Persistence of cholera in the United States. Am. J. Trop. Med. Hyg. *32*, 812–817.

Sharma, C., Nair, G.B., Mukhopadhyay, A.K., Bhattacharya, S.K., Ghosh, R.K. and Ghosh, A. (1997). Molecular characterization of *Vibrio cholerae* O1 biotype E1 Tor strains isolated between 1992 and 1995 in Calcuta, India: evidence for the emergence of a new clone of the El Tor biotype. J. Infect. Dis. *175*, 1134–1141.

Sharma, C., Thungapathra, M., Ghosh, A., Mukhopadhyay, A.K., Basu, A., Mitra, R., Basu, I., Bhattacharya, S.K., Shimada, T., Ramamurthy, T., Takeda, T., Yamasaki, S., Takeda, Y. and Nair, G.B. (1998). Molecular analysis of non-O1, non-O139 *Vibrio cholerae* associated with an unusual upsurge in the incidence of cholera-like disease in Calcutta, India. J. Clin. Microbiol. *36*, 756–763.

Shears, P. (1994). Cholerae. Ann. Trop. Med. Para. *88*, 109–122.

Sheehy, T.W., Sprinz, H., Angerson, W.S. and Formal, S.B. (1966). Laboratory *Vibrio cholerae* infection in the United States. JAMA *197*, 321–326.

Shimada, T., Arakawa, E., Itoh, K., Okitsu, T., Matsushima, A., Asai, Y., Yamai, Sh., Nakazato, T., Nair, G.B., Albert, M.J. and Takeda, Y. (1994). Extended Serotyping Scheme for *Vibrio cholerae*. Curr. Microbiol. *28*, 175–178.

Shimada, T., Arakawa, E. and Itoh, K. (1995). Two strains of *Vibrio cholerae* non-O1 possessing somatic antigen factors in common with *Vibrio cholerae* serogroup O139 synonym 'Bengal'. Curr. Microbiol. 29:1–3.

Shukla, B.N., Singh, D.V. and Sanyal, S.C. (1995). Attachment of non-culturable toxigenic *Vibrio cholerae* O1, and non-O1, and Aeromonas ssp. to the aquatic arthopod Gerris spinolae and plants in the River Ganga, Varanasi. FEMS. Inmmunol. Med. Microbiol. *12*, 113–120.

Singh, D.V., Matte, M.H., Matte, G.R., Jiang, S., Sabeena, F., Shukla, B.N., Sanyal, S.C., Huq, A. and Colwell, R.R. (2001). Molecular analysis of *Vibrio cholerae* O1, O139, non-O1, and non-O139 strains: Clonal relationships between clinical and environmental isolates. Appl. Environ. Microbiol. *67*, 910–921.

Sochard, M.R., Wilson, D.F., Austin, B. and Colwell, R.R. (1979). Bacteria associated with the surface and the gut of marine copepods. Appl. Environ. Microbiol. *37*, 750–759.

Stine, O.C., Sozhamannan, S., Gou, Q., Zheng, S., Morris Jr., J.G. and Johnson, J.A. (2000). Phylogeny of *Vibrio cholerae* based on recA sequence. Infect. Immun. *68*, 7180–7185.

Stroeher, U.H., Jedani, K.E., Dredge, B.K., Morona, R., Brown, M.H., Karageorgos, L.E., Albert, M.J. and Manning, P.A. (1995). Genetic rearrangements in the rfb regions of *Vibrio cholerae* O1 and O139. Proc. Natl. Acad. Sci. USA *92*, 10374–10378.

Stroeher, U.H., Kanageorgos, L.E., Morona, R. and Manning, P.A. (1992). Serotypeconversion in *Vibrio cholerae* O1. Proc. Natl. Sci. USA *89*, 2566–2570.

Taneja, N., Biswal, M., Tarai, B. and Sharma, M. (2005). Emergence of *Vibrio cholerae* O1 biotype E1 Tor serotype Inaba in North India Jpn. J. Infect. Dis. *58*, 238–240.

Taylor, R.K., Miller, V.L., Furlong, D.B. and Mekalanos, J.J. (1987). The use of *phoA* gene fusions to identify a pilus colonization factor coordinately regulated with cholera toxin. Proc. Natl. Acad. Sci. USA *84*, 2833–2837.

Thompson, F.L., Thompson, C.C., Vicente, A.C.P., Theophilo, G.N.D., Hofer, E. and Swings, J. (2003). Genomic diversity of clinical and environmental *Vibrio cholerae* strains isolated in Brazil between 1991 and 2001 as revealed by fluorescent amplified fragment length polymorphism analysis. J. Clin. Microbiol. *41*, 1946–1950.

Tibayrenc, M. (2005). Bridging the gap between molecular epidemiologists and evolutionists. Trends Microbiol. *13*, 575–580.

Wachsmuth, K, Bopp, C., Fields, P. and Carrillo, C. (1991). Difference between toxigenic *Vibrio cholerae* O1 from South America and US gulf coast. Lancet *337*, 1097–1098.

Wachsmuth, K, Evins, G., Fields, P., Olsvik, O., Popovic, T., Bopp, C., Wells, J., Carrillo, C. and Blake, P. (1993). The molecular epidemiology of cholera in Latin America. J. Infect. Dis. *167*, 621–626.

Wachsmuth, K., Olsvik, O., Evins, G. and Popovic, T. (1994). Molecular epidemiology of cholera. In *Vibrio cholerae* and cholera: Molecular to global perspectives. K. Wachsmuth, P. Blake, and O. Olsvik eds., (Washington DC, USA; ASM Press). pp. 357–370.

Waldor, M.K. and Mekalanos, J.J. (1996). Lysogenic conversion by a filamentous phage encoding cholera toxin. Science. *272*, 1910–1914.

Waldor, M.K., Rubin, E.J., Pearson, G.D.N., Kimsey, H. and Mekalanos, J.J. (1997). Regulation, replication and integration functions of the *Vibrio cholerae* CTXϕ are encoded by region RS2. Mol. Microbiol. *24*, 917–926.

Wallace, B. (1981). Basic populations genetics. New York: Columbia University Press. pp. 429–432.

Whittam, T.S. (1986). Methods of multilocus enzyme electrophoresis for bacterial population genetics and systemics. Appl. Environ. Microbiol. *51*, 873–884.

Whittam, T.S. (1996). Genetic variation and evolutionary processes in natural populations of *Escherichia coli*. In *Escherichia coli* and Salmonella: cellular and molecular biology. F. C. Neidhardt, R. Curtiss, J.L. Ingraham., E.C.C. Lin, K.B. Low, B. Magasanik, W.S. Reznikoff, M. Schaechter and H.E. Umbager eds, (Washington DC, USA; ASM Press). pp. 2708–2720.

Yamai, S., Okitsu. T., Shimada, T. and Katsube, Y. (1997). Distribution of serogroups of *Vibrio cholerae* non-O1 non-O139 with specific reference to their ability to produce cholera toxin, and addition of novel serogroups. Kansenshogaku Zasshi *71*, 1037–1045.

Young, J.P.W. (1989). The population genetics of bacteria. In Genetics of bacterial diversity, D.A. Hopwood, and K.E. Charter ed., (London, U.K., Academic Press), pp. 417–438.

Filamentous Bacteriophages in *Vibrio cholerae*: Genetics and Evolution

4

E. Fidelma Boyd

Abstract

The role of filamentous phages, particularly CTXϕ, in *Vibrio cholerae* genetics and evolution are discussed. An overview of the CTXϕ genome structure and interaction with the *V. cholerae* host cell during the various stages of the lifecycle of CTXϕ are outlined. The evolution of CTXϕ is explored, and the co-evolution of pIII^{CTX} and the TcpA, the major pilin protein of TCP the receptor for CTXϕ is examined. Seven diverse filamentous phages have been characterized from *V. cholerae* O1 and O139 serogroup isolates and these phages and their putative roles in *V. cholerae* biology are reviewed.

Introduction

Bacteriophages, through a process called phage conversion, can convert their bacterial host from a non-pathogenic strain to a pathogenic strain by providing the host with bacteriophage-encoded virulence genes (Boyd, 2005). *Vibrio cholerae* isolates that cause cholera carry the A–B type enterotoxin, cholera toxin (CT) encoded by the *ctxAB* genes, which reside in the genome of a lysogenic filamentous bacteriophage CTXϕ (Waldor and Mekalanos, 1996). Only those strains that carry CTXϕ cause epidemic and pandemic cholera. CTXϕ confers a selective advantage to its *V. cholerae* host by producing CT, the action of which results in profuse diarrhoea, promoting dissemination of the bacterium. CTXϕ forms stable lysogens in *V. cholerae*, and replication of the phage does not result in *V. cholerae* cell lysis; therefore, CTXϕ can be both vertically and horizontally transferred (Waldor and Mekalanos, 1996). Bacteriophages, as well as conferring a fitness advantage on their bacterial host, can also be their foe; for example, lytic phages cause bacterial cell death by lysis. The possible therapeutic role of phages in the treatment of cholera patients and their use in the control of *V. cholerae* in the environment are re-emerging areas of research interest (Jensen *et al.*, 2006). As far back as the mid-nineteenth century it was recognized that certain elements (most likely phages) in the water could protect against cholera. The recent recognition that phage blooms coincided with *V. cholerae* decline in waters analysed from endemic areas may point to possible control measures (Faruque *et al.*, 2005; Jensen *et al.*, 2006).

Bacteriophages and their role in *V. cholerae* genetics and evolution are the focus of this chapter, with emphasis on filamentous phages, and CTXϕ, in particular. A brief overview of the CTXϕ genome structure is first presented (Fig.1). Interaction of CTXϕ with the *V. cholerae* host cells is discussed in terms of the phage and *V. cholerae* genes involved in the various stages of the lifecycle of CTXϕ: host attachment and infection, integration into host genome, replication within the host, secretion from the host, and the regulation of these stages (Table 4.1). The prevalence, distribution, and evolution of CTXϕ are also explored. In addition to CTXϕ, at least seven filamentous phages have been characterized from *V. cholerae* O1 and O139 serogroup isolates: phages 493, VSK, VSKK, VGJ, KSF-1, fs-1, and fs-2. These phages and their putative roles in *V. cholerae* biology are investigated. Furthermore, the phylogenetic distribution and

Table 4.1 CTXϕ and *V. cholerae*-encoded proteins required for CTXϕ life cycle

Stage	CTXϕ	*V. cholerae*
Attachment	pIII^CTX	TcpA, TolQRA
Infection	pIII^CTX	TolA
Integration	RstB	XerC, XerD
Replication	RstC, RstA	ToxT
Secretion	Zot	EspD
Regulation		
Repression	RstR	LexA
Activation	RstC	ToxT

evolutionary relationships among filamentous phages have been explored based on RstA and Zot phylogenetic analysis.

CTXϕ genome structure

Cholera toxin (CT), encoded by a filamentous bacteriophage (CTXϕ), is the main cause of the profuse watery diarrhoea associated with cholera, the devastating disease caused by *V. cholerae* (Waldor and Mekalanos, 1996). CT is an A-B toxin type enterotoxin similar to shiga toxin and consists of one active A subunit and five binding B subunits. CT interacts with the human host cell, resulting in the increase of adenylate cyclase, which in turn increases cAMP, thus affecting membrane permeability by disrupting the normal flow of ions in the small intestines. The *ctxAB* genes encoding the A-B subunits share extensive sequence homology with heat-labile enterotoxins produced by enterotoxigenic *Escherichia coli* isolates. The *ctxAB* genes reside in the genome of CTXϕ which exists in a lysogenic form in toxigenic *V. cholerae*, and the phage is similar in size, structure, and gene order to filamentous coliphages such as M13 and f1 from *E. coli* (Fig. 4.1).

CTXϕ has a 6.9-kb genome and two functionally distinct modules: a DNA replication module named RS2 (repeat sequence 2), which encodes three genes, *rstR*, *rstA*, and *rstB*, and a core region encoding *psh*, *cep*, *orfU* (*gIII*), *ace*, *zot*, and *ctxAB* (Fig. 4.1) (Waldor and Mekalanos, 1996). The RstA protein is required for CTXϕ DNA replication, which in *E. coli* Ff filamentous coliphage is performed by pII, the gene for which is located at the same genomic region as the *rstA* gene in CTXϕ (Fig. 4.1) (Waldor *et al.*, 1997). The *rstR* and *rstB* genes encode a repressor protein and a protein required for CTX integration into the host genome, respectively (Waldor *et al.*, 1997). The RstB protein contains a region similar to single-stranded (ss) DNA-binding proteins, which in other filamentous phages stabilize phage ssDNA (Russel, 1995).

The core region genes *cep*, *orfU*, *ace*, and *zot* correspond to genes gVIII, gIII, gVI, and gI of Ff phage (Fig. 4.1). The Psh, Cep, OrfU, and Ace proteins are phage structural proteins and Zot is an assembly protein (Waldor and Mekalanos, 1996). Ace and Zot proteins were also shown to have an enterotoxic effect in humans (Fasano *et al.*, 1991; Trucksis *et al.*, 1993), which is perhaps a side consequence of their expression during bacterial invasion of the human host. The *ctxAB* genes are unique to CTXϕ, but are not essential for phage production, and are located at the same site as gIV in *E. coli* Ff phage (Fig. 4.1). CTXϕ does not encode a pIV protein homologue, which in Ff phage is required for secretion from the bacterial cell.

Associated with CTX prophage in *V. cholerae* O1 El Tor biotype and O139 isolates is a region named RS1 that encodes four proteins, RstR, RstA, and RstB, that are identical to the three RS2 region proteins, and an RctC, an antirepressor that allows CTXϕ gene expression (Davis *et al.*, 2002; Waldor *et al.*, 1997). CTXϕ differs from typical *E. coli* Ff filamentous coliphages in that it integrates into the *V. cholerae* genome to form stable lysogens similar to lysogenic phages, but it does not encode an integrase gene that performs this function in lysogenic phages (Waldor and Mekalanos, 1996).

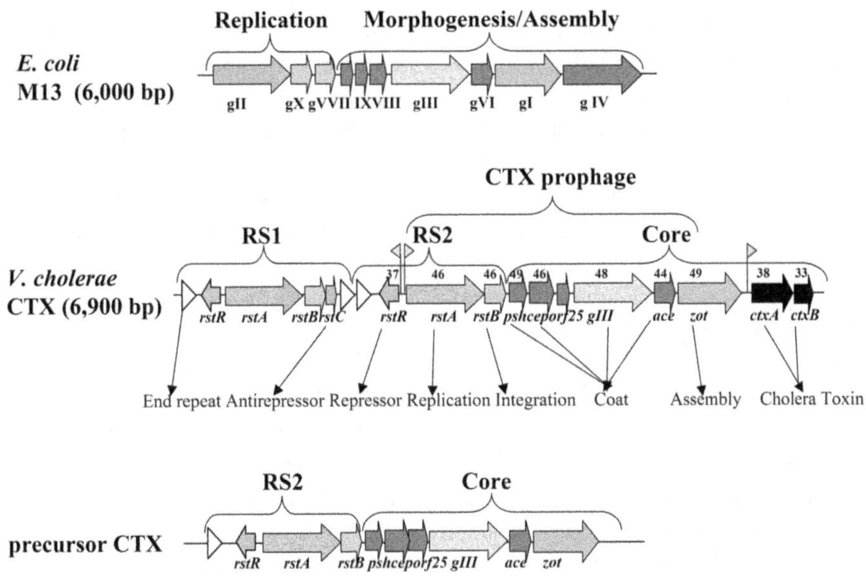

Figure 4.1 Gene organization and gene designation of filamentous phages M13 from *Escherchia coli*, the El Tor-derived CTXϕ and RS1ϕ from *V. cholerae* and the precursor CTX phage (Boyd *et al.*, 2000a; Waldor and Mekalanos, 1996; Waldor *et al.*, 1997). Open arrows represent open reading frames (ORFs) and the direction of transcription of each gene. White triangles indicate end repeat sequences. Vertical flags represent promoter regions. The numbers and lettering underneath the arrows indicate each gene designation. The rstR, rstA, and rstB genes of both RS1ϕ and RS2 regions are almost identical. The striped arrows indicate gene required for replication, hatched arrows are genes required for regulation, solid arrows are genes involved in morphogenesis and black arrows indicate cholera toxin encoding genes.

The life cycle of CTXϕ

In this section, the four main stages in the lifecycle of CTXϕ are described in terms of the phage- and host-encoded proteins required in (i) CTXϕ attachment and entry into *V. cholerae*, (ii) CTXϕ integration into *V. cholerae* genome, (iii) CTXϕ DNA replication and (iv) CTXϕ secretion from *V. cholerae*, as well as the regulation of these stages (Table 4.1).

CTXϕ attachment and entry into *Vibrio cholerae*

The receptor for CTXϕ on the *V. cholerae* cell surface is a type IV pilus, the toxin co-regulated pilus (TCP), and the periplasmic and inner membrane proteins TolQRA (Heilpern and Waldor, 2000; Waldor and Mekalanos, 1996) (Table 4.1). Genetic analyses of mutants of both *tcpA* and *tolQRA* showed a marked decrease in the ability of *V. cholerae* to uptake kanamycin-marked copies of CTXϕ (Heilpern and Waldor, 2000). The OrfU protein of CTXϕ, which has been renamed $pIII^{ctx}$, is hypothesized to bind to the major pilin protein TcpA of TCP, similar to the binding of Ff coliphages to the F pilus mediated by the minor coat protein pIII (Heilpern and Waldor, 2003). The Ff pIII protein contains a leader sequence followed by three domains separated by glycine-rich regions. The first two domains, named g3p-D1 (amino acids 1–86) and g3p-D2 (amino acids 105–236), are involved in phage adsorption and entry, and g3p-D3 (amino acids 270–406) is involved in phage assembly and release (Deng *et al.*, 1999; Deng and Perham, 2002; Rakonjac and Model, 1998; Rakonjac *et al.*, 1999). The g3p-D1 and g3p-D2 domains bind to the host cell receptors TolA and the F pilus in a two-step process: g3p-D2 binds to F, which results in a conformational change that allows g3p-D1 to bind to TolA (Deng *et al.*, 1999; Holliger and Riechmann, 1997; Holliger *et al.*, 1999; Riechmann and Holliger, 1997). Although there is limited sequence homology between pIII from Ff-related phages and $pIII^{CTX}$, magnetic resonance spectroscopy has predicted that amino acid residues 18–86 of $pIII^{CTX}$ may fold into a barrel similar to that of g3p-D1 (Holliger *et al.*, 1999).

Heilpren and Waldor (2003) further tested this hypothesis by displaying $pIII^{CTX}$ on the surface of an Ff phage as a $pIII^{CTX}$–$pIII^{Ff}$ fusion protein (Heilpern and Waldor, 2003). The resultant construct was found to infect both F pilus positive *E. coli* and TCP positive *V. cholerae* isolates. They found that different segments of $pIII^{CTX}$ mediate different steps in CTX infection, similar to pIII from fd. The $pIII^{CTX}$ amino acid residues 15–138 were sufficient for entry of fd into *V. cholerae* cells, which was dependent on TolA but independent of TCP, thus identifying the g3p-D1 penetration domain. The $pIII^{CTX}$ residues 15–274 increased CTXϕ infection by up to three orders of magnitude in a TCP-dependent manner, suggesting that this region contained the TCP binding domain, the g3p-D2 adsorption domain (Heilpern and Waldor, 2003). In addition, it was demonstrated that $pIII^{CTX}$ is required for *V. cholerae* CTXϕ assembly and release, but they could not locate the precise domain(s) involved (Heilpern and Waldor, 2003).

To explore further and demonstrate directly the proposed interaction between CTXϕ pIII and TcpA, the TCP major pilin protein, and to investigate the ability of CTXϕ pIII alleles to interact with TcpA/TolA proteins, we used the CytoTrap yeast two-hybrid system, which identifies interacting proteins through the functional restoration of the yeast GAL4 transcriptional activator *in vivo* (Reen and Boyd, unpublished data). Significant sequence divergence has been found in both CTXϕ pIII and TcpA proteins from *V. cholerae* classical and El Tor biotype strains, which otherwise show high sequence similarity (Boyd *et al.*, 2000a; Boyd and Waldor, 2002; O'Shea *et al.*, 2004). We used variant TcpA proteins as bait and diverse $pIII^{ctx}$ proteins as prey and demonstrated TcpA–$pIII^{ctx}$ interaction with one another using the yeast two-hybrid system (Reen and Boyd, unpublished data). We found for the first time that CTXϕ pIII interacts directly with TcpA. Our results revealed that the divergence at the *tcpA* locus is not reflected in the absolute ability of the pilin to bind the $pIII^{CTX}$ protein. We also found that the co-receptor TolA was unable to bind CTXϕ pIII in the absence of TcpA, confirming the requirement for a conformational change. To characterize the effect that TcpA sequence divergence has on colonization efficiency and frequency of CTX binding, we have introduced diverse *tcpA* alleles into a classical *V. cholerae* strain O395 background. Candidates that maintain the O395 endogenous promoter, as well as those that acquired the promoter of the introduced allele, will be selected and assayed for colonization efficiency in an infant mouse model.

CTXϕ integration into *Vibrio cholerae* genome

CTXϕ integration into the *V. cholerae* host genome requires a number of phage-encoded and host-encoded factors (Table 1). After CTXϕ DNA adsorption to the *V. cholerae* cell wall, the viral ssDNA is injected to the cell cytoplasm. Upon entry into the cell cytoplasm, the linear ssDNA circularizes, forming pCTX, which can remain as an extra chromosomal element or more usually integrate into the *V. cholerae* genome at site-specific attachment sites (Huber and Waldor, 2002; Waldor and Mekalanos, 1996). CTXϕ does not encode an integrase, which usually catalyzes integration in other phages; instead CTXϕ depends on two host-encoded tyrosine recombinases, XerC and XerD (Huber and Waldor, 2002).

The XerCD proteins are conserved among eubacteria, as they serve to resolve chromosome dimers at cell division (Huber and Waldor, 2002). In *E. coli*, XerCD proteins bind and catalyse recombination at a *dif* site located near the bacterial replication terminus (Blakely *et al.*, 1993). Integration of CTXϕ at *dif1*, a 28-bp site on chromosome I of *V. cholerae*, requires binding of XerC and XerD separated by a 6-bp central region, and, unlike in *E. coli*, FtsK, a cell division protein, is not required and integration is irreversible (Huber and Waldor, 2002). The integration site *attP* on the CTXϕ genome, a 200-bp CTXϕ intergenic region (Ig-1) was identified, and deletion of this region resulted in loss of integration (Huber and Waldor, 2002). It was proposed that XerC and XerD bind at two sites within *attP*, the core XerCD binding site that spans the recombination region and a second site 80 to 90 bp away shown to be required for integration, and suggested to serve as an architectural or stabilizing factor (McLeod and Waldor, 2004)

An alternative model for CTXϕ integration was recently proposed, which suggests that the ssDNA CTXϕ genome is used in integration into the *V. cholerae* genome (Val *et al.*, 2005). Val and colleagues (2005) uncovered on the single-stranded pCTX genome two pairs of binding sites in inverted orientation for XerC and XerD binding that fold into a hairpin structure, creating a recombination target for XerCD. XerC can catalyse a single pair of strand exchanges between this target and *dif1* in the presence of XerD, resulting in CTXϕ integration lacking functional Xer sites (Val *et al.*, 2005). In contrast to the classical model of phage integration of ssDNA phage being converted to dsDNA prior to integration, Val and colleagues proposed that Xer recombinases promote the direct insertion of the ssDNA genome into the *dif1* site of *V. cholerae.* The ssDNA forms a stem–loop structure, producing a region of duplex DNA upon which Xer recombinase acts. This model appears to explain many of the unusual aspects of CTXϕ integration compared with XerCD recombination at the *E. coli dif* site and why and how integration is irreversible (Blakely, 2004; McLeod and Waldor, 2004; Val *et al.*, 2005). The wide distribution of XerCD recombinase system among bacteria and the prevalence of *dif* sites among filamentous phages suggest that this system may be the paradigm for filamentous phage integration into the bacterial genome.

CTXϕ DNA replication

CTX integration into the *V. cholerae* genome is not reversible; therefore, once acquired, CTXϕ is vertically transmitted along with the rest of the genome. However, chromosomally integrated CTX prophage can act as template for the production of ssDNA phage, and this process requires the presence of either a tandem array of CTX prophages or a CTX-RS1 array (Davis *et al.*, 2000b; Davis and Waldor, 2000; Moyer *et al.*, 2001). The plasmid form of CTX is produced by rolling circle replication of tandemly repeated CTX-CTX or CTX-RS1 (Moyer *et al.*, 2001). Moyer and coworkers demonstrated that RstA

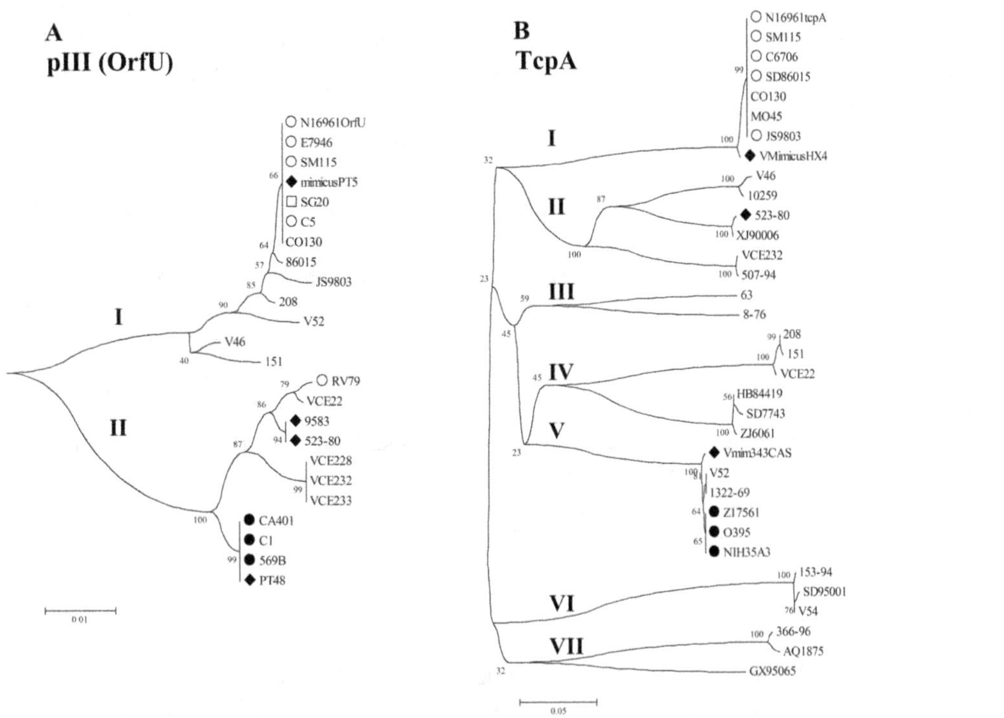

Figure 4.2 Unrooted Neighbor-joining trees (poisson correction), bootstrap 1000, constructed from $pIII^{CTX}$ (A) and TcpA (B) amino acid sequences. Open circles represent *V. cholerae* O1 biotype El Tor isolates and closed circles represent *V. cholerae* O1 biotype classical isolates. Open square represents *V. cholerae* O130 serogroup isolate. Closed diamonds represent *V. mimicus* isolates. A. Two main lineages of pIII are designated I and II. B. Seven lineages of TcpA are designated I to VII.

is required for pCTX replication, and they suggested that RstA makes a single-stranded nick in the CTX prophage in Ig-1 (Moyer *et al.*, 2001). The resulting 3'-end serves as a template for new DNA synthesis, presumably by host-encoded DNA polymerase. During DNA replication, strand displacement occurs and replication continues into downstream CTX prophage until Ig-1 is reached. The RstA protein nicks the displaced strand, yielding single stranded CTX genome (Moyer *et al.*, 2001). This model explains the observation that classical CTXϕ do not produce infectious phage particles because these isolates do not contain complete CTX arrays or RS1 prophages, the substrate for ssDNA replication (Davis *et al.*, 2000b).

CTXϕ secretion from *Vibrio cholerae*

As stated previously, the genome of CTXϕ differs from that of filamentous phages from *E. coli* in that it lacks the gene for a putative bacteriophage export or secretin protein (a homologue of Ff gIV) (Fig. 4.2). However, similar to other filamentous phages, CTX is secreted from the cell without *V. cholerae* cell lysis. Davis and colleagues (2000) demonstrated that CTXϕ uses the EspD secretin, part of a type II extracellular protein secretion system, the same pathway for export out of the cell of CT, haemagglutinin-protease, chitinase, and other proteins (Davis *et al.*, 2000a; Johnson *et al.*, 2006; Sandkvist, 2001). The EspD protein, an outer membrane component, appears to be the only component of the type II secretion system required for secretion of CTXϕ (Davis *et al.*, 2000a).

CTXϕ regulation

The RstR protein is the key repressor of CTX genome replication, which in the phage lysogenic state is produced at high levels, repressing virion production (McLeod *et al.*, 2005). RstR binds to the *rstA* promoter (*PrstA*), thereby repressing the expression of genes in the RS2 and core regions. Interestingly, the *OPrstA* (operator/promoter region) and *PrstR* overlap one another, indicating that RstR is autorepressed (Fig. 4.1). It has been hypothesized that CTX prophage may occasionally produce virions (McLeod *et al.*, 2005). For example, treatment of CTX lysogens with mitomycin C or UV light, DNA damaging agents that induce the SOS response, also results in increased virion production. The SOS induction of CTXϕ production required RecA-stimulated autocleavage of LexA, a global regulator in *E. coli* (Quinones *et al.*, 2005; Waldor and Mekalanos, 1996). Quinones and coworkers (2005) proposed that both LexA and RstR repress transcription from *PrstA* by binding to the promoter region and that LexA autoproteolysis occurs upon activation of RecA and DNA damage (Quinones *et al.*, 2005). It is hypothesized that removal of LexA from *PrstA* allows RstR to bind *PrstR*, resulting in suppression of RstR synthesis, which enables expression of CTX genes (Quinones *et al.*, 2005). In addition, expression of *rstC* from RS1, which is under the control of *PrstA*, also allows increased CTX production. Davis and colleagues (2002) showed that RstC binds directly to RstR, thereby preventing it from binding to *PrstA* (Davis *et al.*, 2002).

Skorupski and Taylor demonstrated that expression of *ctxAB* from CTX prophages depends on the transcriptional activators ToxR and ToxT (Skorupski and Taylor, 1997). However, *in vitro* and *in vivo* expression of CT by the CTXϕ replicative form does not require ToxR and ToxT, suggesting that induction of CTX prophage may up-regulate toxin production during intestinal infection by increasing the *ctxAB* copy number and by eliminating the requirement for ToxR and ToxT (Wagner and Waldor, 2002). More recently, it has been found that in the infant mouse model of cholera, intraintestinal *ctxA* expression does not depend on *PrstA* and that the mouse intestine is not a significant inducer of the SOS response in *V. cholerae* (Quinones *et al.*, 2006).

Evolutionary genetics of CTXϕ

Precursor of CTX (preCTX)

A number of features of the CTXϕ genome and the *ctxAB* genes suggest that *ctxAB* were acquired by CTXϕ from an unknown source (Boyd *et al.*, 2000a; Waldor and Mekalanos, 1996). The *ctxAB* genes are located in the 3' region of the CTXϕ genome at the *attR* attachment site (Fig. 4.1). This is the site in many phages of insertion of phage-encoded bacterial virulence genes that were likely acquired by imprecise excision from the bacterial host genome

(Boyd, 2005). As stated previously, two promoters are known to promote CT production: the P*rstA* promoter, which is ~ 5 kb upstream of *ctxA* and is required for CTXϕ virion synthesis and can extend into *ctxAB*; and P*ctxAB*, a promoter found immediately upstream of *ctxAB*, indicating that the *ctxAB* genes are regulated differently than other core genes. The percent GC of the *ctxAB* genes, with a mean of 38%, is different from that of the rest of the CTX genome, with a mean of 40%, suggesting a foreign origin compared to other core genes (Fig. 4.1). The location of *ctxAB* at the same site as gIV of F phages from *E. coli* initially suggested that *ctxAB* may have replaced the phage secretion gene. However, the identification of three *V. cholerae* strains that contained CTX prophage sequences that neither encoded the *ctxAB* genes nor a gIV homologue indicates that this is not the case (Fig. 4.1) (Boyd *et al.*, 2000a). These precursor CTX (preCTX) prophages can produce the replicative form, whose genome is very similar to that of the pCTX replicative form, with the exception of missing *ctxAB* (Boyd *et al.*, 2000a). Molecular analysis of these preCTX plasmids revealed that they lacked the upstream control region normally found 5′ of *ctxA*, as well as the *ctxAB* promoter region and coding sequences. Thus, it appears that a preCTX simultaneously acquired the toxin genes and their regulatory sequences (Boyd *et al.*, 2000a). Subsequent to this finding, a number of studies have identified preCTX among environmental *V. cholerae* isolates (Bhattacharya *et al.*, 2006; Maiti *et al.*, 2006). More recently, an environmental isolate of *V. cholerae* was shown to encode preCTX phages tandemly arranged on chromosome I at the CTX integration site (Maiti *et al.*, 2006). A hypothetical scenario for the acquisition of *ctxAB* is that integration of a preCTX phage occurred next to the *ctxAB* genes from an unknown source and because CTX virion production is by rolling circle replication from the chromosome, *ctxAB* was acquired by imprecise prophage replication (Boyd, 2005).

Molecular evolutionary genetic analysis of CTXϕ

The CTXϕ genome has a modular structure consisting of RS2 and core regions. The RS2 region genes *rstA* and *rstB* show a high level of sequence conservation among *V. cholerae* isolates (>99% sequence similarity). In contract, the *rstR* gene can vary from 100% identity among El Tor isolates to <30% amino acid similarity among classical, El Tor, and O139 isolates. Thus, CTXϕ is classified based on *rstR* sequence variation: CTXϕ from classical isolates carry $rstR^{class}$ and are designated $CTX^{class}ϕ$; whereas that from El Tor isolates usually carry $rstR^{ET}$, and are designated $CTX^{ET}ϕ$. *V. cholerae* O139 carry either $CTX^{ET}ϕ$ or $CTX^{Calc}ϕ$, which has a distinct rstR designated $rstR^{Calc}$ (Davis *et al.*, 1999; Kimsey *et al.*, 1998). *V. cholerae* O139 strains may also carry both $CTX^{ET}ϕ$ or $CTX^{Calc}ϕ$ prophages simulataneously (Faruque *et al.*, 2003d). A number of additional novel *rstR* sequences have been identified in *V. cholerae* non-O1/non-O139 isolates (Bhattacharya *et al.*, 2006; Li *et al.*, 2002; Li *et al.*, 2003). Within most core region genes, limited sequence variation is found; however, at the $gIII^{CTX}$ locus, significant biotype-specific sequence diversity was uncovered (Boyd *et al.*, 2000a; Boyd and Waldor, 2002). Molecular evolutionary analysis of CTXϕs derived from a variety of *V. cholerae* natural isolates based on sequence analysis of $gIII^{CTX}$ revealed the presence of two phylogenetic-distinct lineages of CTXϕs between biotype El Tor and classical isolates (Bhattacharya *et al.*, 2006; Boyd *et al.*, 2000a; Boyd and Waldor, 2002). Bhattacharya *et al.* (2006) also performed phylogenetic sequence analysis on *rstR* sequences from a similar cohort of strains and found four major clades: (i) *rstR* sequences from El Tor isolates, (ii) *rstR* sequences from classical isolates similar to the gIII gene tree, (iii) *rstR* sequences from O139 serogroup isolates, and (iv) *rstR* sequences from nonO1/nonO139 isolates (Bhattacharya *et al.*, 2006). These data taken together indicate independent acquisition of CTXϕ among *V. cholerae* isolates and that classical, El Tor and O139 isolates each obtained novel CTX (Boyd *et al.*, 2000a, 2005).

$pIII^{CTX}$ and TcpA sequence diversity

A neighbour-joining phylogenetic tree based on $pIII^{CTX}$ sequences from 20 *V. cholerae* and four *V. mimicus* isolates was constructed using poisson correction (Fig. 4.2A). This tree represents the diversity of $pIII^{CTX}$ sequences in the databases to date. Similar to previous analyses, two divergent $pIII^{CTX}$ lineages are present. Lineage I consists of identical $pIII^{CTX}$ sequences from *V. cholerae* O1 El Tor biotype and O139 serogroup isolates, as well as an O37 serogroup isolate CO130 and a *V. mimicus* isolate PT5. Clustering with these pIII sequences are sequences from a number of *V. cholerae* non-O1/non-O139 isolates (151, 208, V46 and V52) (Fig. 4.2A). The most divergent sequences within this lineage are pIII from *V. cholerae* strains V52 (O37 serogroup), V46 (O141 serogroup) and 151 (O35 serogroup). Within lineage II are $pIII^{CTX}$ sequences from classical biotype isolates, which are all identical and include $pIII^{CTX}$ from a *V. mimicus* isolate PT48 (Fig. 4.2A). On a separate branch within lineage II are $pIII^{CTX}$ sequences from a number of non-O1/non-O139 isolates and two *V. mimicus* isolates (9583 and 523–80). The $pIII^{CTX}$ from strain RV79 (a derivative of MAK757), an El Tor isolate recovered in the 1930s, clustered with these sequences in lineage II (Fig. 4.2A). This suggests that this El Tor isolate had a classical CTXϕ and this was the only exception noted.

A phylogenetic tree based on TcpA, the major pilin protein that directly interacts with $pIII^{CTX}$, using Poisson correction is shown in Fig. 4.2B. The TcpA tree represents the diversity of TcpA present in the databases. TcpA from 31 *V. cholerae* and three *V. mimicus* isolates were examined. There are 12 *V. cholerae* isolates in common to both the pIII and TcpA tree. The TcpA tree differs from the pIII tree in several aspects. Firstly, the diversity of TcpA sequences in the databases is much greater than the diversity of $pIII^{CTX}$. The TcpA sequences are highly divergent from one another (long branch lengths) and seven distinct lineages are noted. The internal node bootstrap values are very high indicating that the relationships based on TcpA among the strains within each lineage is robust. However, the bootstrap values at the base of the tree are very weak indicating that the relationships between the lineages are not resolved. Similar to the $pIII^{CTX}$ tree, TcpA from all *V. cholerae* El Tor and O139 isolates are identical and cluster together on the same branch of the tree with *V. cholerae* strain CO130 (O37) and a *V. mimicus* isolate (Fig. 4.2B). Similarly, within a separate divergent lineage, the TcpA sequences from *V. cholerae* classical isolates have identical sequence and clustered with a *V. mimicus* isolate (Fig. 4.2B). In contrast to the $pIII^{CTX}$ sequences from nonO1/non-O139 isolates, the TcpA sequences are found on five separate divergent lineages indicating greater heterogeneity in TCP among *V. cholerae* isolates. Inconsistencies between the $pIII^{CTX}$ and TcpA trees are noted. For example, *V. cholerae* strains 208 and 151 on the $pIII^{CTX}$ tree grouped with the El Tor–like sequences, whereas on the TcpA tree they form a separate divergent lineage and group with strain VCE22 (Fig. 4.2B). Also *V. cholerae* strain V52 (O37) groups with the El Tor-like sequences on the $pIII^{CTX}$ tree but with the classical-like sequences on the TcpA tree.

The interbiotype sequence diversity at both $pIII^{CTX}$ and TcpA loci is all the more striking since these strains are otherwise highly similar when additional genetic markers such as housekeeping genes are examined. For example, El Tor and classical biotype strains show 100% sequence identity at the Mdh, GroEL-1 and GroEL-2 loci, whereas at the TcpA locus between the biotypes share less than 85% sequence similarity. The most divergent TcpA sequence from *V. cholerae* strains V54, SD95001 and 153–94 share less than 70% sequence similarity with classical biotype strains (Fig. 4.2B). Monomers of the TcpA protein make up the TCP pilus, which is a surface exposed structure that interacts directly with the environment. The extreme genetic diversity uncovered at TcpA suggests positive Darwinian selection acting at this locus perhaps playing a role in host immune avoidance in the case of pathogenic interactions with humans and/or TCP maybe involved in interactions with a range of zooplankton in the environment perhaps conferring specificity.

The correlation between divergence at the TcpA and the pIII loci from classical and El Tor isolates is intriguing and suggests selective pressure for change; that is, the $pIII^{CTX}$ sequence has changed in order to bind to the respective TcpA proteins. However, our *in vitro* yeast two-model system did not uncover a qualitative difference in pIII and TcpA binding regardless of the different alleles examined indicating that the relationship may be very subtle, which will require further experimentation.

Distribution of CTXϕ

Apart from *V. cholerae* O1 and related strains i.e. *V. cholerae* O139, the distribution of CTXϕ is sporadic among nonO1/nonO139 serogroup strains (Faruque *et al.*, 1998a; Faruque *et al.*, 1998b; Faruque *et al.*, 2003b; O'Shea *et al.*, 2004). However, CTXϕ is readily mobile via transduction, which is detectable *in vitro* and *in vivo* (Waldor and Mekalanos, 1996; Waldor *et al.*, 1997). The limited distribution is likely due to the limited distribution of TCP, which is the host receptor for CTXϕ. TCP is encoded on the TCP pathogenicity island, alternatively known as Vibrio pathogenicity island (VPI), and the bacterium must acquire this region before it can uptake CTXϕ (Waldor and Mekalanos, 1996). CTXϕ has been isolated from a range of *V. mimicus* isolates and these isolates also encode TCP, indicating a similar method of phage infection. The presence of CTXϕ and VPI among *V. mimicus* isolates represents an additional reservoir of *V. cholerae* virulence factors (Boyd *et al.*, 2000b). Comparative sequence analysis of CTXϕ genes from *V. cholerae* and *V. mimicus* showed high sequence similarity, indicating recent horizontal transfer of CTX between these species (Boyd *et al.*, 2000b). From the genome sequence of *V. fischeri*, homologues of CTX as well as the CTX host receptor TCP have been identified; however, it is not known whether the TCP sequence is functional. The CTX prophage from *V. fischeri* lacks the *ctxAB* genes. Examination of more than 15 *V. fischeri* isolates recovered from different sources for the presence of both CTX and TCP genes indicated that they have a limited distribution, but, similar to *V. cholerae*, the presence of CTX always correlated with the presence of TCP (Boyd, unpublished data).

CTXϕ host receptor: the toxin co-regulated pilus (TCP)

TCP, a type IV pilus, is the receptor for CTXϕ and an essential human intestinal colonization factor (Taylor *et al.*, 1987; Waldor and Mekalanos, 1996). TCP is encoded by the *tcp* gene cluster found on a 41-kb region that encodes additional virulence gene clusters, as well as a phage-like integrase and the region inserts at a tRNA site (*ssrA*) (Karaolis *et al.*, 1998; Kovach *et al.*, 1996; Taylor *et al.*, 1987). The 41-kb region was defined as a pathogenicity island (PAI), the TCP pathogenicity island, or *Vibrio* pathogenicity island (VPI) (Karaolis *et al.*, 1998; Kovach *et al.*, 1996). Subsequently, Karaolis and colleagues proposed that TCP is encoded by a filamentous phage named VPIϕ and that the major pilus protein (TcpA), is the VPIϕ's coat protein (Karaolis *et al.*, 1999). However, more recent data show that the initial study was incorrect and that VPI, although it can excise from the cell and form circular intermediates (CIs), is not itself a phage, nor does it encode a phage (Faruque and Mekalanos, 2003; Faruque *et al.*, 2003c; Rajanna *et al.*, 2003). It is more likely transmitted by phage transduction. For example, O'Shea and Boyd (2002) demonstrated interstrain transfer of VPI mediated by generalized transducing phage CP-T1. They used a kanamycin-labelled copy of the VPI to show transfer from a *V. cholerae* El Tor biotype strain to four VPI-negative *V. cholerae* O1 serogroup strains via CP-T1 (O'Shea and Boyd, 2002). Previous work had shown that CP-T1 phage preferential packaged the circularized form of CTXϕ for transfer to CTX negative isolates (Boyd and Waldor, 1999). The observation that VPI can excise and form a CI also lends support to the horizontal transfer of this region and is likely a first step in its transfer from cell to cell. Excision and formation of CIs is also a property of other *V. cholerae* encoded PAIs such as VPI-2, VSP-I and VSP-II (Murphy and Boyd, 2008). VPI-I, VPI-2 and VSP-II each encode a P4-like integrase adjacent to a tRNA locus and each PAI is flanked by direct repeats. All four regions appear to have a mechanism of excision similar to phages and in the case of VPI-2 and VSP-II the cognate P4-like integrase was required for excision (Murphy and Boyd, 2008). The recent discovery that *V. cholerae* becomes naturally competent for the uptake of naked DNA in the present of chitin suggests an alternative mechanism of horizontal transfer of these regions among *V. cholerae*, which needs to be investigated further (Meibom *et al.*, 2005; Miller *et al.*, 2007).

The RS1ϕ

RS1ϕ is a satellite filamentous phage that encodes four open reading frames (ORFs) RstR, RstA, RstB, and RstC. RstR, RstA, and RstB are homologous to the RS2 region genes of CTXf. An additional ORF RstC, is an antirepressor protein that causes the RstR repressor to aggregate and prevent binding to the P*rstA* promoter as discussed previously (Davis *et al.*, 2002; Faruque *et al.*, 2002). The RstC protein promotes transcription of CTX genes required for phage production, supporting the transmission of both RS1ϕ and CTXϕ (Davis *et al.*, 2002). RS1 is a satellite filamentous phage that depends on the CTXϕ for capsid proteins, packaging, and transmission (Faruque *et al.*, 2002). Filamentous bacteriophages such as VGJϕ can also supply capsid, assembly, and packaging proteins for RS1 that are required to exit the bacterial host cell (Campos *et al.*, 2003a). Faruque and colleagues found that a kanamycin resistance (Kn^r)-marked copy of the RF of RS1, pRS1-Kn, transduced *V. cholerae* isolates to Kn^r and similar to CTXϕ, integrated site-specifically into the *V. cholerae* genome at the 20-bp *attRS* sequence (Faruque *et al.*, 2002).

Genetic organization of CTX prophages

Many different CTX and RS1 arrangements have been observed among *V. cholerae* pandemic isolates (Mekalanos and, 1983). For example, El Tor strain E7946 was shown to contain an RS1-CTX^{ET}-RS1-CTX^{ET}-RS1 arrangement, whereas strain C6709, isolated in Peru in 1991, contained a CTX^{ET}-RS1 arrangement (Mekalanos and, 1983; Waldor and Mekalanos, 1994). The arrangement of CTX prophage on the genome of *V. cholerae* varies depending on whether (i) it is integrated at one or two sites and (ii) the biotype or serogroup of the isolate examined. In *V. cholerae* classical biotype isolates, CTX is integrated on both chromosomes and never contains an RS1 element (Davis *et al.*, 2000b; Sharma *et al.*, 1997). In *V. cholerae* classical strains on chromosome I, CTX is present in the intergenic region between Toxin Linked Cryptic (TLC) plasmid (ORF VC1465) and Repeat in toxin (RTX) (ORF VC1451) gene clusters, designated as the El Tor site, and on chromosome II in the intergenic region between VCA0569 and VCA0570, designated as the classical site (Davis *et al.*, 2000b). Within classical strains O395, GP12, and C33, at the El Tor integration site on chromosome I CTX^{class} is present as an array of truncated, fused prophages, whereas at the classical integration site on chromosome II, a solitary CTX^{class} prophage is present (Davis *et al.*, 2000b). In classical strains CA401, C1, C14, C21, and C34, the opposite genetic arrangement is found. At the El Tor site, a single CTX^{class} prophage is present and at the classical site, a truncated CTX^{class} array is found (Davis *et al.*, 2000b). Davis and Waldor found that production of infectious CTX particles requires the presence of either two CTX-CTX prophages or CTX-RS1 array and that the DNA in CTXϕ is a hybrid sequence that is derived from these elements (Davis and Waldor, 2000). Thus, the classical prophage arrangements cannot generate pCTX^{class} and CTX^{class}ϕ (Davis *et al.*, 2000b). Among *V. cholerae* El Tor isolates, CTX^{ET} is present at the El Tor site on chromosome I either as a CTX^{ET}-CTX^{ET} array or a CTX^{ET}-RS1 array (Davis and Waldor, 2000; Mekalanos, 1983; Waldor and Mekalanos, 1994).

Within *V. cholerae* O139 serogroup strains, a number of CTX prophage arrangements have been described (Bhadra *et al.*, 1995; Davis *et al.*, 1999; Davis *et al.*, 2000b; Davis and Waldor, 2000; Khetawat *et al.*, 1999; Kimsey *et al.*, 1998). For example, strain SG24 contains a CTX^{ET}-RS1-CTX^{ET} array, and strains AS197 and AS209 contain a RS1-CTX^{ET}-CTX^{Calc}-CTX^{Calc} array. Since these initial studies of CTX prophage arrangements, several variant arrays have been described. For example, a *V. cholerae* El Tor isolate recovered before 1992 carried a CTX-CTX array on chromosome II, which includes a CTX similar CTX^{ET}, and a novel CTX similar to CTX^{Calc} isolated from O139 isolates recovered after 1992 (Nandi *et al.*, 2003). Genomic analysis of an El Tor *V. cholerae* strain B-33 isolated in Mozambique in 2004 revealed the presence of CTX^{class} with a CTX^{class}-CTX^{class} array at the classical insertion site (Faruque *et al.*, 2007; Lee *et al.*, 2006). These data indicate that CTX prophage arrangement is dynamic and diverse, that is continually giving rise to new variant strains.

Genomics of other *Vibrio cholerae* filamentous phages

Along with CTX and RS1, there are at least seven well described filamentous phages from *V. cholerae* isolates: VGJ, KSF-1, fs-1, VSK, VSKK,

493, and fs-2 (Fig. 4.3 and Table 4.2). As more *V. cholerae* genome sequences become available (six O1, three O139, and three non-O1/non-139 genome sequences are in progress), the number of filamentous phage whole genome sequences should increase dramatically and with it our knowledge of the evolution of these phages.

VGJϕ

The filamentous phage VGJϕ was shown to integrate at the same chromosomal site as CTXϕ (Campos *et al.*, 2003b). VGJϕ can infect O1 (classical and El Tor) and O139 strains. The genome of VGJϕ is 7542 bp with a GC content of 43% encoding 10 ORFs and is known to use the type IV pilus mannose-sensitive haemagglutinin (MSHA) as a host cell receptor (Fig. 4.3 and Table 4.2)(Campos *et al.*, 2003b). The VGJϕ genome sequence showed the presence of two sites homologous to functional *att* sequences from other filamentous phages such as CTXϕ and VSK from *V. cholerae*, and f237 from *V. parahaemolyticus*. These sites mapped to similar relative positions in the genome. The VGJϕ *att* sites overlapped in opposite directions, similar to VSK and f237 (Campos *et al.*, 2003b). VGJ integrates via its own *attP* sequence (CTTACTTCGCATTATGT) into the same chromosomal *attB* site (CTTAGTGCGTATTATGT) where CTXϕ integrates. It is suggested that integration of VGJϕ and other related filamentous phages depends on the XerCD recombinases of their host, similar to CTXϕ and f237 (Campos *et al.*, 2003b; Huber and Waldor, 2002; Iida *et al.*, 2002). It has also been demonstrated that VGJϕ provides functions required for production and transmission of both CTXϕ and RS1ϕ (Campos *et al.*, 2003a). The significance of this finding is the ability of VGJϕ to transfer CTXϕ to *V. cholerae* cells by a TCP-independent mecha-

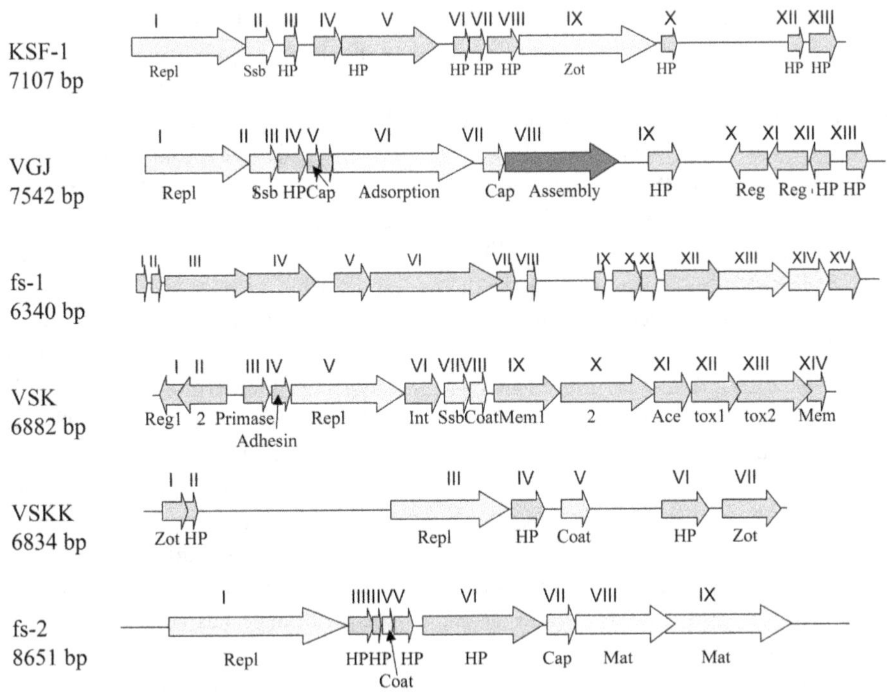

Figure 4.3 Schematic representation of ssDNA filamentous bacteriophages from *V. cholerae*. Arrows indicate ORFs, and the direction of the arrows indicates the direction of transcription. The roman numerals above the ORFs indicate ORF number and underneath are indicate protein functions. Abbreviations: Rep1, replicase, Ssb, single strand binding protein, HP, hypothetical protein, Zot, zonula occludens toxin, Cap, capsid protein, Reg, regulation, Mem, membrane protein, Mat, maturation protein. Genes encoding similar proteins have identical pattern arrows among the phages. DNA replication genes are represented as striped arrows, single strand binding proteins are represented as bubbled arrows, phage morphogenesis and assembly genes are represented as hatched and dotted arrows.

Table 4.2 Filamentous bacteriophages identified in *V. cholerae*

Phage	Size (bp)	GC (%)	Receptor
CTXϕ	6900	44	TCP
RS1ϕ	3000	44	TCP/MSHA
KSF-1ϕ	7107	44	MSHA
VGJϕ	7542	43	MSHA
fs-1	6340	43	MSHA
VSKϕ	6882	43	Unknown
VSKK	6834	43	Unknown
fs-2	8651	44	MSHA

nism, because the host receptor for VGJϕ is the MSHA pilus thus bypassing the requirement for TCP. Campos and colleagues (2002b) described a novel type of specialized transduction in which the site-specific co-integration of VGJϕ and CTXϕ (or RS1ϕ) RF genome results in single hybrid molecules that generate ssDNA hybrid genomes that are packaged into hybrid phages (VGJϕ/CTXϕ hybrid or VGJϕ/RS1ϕ hybrid) (Campos *et al.*, 2003a).

KSF-1ϕ

Similar to VGJϕ, KSF-1ϕ is a viable infectious phage that integrates into the host genome and contains an 18bp *att*-like sequence (CAAGCCGATACTGCGCGA) (Faruque *et al.*, 2003a). The host cell receptor for KSF-1ϕ is also the MSHA pilus (Faruque *et al.*, 2005). KSF-1ϕ has a 7,107-bp genome with a GC content of 44% encoding 14 ORFs (Fig. 4.3) (Faruque *et al.*, 2005). KSF-1ϕ, similar to VGJϕ, was shown to produce RS1 phage independently of CTXϕ (Faruque *et al.*, 2003a).

VSK, 493, VSKK, fs-1 and fs-2 phages

Five other filamentous phages that also use MSHA as their host cell receptor are phage VSK, 493, VSKK, fs-1, and fs-2 (Kar *et al.*, 1996; Ehara *et al.*, 1997; Honma *et al.*, 1997; Ikema and Honma, 1998; Jouravleva *et al.*, 1998a,b). VSKϕ is a 6880-bp phage with a GC content of 43% that was isolated from a *V. cholerae* O139 isolate in 1996, and it integrated into the genome (Kar *et al.*, 1996). VSKϕ shows high sequence similarity of 83% at the nucleotide level to VGJϕ and 87% to fs-1 (Kar *et al.*, 1996). The genome sequence of phage 493 is not available but its approximate size is 9,300 bp with a 2300-bp stem-loop structure, which may represent a transposable element. Phage 493 is of interest as it was isolated from a 1994 isolate of *V. cholerae* O139, and was shown to inhibit pre-O139 but not post-O139 El Tor biotype *V. cholerae* strains in plaque assays (Jouravleva *et al.*, 1998a). In addition, the emergence and decline of O139 in India and Bangladesh were correlated with the susceptibility and resistance of El Tor strains to 493ϕ (Jouravleva *et al.*, 1998b). Phages fs-1 and fs-2 were isolated from *V. cholerae* O139 strains and were found to form infectious phage particles and also could integrate into the host genome. Phage fs-1 was also isolated from a number of clinical *V. cholerae* O1 isolates (Ehara *et al.*, 1997). Phage fs-1 has a 6,340-bp ssDNA genome that encodes 15 ORFs; the fs-2 genome size is 8,651 bp, containing nine ORFs and a large stem loop region (Fig. 4.3) (Honma *et al.*, 1997; Ikema and Honma, 1998). Almost all *V. cholerae* O139 and O1 El Tor biotype strains tested were sensitive to fs-2, whereas most O1 classical biotype strains and non O1/nonO139 strains were resistant, demonstrating a limited host range (Ikema and Honma, 1998). Although the fs-2 phage has conserved gene content and gene order, it shares very little sequence similarity to other *V. cholerae* filamentous phages (Fig. 4.3).

Common themes

Some key themes emerge from the characterization and molecular analysis of filamentous phages from *V. cholerae*. From a structural and gene content perspective, most of the filamentous phages discussed here have genomes of a similar size (range 6–8.7 kb) and encode a similar number of ORFs (range 7–15). Also, distinct replication and morphogenesis modules are present in most of these phages. From a mechanistic perspective also, common features are found, for example most of the filamentous phages studied integrate into the host genome. In addition, since most of these phages contain *att*-like sites similar to CTXϕ, it is likely that most use a similar XerD and XerC mechanism of integration. Interestingly, outside of *V. cholerae*, the *att*-like sites and *dif* sequence are also found in a number of filamentous phages from a range of species, suggesting similar pathways to integration (Huber and Waldor, 2002; Iida *et al.*, 2002).

Evolutionary phylogenetic analysis of *Vibrio cholerae* filamentous phages

To understand better the diversity and evolutionary relationships among filamentous phages, phylogenetic trees were reconstructed using minimum evolution based on two loci; the RstA (representing the replication module) and Zot (representing the core module) homologues in the databases (Fig. 4.4).

RstA sequence tree

All of the RstA sequences from *V. cholerae* serogroup O1 and O139 strains were highly similar, forming a tight lineage on the RstA tree represented by three sequences from *V. cholerae* CTX^{ElTor}, CTX^{class}, and CTX^{Calc} and RstA from *V. mimicus* (Fig. 4.4A). The RstA CTX sequence from *V. fischeri* was the most closely related to RstA from *V. cholerae* CTX phage, and clustered with these sequences on a separate divergent branch (Fig. 4.4). The RstA sequences from VGJ, KSF-1, VSK, VSKK, and fs-1 all clustered together on the tree forming a closely related lineage. These RstA sequences were the most closely related sequences to the RstA from CTX phage (Fig. 4.4A). On a separate clade, the RstA sequences from *V. parahaemolyticus* filamentous phages of Vf33 and VfO3:K6 clustered with a sequence from *Vibrio* spp. strain Ex25, and these sequences clustered with a sequence from *V. cholerae* strain MAK757, an El Tor isolate recovered in the 1930s. This clade also included sequences from *V. vulnificus* strain YJ016 and *V. cholerae* strain MZO-3, which both clustered together (Fig. 4.4A). In summary, the RstA tree contained four main lineages; lineage I represents sequences from CTX from *V. cholerae*, *V. mimicus*, and *V. fischeri*; lineage II represents sequences encoded on VSK, VSKK, VGJ, fs-1, and KSF-1 from *V. cholerae* strains; lineage III encompasses phage Vf33 and VfO3:K6 from *V. parahaemolyticus*, sequences from *Vibrio* spp. Ex25 and *V. cholerae* strain MAK757; lineage IV represents a sequence from a phage-like region from *V. vulnificus* strain YJ016 and a sequence from *V. cholerae* strain MZ0–3.

Zot sequence tree

The phylogenetic tree based on the Zot sequences gave a similar clustering of CTX phages from the various *Vibrio* species as the RstA sequences (Fig. 4.4B). For example, all the Zot sequences of CTX from *V. cholerae* isolates clustered together on a divergent branch of the tree, which also included the Zot sequences of CTX from *V. mimicus* and *V. fischeri*. On a separate lineage, the Zot sequences from KSF-1 and sequences from a number of *V. cholerae* isolates branched together. This lineage also included a Zot sequence from *V. alginolyticus*. On a divergent branch from this cluster are Zot-like proteins from *V. parahaemolyticus* phage VfO3:K6 and *Vibrio* sp strain Ex25. Zot sequences from *Colwellia psychrerythraea* and *Shewanella baltica* formed separate divergent branches from the others and each other. Zot sequences from all the *Pseudomonas* species clustered together; for example, Zot from phage pf1 from *Pseudomonas aeruginosa* and sequences from two additional *Pseudomonas* species formed a separate divergent branch on the tree (Fig. 4.4B). The Zot-like sequences encoded on VSK, VSKK, fs-1 and VGJ do not share significant homology to the Zot proteins from CTX or the other Zot proteins discussed above. These were not included in the analysis. The Zot-like proteins from these phages cluster together and are related to Zot-like proteins from *Shewanella baltica*, Vf33 and a filamentous phage from *Photobacterium profundum*.

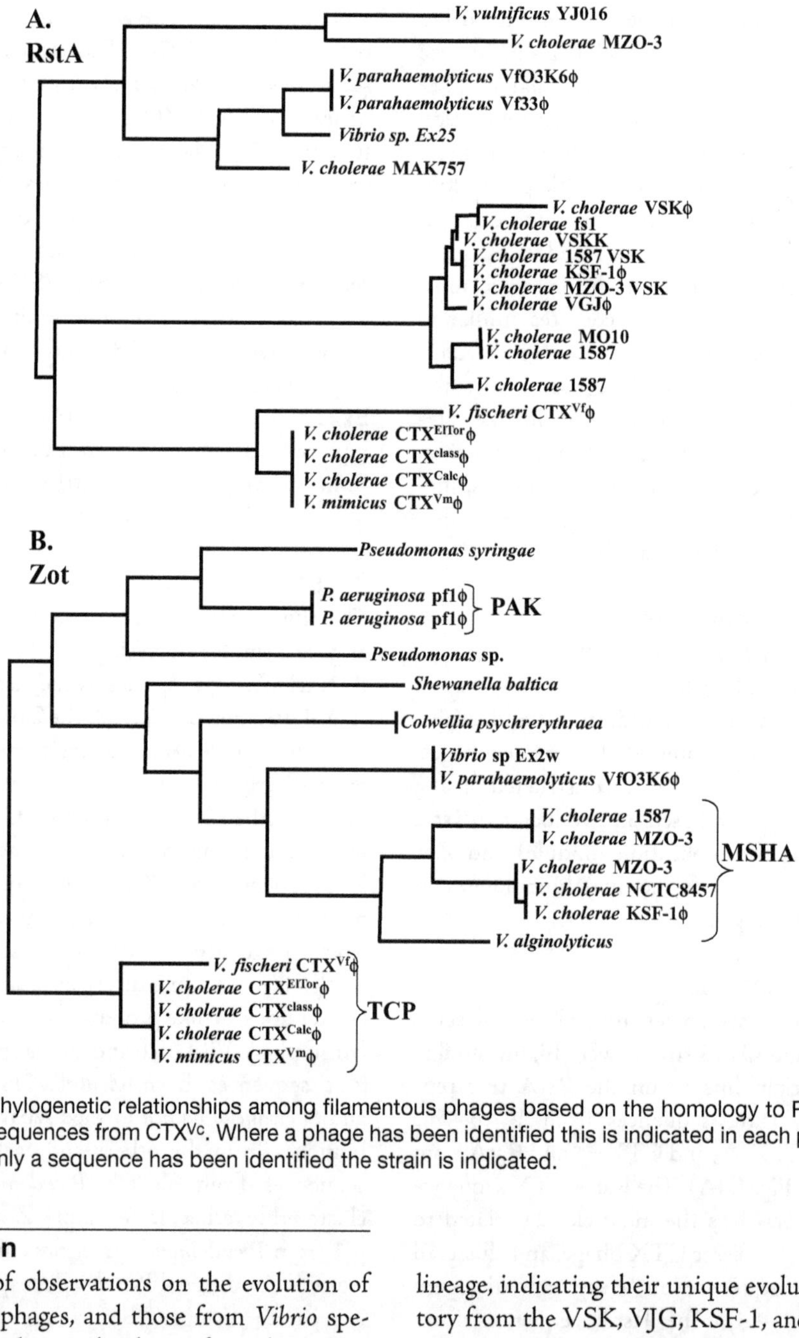

Figure 4.4 Phylogenetic relationships among filamentous phages based on the homology to RstA and Zot amino acid sequences from CTX^{Vc}. Where a phage has been identified this is indicated in each phylogenetic tree. When only a sequence has been identified the strain is indicated.

Conclusion

A number of observations on the evolution of filamentous phages, and those from *Vibrio* species in particular, can be drawn from these data. First, it appears that the RstA sequences are conserved (short branch lengths) among VSK, VJG, KSF-1, and f1 phages within *V. cholerae* strains relative to the Zot sequences. RstA, and by extension the replication module, has a phylogenetically limited distribution since it was only identified in *Vibrio* species. Second, the CTX phages from *V. cholerae, V. mimicus*, and *V. fischeri* are clustered together forming a separate lineage, indicating their unique evolutionary history from the VSK, VJG, KSF-1, and f1 phages. In contrast, the analysis of Zot indicates that these proteins are more divergent (long branch lengths) from one another and are phylogenetically more widely distributed than RstA. Unlike, RstA which is confined to aquatic *Vibrio* species, Zot is found not only among *Vibrio* species and other aquatic species but also among a number of *Pseudomonas* species.

As discussed in an earlier section, the host cell receptor for CTX from *V. cholerae* and *V.*

mimicus is TCP. A gene cluster homologous to TCP is also present in *V. fischeri*, as are the *tolQRA* genes required for CTXϕ infection in *V. cholerae*; it therefore seems likely that TCP is the receptor for $CTX^{VF}ϕ$. Similar to CTX, TCP is limited to subsets of strains within three species, *V. cholerae*, *V. mimicus* and *V. fischeri*, thus explaining the limited distribution of CTX. The cell receptor for VSK, VJG, KSF-1, and f1 phages is MSHA, a IV pilus that is widely present among *Vibrio* species. Of the four completed *Vibrio* genomes and the 20 genomes in progress encompassing 10 different *Vibrio* species, all encode the MSHA gene cluster, indicating the potential broad host range of these filamentous phages and their potential role in horizontal gene transfer across species boundaries. The receptor for Pf1 from *P. aeruginosa* is the type IV PAK pilus and for Pf3 the conjugative RP4 pilus (Holland *et al.*, 2006). The receptor for Vf33 is unknown but is likely to be a type IV pilus based on the what appears to be a conserved pathway of infection among filamentous phages discussed in this section.

Acknowledgements

I thank past and present members of my laboratory for their enthusiasm, support, and hard work, especially Ana Luisa Cohen, Salvador Almagro Moreno, Ronan A. Murphy, Lynn M. Naughton, Jennifer Moynihan, David Woods and Noelle McCarthy. I thank the Department of Biological Sciences, University of Delaware, for continued support and University of Delaware Research Foundation grants. Literature citations have been limited, in some cases to recent relevant reviews, due to space limitations. Therefore, I apologize to those authors whose research is not included or cited.

References

Bhadra, R., Roychoudhury, S., Banerjee, R., Kar, S., Majumdar, R., Sengupta, S., Chatterjee, S., Khetawat, G. and Das, J. (1995). Cholera toxin (CTX) genetic element in *Vibrio cholerae* o139. Microbiology *141*, 1977–1983.

Bhattacharya, T., Chatterjee, S., Maiti, D., Bhadra, R., Takeda, Y., Nair, G. and Nandy, R. (2006). Molecular analysis of the *rstR* and *orfU* genes of the ctx prophages integrated in the small chromosomes of environmental *Vibrio cholerae* non-o1, non-o139 strains. Environ. Microbiol. *8*, 526–634.

Blakely, G., May, G., McCulloch, R., Arciszewska, L.K., Burke, M., Lovett, S.T. and Sherratt, D.J. (1993). Two related recombinases are required for site-specific recombination at dif and cer in *E. coli* K12. Cell *75*, 351–361.

Blakely, G. (2004). Smarter than the average phage. Mol. Microbiol. *54*, 851–854.

Boyd, E.F. and Waldor, M.K. (1999). Alternative mechanism of cholera toxin acquisition by *Vibrio cholerae*: generalized transduction of CTXϕ by bacteriophage CP-T1. Infect. Immun. *6*, 5898–5905.

Boyd, E.F., Heilpern, A.J. and Waldor, M.K. (2000a). Molecular analysis of a putative CTXϕ precursor and evidence for independent acquistion of distinct CTXϕs by toxigenic *Vibrio cholerae*. J. Bacteriol. *182*, 5530–5538.

Boyd, E.F., Moyer, K.L., Shi, L. and Waldor, M.K. (2000b). Infectious CTXϕ and the *Vibrio* pathogenicity island prophage in *Vibrio mimicus*: Evidence for recent horizontal transfer between *V. mimicus* and *V. cholerae*. Infect. Immun. *68*, 1507–1513.

Boyd, E.F. and Waldor, M.K. (2002). Evolutionary and functional analyses of variants of the toxin-coregulated pilus protein tcpa from toxigenic *Vibrio cholerae* non-O1/non-O139 serogroup isolates. Microbiology *148*, 1655–1666.

Boyd, E.F. (2005). Bacteriophages and bacterial virulence. In Bacteriophages: Biology and applications, pp. 223–266. Edited by E. Kutter and A. Sulakvelidze: CRC Press.

Campos, J., Martinez, E., Marrero, K., Silva, Y., Rodriguez, B., Suzarte, E., Ledon, T. and Fando, R. (2003a). Novel type of specialized transduction for CTXphi or its satellite phage RS1 mediated by filamentous phage VGJphi in *Vibrio cholerae*. J. Bacteriol. *185*, 7231–7240.

Campos, J., Martinez, E., Suzarte, E., Rodriguez, B.L., Marrero, K., Silva, Y., Ledon, T., del Sol, R. and Fando, R. (2003b). VGJphi, a novel filamentous phage of *Vibrio cholerae*, integrates into the same chromosomal site as CTXphi. J. Bacteriol. *185*, 5685–5696.

Davis, B.M., Kimsey, H.H., Chang, W. and Waldor, M.K. (1999). The *Vibrio cholerae* o139 Calcutta CTXϕ is infectious and encodes a novel repressor. J. Bacteriol. *181*.

Davis, B.M., Lawson, E.H., Sandkvist, M., Ali, A., Sozhamannan, S. and Waldor, M.K. (2000a). Convergence of the secretory pathways for cholera toxin and the filamentous phage, CTXphi. Science, *288*, 333–335.

Davis, B.M., Moyer, K.E., Boyd, E.F. and Waldor, M.K. (2000b). CTX prophages in classical biotype *Vibrio cholerae*: Functional phage genes but dysfunctional phage genomes. J. Bacteriol. *182*, 6992–6998.

Davis, B.M. and Waldor, M.K. (2000). CTXϕ contains a hybrid genome derived from tandemly integrated elements. Proc. Natl. Acad. Sci. USA *97*, 8572–8577.

Davis, B.M., Kimsey, H.H., Kane, A.V. and Waldor, M.K. (2002). A satellite phage-encoded antirepressor induces repressor aggregation and cholera toxin gene transfer. EMBO J. *21*, 4240–4249.

Deng, L.W., Malik, P. and Perham, R.N. (1999). Interaction of the globular domains of pIII protein of filamentous bacteriophage fd with the F-pilus of *Escherichia coli*. Virology *253*, 271–277.

Deng, L.W. and Perham, R.N. (2002). Delineating the site of interaction on the piii protein of filamentous bacteriophage fd with the f-pilus of *Escherichia coli*. J. Mol. Biol. *319*, 603–614.

Ehara, M., Shimodori, S., Kojima, F., Ichinose, Y., Hirayama, T., Albert, M.J., Supawat, K., Honma, Y., Iwanaga, M. and Amako, K. (1997). Characterization of filamentous phages of *Vibrio cholerae* O139 and O1. FEMS Microbiol. Lett. *154*, 293–301.

Faruque, S.M., Kamruzzaman, M., Asadulghani, Sack, D., Mekalanos, J.J. and Nair, G.B. (2003a). CTXΦ-independent production of the RS1 satellite phage by *Vibrio cholerae*. Proc. Natl. Acad. Sci. USA *100*, 1280–1285.

Faruque, S.M., Naser Bin, I., Fujihara, K., Diraphat, P., Chowdhury, N., Kamruzzaman, M., Qadri, F., Yamasaki, S., Ghosh, A.N. and Mekalanos, J.J. (2005). Genomic sequence and receptor for the *Vibrio cholerae* phage KSF-1phi: Evolutionary divergence among filamentous vibriophages mediating lateral gene transfer. J. Bacteriol. *187*, 4095–4103.

Faruque, S.M., Tam, V., Chowdhury, N., Diraphat, P., Dziejman, M., Heidelberg, J., Clemens, J., Mekalanos, J.J. and Nair, G. (2007). Genomic analysis of the mozambique strain of *Vibrio cholerae* O1 reveals the origin of el tor strains carrying classical CTX prophage. Proc. Natl. Acad. Sci. USA *104*, 5151–5156.

Faruque, S.M., Albert, M.J. and Mekalanos, J.J. (1998a). Epidemiology, genetics, and ecology of toxigenic *Vibrio cholerae*. Microbiol. Mol. Biol. Rev. *62*, 1301–1314.

Faruque, S.M., Asadulghani, Saha, M.N., Alim, A.R., Albert, M.J., Islam, K.M. and Mekalanos, J.J. (1998b). Analysis of clinical and environmental strains of nontoxigenic *Vibrio cholerae* for susceptibility to CTXϕ: Molecular basis for origination of new strains with epidemic potential. Infect. Immun, *66*, 5819–5825.

Faruque, S.M., Asadulghani, Kamruzzaman, M., Nandi, R.K., Ghosh, A.N., Nair, G.B., Mekalanos, J.J. and Sack, D.A. (2002). RS1 element of *Vibrio cholerae* can propagate horizontally as a filamentous phage exploiting the morphogenesis genes of CTXphi. Infect Immun. *70*, 163–170.

Faruque, S.M. and Mekalanos, J.J. (2003). Pathogenicity islands and phages in *Vibrio cholerae* evolution. Trends Microbiol. *11*, 505–510.

Faruque, S.M., Sack, D.A., Sack, R.B., Colwell, R.R., Takeda, Y. and Nair, G.B. (2003b). Emergence and evolution of *Vibrio cholerae* O139. Proc. Natl. Acad. Sci. USA *100*, 1304–1309.

Faruque, S.M., Zhu, J., Asadulghani, Kamruzzaman, M. and Mekalanos, J.J. (2003c). Examination of diverse toxin-coregulated pilus-positive *Vibrio cholerae* strains fails to demonstrate evidence for Vibrio pathogenicity island phage. Infect. Immun. *71*, 2993–2999.

Faruque, S.M., Chowdhury, N., Kamruzzaman, M., Ahmad. Q.S., Faruque. A.S., Salam, M.A., Ramamurthy, T., Nair, G.B., Weintraub, A., Sack, D.A. (2003d). Reemergence of epidemic *Vibrio cholerae* O139, Bangladesh. Emerg. Infect. Dis. *9*, 1116–1122.

Faruque, S.M., Naser, I.B., Islam, M.J., Faruque, A.S., Ghosh, A.N., Nair, G.B., Sack, D.A. and Mekalanos, J.J. (2005). Seasonal epidemics of cholera inversely correlate with the prevalence of environmental cholera phages. Proc. Natl. Acad. Sci. USA *102*, 1702–7.

Fasano, A., Baudry, B., Pumplin, D., Wasserman, S., Tall, B., Ketley, J. and Kaper, J. (1991). *Vibrio cholerae* produces a second enterotoxin, which affects intestinal tight junctions. Proc. Natl. Acad. Sci. USA *88*, 5242–5246.

Heilpern, A. and Waldor, M. (2003). PIIICTX, a predicted CTXphi minor coat protein, can expand the host range of coliphage fd to include *Vibrio cholerae*. J. Bacteriol. *185*, 1037–1044.

Heilpern, A.J. and Waldor, M.K. (2000). CTXphi infection of *Vibrio cholerae* requires the tolqra gene products. J. Bacteriol. *182*, 1739–1747.

Holland, S.J., Sanz, C. and Perham, R.N. (2006). Identification and specificity of pilus adsorption proteins of filamentous bacteriophages infecting pseudomonas aeruginosa. Virology *345*, 540–548.

Holliger, P. and Riechmann, L. (1997). A conserved infection pathway for filamentous bacteriophages is suggested by the structure of the membrane penetration domain of the minor coat protein g3p from phage fd. Structure *5*, 265–275.

Holliger, P., Riechmann, L. and Williams, R.L. (1999). Crystal structure of the two n-terminal domains of g3p from filamentous phage fd at 1.9 a: Evidence for conformational lability. J. Mol. Biol. *288*, 649–657.

Honma, Y., Ikema, M., Toma, C., Ehara, M. and Iwanaga, M. (1997). Molecular analysis of a filamentous phage (fsl) of *Vibrio cholerae* O139. Biochim. Biophys. Acta *1362*, 109–115.

Huber, K.E. and Waldor, M.K. (2002). Filamentous phage integration requires the host recombinases XerC and XerD. Nature *417*, 656–659.

Iida, T., Makino, K., Nasu, H., Yokoyama, K., Tagomori, K., Hattori, A., Okuno, T., Shinagawa, H. and Honda, T. (2002). Filamentous bacteriophages of vibrios are integrated into the dif-like site of the host chromosome. J. Bacteriol. *184*, 4933–4935.

Ikema, M. and Honma, Y. (1998). A novel filamentous phage, fs-2, of *Vibrio cholerae* O139. Microbiology *144*, 1901–1906.

Jensen, M., Faruque, S., Mekalanos, J. and Levin, B. (2006). Modeling the role of bacteriophage in the control of cholera outbreaks. Proc. Natl. Acad. Sci. USA *103*, 4652–4657.

Johnson, T., Abendroth, J., Hol, W.G. and Sandkvist, M. (2006). Type II secretion: From structure to function. FEMS Microbiol. Lett. *255*, 175–186.

Jouravleva, E.A., McDonald, G.A., Garon, C.F., Boesman-Finkelstein, M. and Finkelstein, R.A. (1998a). Characterization and possible functions of a new filamenous bacteriophage from *Vibrio cholerae* O139. Microbiology *144*, 315–324.

Jouravleva, E.A., McDonald, G.A., Marsh, J.W., Taylor, R.K., Boesman-Finkelstein, M. and Finkelstein, R.A. (1998b). The *Vibrio cholerae* mannose-sensitive aemaglutinin is the receptor for a filamentous bacteriophage from *Vibrio cholerae* O139. Infect. Immun. *66*, 2535–2539.

Kar, S., Ghosh, R.K., Ghosh, A.N. and Ghosh, A. (1996). Integration of the DNA of a novel filamentous bacteriophage vsk from *Vibrio cholerae* O139 into the host chromosomal DNA. FEMS Microbiol. Lett. *145*, 17–22.

Karaolis, D.K., Johnson, J.A., Bailey, C.C., Boedeker, E.C., Kaper, J.B. and Reeves, P.R. (1998). A *Vibrio cholerae* pathogenicity island associated with epidemic and pandemic strains. Proc. Natl. Acad. Sci. USA *95*, 3134–3139.

Karaolis, D.K., Somara, S., Maneval, D.R., Johnson, Jr. J.A. and Kaper, J.B. (1999). A bacteriophage encoding a pathogenicity island, a type-IV pilus and a phage receptor in cholera bacteria. Nature. *399*, 375–379.

Khetawat, G., Bhadra, R., Nandi, S. and Das, J. (1999). Resurgent *Vibrio cholerae* O139: Rearrangement of cholera toxin genetic elements and amplification of rrn operon. Infect. Immun. *67*, 148–154.

Kimsey, H.H., Nair, G.B., Ghosh, A. and Waldor, M.K. (1998). Diverse CTXphis and evolution of new pathogenic *Vibrio cholerae*. Lancet *352*, 457–458.

Kovach, M.E., Shaffer, M.D. and Peterson, K.M. (1996). A putative integrase gene defines the distal end of a large cluster of toxr-regulated colonization genes in *Vibrio cholerae*. Microbiology *142*, 2165–2174.

Lee, J., Han, K., Choi, S., Lucas, M., Mondlane, C., Ansaruzzaman, M., Nair, G., Sack, D., von Seidlein, L., Clemens, J., Song, M., Chun, J., Kim, D. and Mozambique Cholera Vaccine Demonstration Project Coordination Group. (2006). Multilocus sequence typing (mlst) analysis of *Vibrio cholerae* o1 el tor isolates from mozambique that harbour the classical CTX prophage. J. Med. Microbiol. *55*, 165–170.

Li, M., Shimada, T., Morris, J.G., Sulakvelidze, Jr. A. and Sozhamannan, S. (2002). Evidence for the emergence of non-O1 and non-O139 *Vibrio cholerae* strains with pathogenic potential by exchange of o-antigen biosynthesis regions. Infect. Immun. *70*, 2441–2453.

Li, M., Kotetishvili, M., Chen, Y. and Sozhamannan, S. (2003). Comparative genomic analyses of the vibrio pathogenicity island and cholera toxin prophage regions in nonepidemic serogroup strains of *Vibrio cholerae*. Appl. Environ. Microbiol. *69*, 1728–1738.

Maiti, D., Das, B., Saha, A., Nandy, R., Nair, G. and Bhadra, R. (2006). Genetic organization of pre-CTX and CTX prophages in the genome of an environmental *Vibrio cholerae* non-O1, non-O139 strain. Microbiology *152*, 3633–3641.

McLeod, S., Kimsey, H., Davis, B. and Waldor, M. (2005). CTXphi and *Vibrio cholerae*: Exploring a newly recognized type of phage-host cell relationship. Mol. Microbiol. *57*, 347–356.

McLeod, S.M. and Waldor, M.K. (2004). Characterization of xerC- and xerD-dependent CTX phage integration in *Vibrio cholerae*. Mol. Microbiol. *54*, 935–947.

Meibom, K.L., Blokesch, M., Dolganov, N.A., Wu, C.Y. and Schoolnik, G.K. (2005). Chitin induces natural competence in *Vibrio cholerae*. Science *310*, 1824–1827.

Mekalanos, J.J. (1983). Duplication and amplification of toxin genes in *Vibrio cholerae*. Cell 35, 253–263.

Miller, M.C., Keymer, D.P., Avelar, A., Boehm, A.B. and Schoolnik, G.K. (2007). Detection and transformation of genome segments variable within a coastal population of *Vibrio cholerae*. Appl. Environ. Microbiol.

Moyer, K.E., Kimsey, H.H. and Waldor, M.K. (2001). Evidence for a rolling-circle mechanism of phage DNA synthesis from both replicative and integrated forms of CTXphi. Mol. Microbiol. *41*, 311–323.

Murphy, R.A. and Boyd, E.F. (2008). Three pathogenicity islands of *Vibrio cholerae* excise from the chromosome and form circular intermediates. J. Bacteriol. *190*, 636–647.

Nandi, S., Maiti, D., Saha, A. and Bhadra, R. (2003). Genesis of variants of *Vibrio cholerae* o1 biotype el tor: Role of the CTXphi array and its position in the genome. Microbiology. *149*, 89–97.

O'Shea, Y.A. and Boyd, E.F. (2002). Mobilization of the Vibrio pathogenicity island between *Vibrio cholerae* isolates mediated by CP-T1 generalized transduction. FEMS Microbiol. Lett. *214*, 153–157.

O'Shea, Y.A., Reen, F.J., Quirke, A.M. and Boyd, E.F. (2004). Evolutionary genetic analysis of the emergence of epidemic *Vibrio cholerae* isolates based on comparative nucleotide sequence analysis and multilocus virulence gene profiles. J. Clin. Microbiol. *42*, 4657–4671.

Quinones, M., Kimsey, H.H. and Waldor, M.K. (2005). Lexa cleavage is required for CTX prophage induction. Mol Cell. *17*, 291–300.

Quinones, M., Davis, B.M. and Waldor, M.K. (2006). Activation of the *Vibrio cholerae* sos response is not required for intestinal cholera toxin production or colonization. Infect. Immun. *74*, 927–930.

Rajanna, C., Wang, J., Zhang, D., Xu, Z., Ali, A., Hou, Y.M. and Karaolis, D.K. (2003). The vibrio pathogenicity island of epidemic *Vibrio cholerae* forms precise extrachromosomal circular excision products. J. Bacteriol. *185*, 6893–6901.

Rakonjac, J. and Model, P. (1998). Roles of pIII in filamentous phage assembly. J Mol. Biol. *282*, 25–41.

Rakonjac, J., Feng, J. and Model, P. (1999). Filamentous phage are released from the bacterial membrane by a two-step mechanism involving a short c-terminal fragment of pIII. J. Mol. Biol. *289*. 1253–1265.

Riechmann, L. and Holliger, P. (1997). The c-terminal domain of tola is the coreceptor for filamentous phage infection of *E. coli*. Cell. *90*, 351–360.

Russel, M. (1995). Moving through the membrane with filamentous phages. Trends Microbiol. 3, 223–228.

Sandkvist, M. (2001). Type II secretion and pathogenesis. Infect. Immun. 69: 3523–3535.

Sharma, C., Nair, G.B., Mukhopadhyay, A.K., Bhattacharya, S.K., Ghosh, R.K. and Ghosh, A. (1997). Molecular characterization of *Vibrio cholerae* O1 biotype El Tor strains isolated between 1992 and 1995 in Calcutta, India: Evidence for the emergence of a new clone of the el tor biotype. J. Infect. Dis. *175*, 1134–1141.

Skorupski, K. and Taylor, R.K. (1997). Control of the toxr virulence regulon in *Vibrio cholerae* by environmental stimuli. Mol. Microbiol. *25*, 1003–1009.

Taylor, R.K., Miller, V.L., Furlong, D.B. and Mekalanos, J.J. (1987). Use of *phoA* gene fusions to identify a pilus colonization factor coordinately regulated with cholera toxin. Proc. Natl. Acad. Sci. USA *84*, 2833–2837.

Trucksis, M., Galen, J.E., Michalski, J., Fasano, A. and Kaper, J.B. (1993). Accessory cholera enterotoxin (ace), the third toxin of a *Vibrio cholerae* virulence cassette. Proc. Natl. Acad. Sci. USA *90*, 5267–5271.

Val, M., Bouvier, M., Campos, J., Sherratt, D., Cornet, F., Mazel, D. and Barre, F. (2005). The single-stranded genome of phage CTX is the form used for integration into the genome of *Vibrio cholerae*. Mol Cell. *19*, 559–566.

Wagner, P.L. and Waldor, M.K. (2002). Bacteriophage control of bacterial virulence. Infect. Immun. *70*, 3985–3993.

Waldor, M.K. and Mekalanos, J.J. (1994). Emergence of a new cholera pandemic: Molecular analysis of virulence determinants in *Vibrio cholerae* O139 and development of a live vaccine prototype. J. Infect. Dis, *170*, 278–283.

Waldor, M.K. and Mekalanos, J.J. (1996). Lysogenic conversion by a filamentous phage encoding cholera toxin. Science *272*, 1910–1914.

Waldor, M.K., Rubin, E.J., Pearson, G.D., Kimsey, H. and Mekalanos, J.J. (1997). Regulation, replication, and integration functions of the *Vibrio cholerae* CTXϕ are encoded by region RS2. Mol. Microbiol. *24*, 917–926.

Genetics of *Vibrio cholerae* Colonization and Motility

5

Brooke A. Jude and Ronald K. Taylor

Abstract

Survival of *Vibrio cholerae*, either in the aquatic environment or in the human host, is mediated by appropriate expression of factors that control motility, colonization, production of virulence factors, as well as sensing the cell density (quorum sensing). Successful transition of the organism between the aquatic and the host intestinal environments thus depends on the coordinated activity of a number of genes and regulatory circuits. Recent developments in our understanding of the complex gene regulation supporting the cyclic transitions and the dual lifestyle of *V. cholerae* has been presented and discussed.

Introduction

Survival and colonization of *Vibrio cholerae* in both the aquatic environment and the epithelium of the human small intestine is mediated by the appropriate regulation of attachment, colonization, motility and additional virulence factors. *V. cholerae* cyclically transitions between these environments. Within its primary aquatic reservoir, *V. cholerae* binds to abiotic, aquatic surfaces. Upon ingestion by humans, there is a transition to its second environment where bacteria undergo a stepwise colonization of the intestinal epithelium of humans with concomitant expression of cholera toxin, leading to the manifestation of cholera. This is followed by bacterial detachment, coincident with excretion of the bacteria back into the aquatic environment. These transitions are accomplished through successful flagellar-mediated motility to the site of intestinal colonization, a coordinated expression of appropriate attachment factors, induction of genes encoding the toxin co-regulated pilus (TCP), expression of the cholera toxin genes, and recognition of the surrounding bacterial population density through quorum sensing. To successfully execute the intestinal program of colonization, *V. cholerae* responds to changes in its surroundings. This ability to respond to environmental changes continues during late stages of cholera infection, when the bacteria apparently signals to exit the intestine and return to the aquatic environment. These cyclic transitions and the complex gene regulation for this dual lifestyle will be explored in detail in this chapter, with an emphasis on pilus and flagellar mediated functions.

Toxin co-regulated pilus (TCP)

The toxin co-regulated pilus (TCP) is the most critical intestinal colonization factor of *V. cholerae*. TCP is classified as a type 4b pilus (Strom and Lory, 1993). Type 4a pili have been identified for a variety of gram-negative bacteria, whereas type 4b pili have only been identified in enteric bacteria thus far. Type 4b pilins are further distinguished from the 4a subclass by possessing a longer leader sequence, numbering 15–30 residues as compared with five or six observed for type 4a pilins. In addition, while the N-terminal amino acid is methylated for both subclasses, this is a conserved phenylalanine for type 4a pilins, while this residue identity varies for type 4b, and thus far includes leucine, valine, or methionine, the last of which is the case for TcpA (Strom and Lory, 1993). The gene for the pilin subunit, *tcpA*, was identified in a genetic screen using Tn*phoA* to identify genes encoding *V. cholerae* surface or secreted factors involved in virulence (Taylor *et al.*, 1987). A *tcpA* mutant has a 5-log defect in the *in vivo* infant mouse model for cholera colonization (Taylor *et al.*, 1987). Likewise, *tcpA* mutants of O1 classical biotype as well as O139 strains have been shown to be essentially completely deficient for colonization in humans (Herrington *et al.*, 1988; Tacket *et al.*, 1998). Polymers of TcpA subunits form pilus filaments that interact to form the bundled TCP structure on the bacterial cell surface, which is readily observed via electron microscopy (Fig. 5.1A) (Taylor *et al.*, 1987). TCP structures protruding from the surface of *V. cholerae* mediate bacteria–bacteria interactions. This association results in the formation of microcolony structures, an essential step for intestinal colonization (Kirn *et al.*, 2000). The crystal structure of the globular head domain of TcpA has been solved and a model of the structure of the pilus filament has been established using this crystal structure (Craig *et al.*, 2003). TcpA pilin subunits align into pilus filaments that form a three-start helix. Repeating surface exposed patches of residues on the globular head domain of TcpA are also positioned on the outwardly facing surface of the pilus filaments. Recent evidence has lead to a model of individual filaments interacting with one another to form the pilus twists, as seen in Fig. 5.1A. These twists can progress to higher order supertwist structures, also seen in Fig. 5.1A, when twists successfully interact with each other (Jude and Taylor, unpublished). This supertwisting step is essential for successful pilus mediated autoagglutination and microcolony formation, resulting in intestinal colonization. Residues within the TcpA subunit have been identified as essential for progression to supertwisting. Alterations of these residues allow for the formation of individual pilus twists, observable via electron microscopy, but pilus interaction arrests at this stage, and supertwisting and autoagglutination

A.

B.

C.

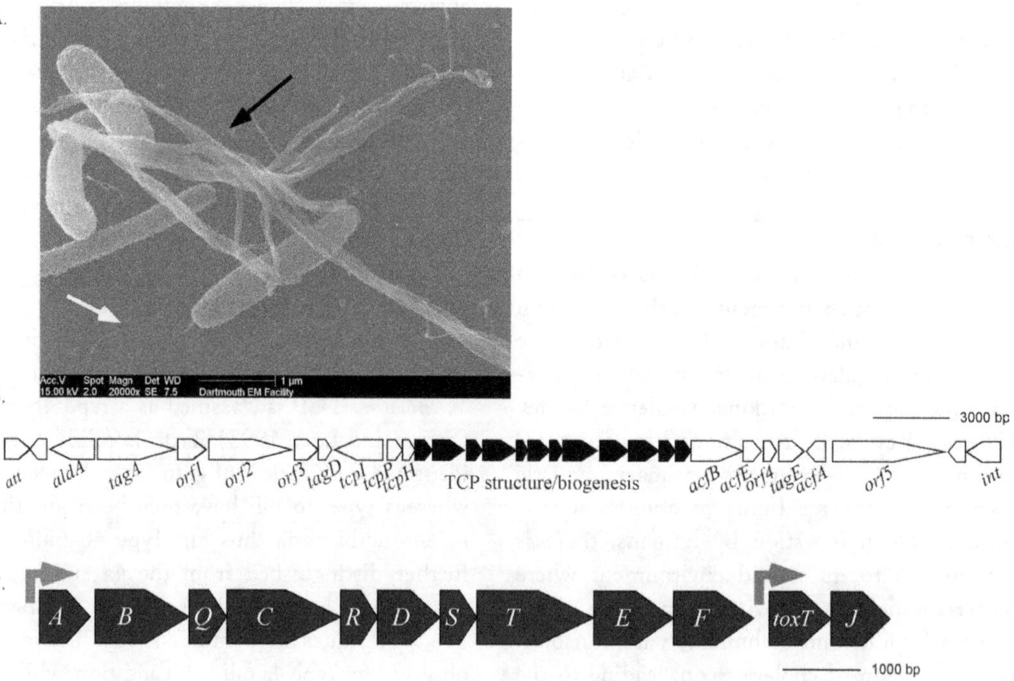

Figure 5.1 FESEM and organization of the *Vibrio cholerae* Pathogenicity Island (VPI) and *tcp* related genes. (A) Field emission scanning electron micrograph (FESEM) of *V. cholerae*. Black arrow indicates TCP structures, white arrow indicates polar flagellum. (B) Order and directionality of the genes encoded on the VPI, as annotated by TIGR. Shaded in black are genes that make up the toxin co-regulated pilus (TCP) biogenesis portion of the *tcp* operon. (C) Nomenclature of the *tcp* biogenesis operon genes, including direction and location of open reading frames. The *tcpA* and *toxT* promoter regions are designated by gray arrows. Although there are two promoters, transcription that initiates from the promoter upstream of *tcpA*, proceeds into *toxT* and beyond.

do not occur. The result of this arrest is that strains expressing this altered pilin subunit are avirulent (Kirn *et al.*, 2000). This observation delineates potential protein regions and residues as targets to block pilus supertwisting as a strategy for vaccine development.

Biogenesis of the pilus structure is mediated by the production of a membrane spanning apparatus composed of proteins encoded by genes downstream of *tcpA* in the *tcp* operon (Table 5.1 and Fig. 5.1B and C). TcpC is the lipoprotein outer membrane secretin pore from which the pilus structure is extracellularly translocated, and requires the presence of an associated periplasmic protein, TcpQ, for localization and stability (Bose and Taylor, 2005). TcpR, D, and E form the inner membrane platform of the biogenesis apparatus, and TcpS, a periplasmic protein, is predicted to associate with the periplasmic side of this inner membrane platform and interacts with TcpQ to form a periplasmic bridge (Taylor *et al.*, 2004). TcpT is the cognate ATPase that

Table 5.1 Genes located on the *Vibrio cholerae* pathogenicity island

VC no.	Gene designation	Role/function, if known
VC0817	*att*	Transposase
VC0818	–	Transposase, point mutant
VC0819	*aldA-1*	Aldehyde dehydrogenase
VC0820	*tagA*	ToxR-regulated gene
VC0821	*orf1*	Hypothetical protein
VC0822	*orf2*	Putative IM protein
VC0823	*orf3*	Hypothetical protein
VC0824	*tagD*	ToxR-regulated gene
VC0825	*tcpI*	Methyl-accepting protein
VC0826	*tcpP*	Transcriptional activator
VC0827	*tcpH*	Stabilizer of TcpP
VC0828	*tcpA*	Type 4b pilin
VC0829	*tcpB*	Unknown; type 4b pilin-like processing site
VC0830	*tcpQ*	Periplasm, in complex with TcpC
VC0831	*tcpC*	OM lipoprotein pore, secretin superfamily
VC0832	*tcpR*	IM platform
VC0833	*tcpD*	IM platform
VC0834	*tcpS*	Periplasm, in complex with IM platform
VC0835	*tcpT*	ATPase for pilus biogenesis, in complex with IM platform
VC0836	*tcpE*	IM platform
VC0837	*tcpF*	Secreted colonization factor
VC0838	*toxT*	AraC family member transcriptional regulator
VC0839	*tcpJ*	Type 4 prepilin peptidase
VC0840	*acfB*	Accessory colonization factor
VC0841	*acfE*	Accessory colonization factor
VC0842	*orf4*	Conserved hypothetical protein
VC0843	*tagE*	ToxR-regulated gene
VC0844	*acfA*	Accessory colonization factor
VC0845	*orf5*	Putative lipoprotein
VC0846	–	Integrase – degenerate
VC0847	*int*	Integrase

provides the energy required for pilus formation. It is a cytoplasmic enzyme that is associated with the cytoplasmic side of the inner membrane platform via direct interaction with TcpR and perhaps additional inner membrane platform proteins (Tripathi and Taylor, 2007). TcpF is a soluble protein essential for intestinal colonization in infant mice by *V. cholerae*, and is secreted via the TCP biogenesis apparatus (Kirn *et al.*, 2003, 2005). Deletion of *tcpF* results in a 5-log decrease in colonization, similar to deletion of *tcpA* (Kirn and Taylor, 2005); yet the mechanism of TcpF function remains unknown. TcpJ, the type 4 prepilin peptidase, an aspartyl protease, is required for cleaving the leader peptide from precursor TcpA, making it competent for assembly into the pilus filament (Kaufman *et al.*, 1991; LaPointe and Taylor, 2000). The role and localization of TcpB are not known. TcpB has a short leader peptide, typical of type 4a pilins, but has a N-terminal valine residue, which can occur with type 4b pilins. The predicted molecular weight of mature TcpB is approximately 46 kDa, much larger than that of a typical type 4 pilin. Non-polar deletion of any of these genes, with the exception of *tcpF*, results in the failure to produce TCP (Kirn *et al.*, 2003). The pilus biogenesis machinery and the resulting TCP structure is an essential element for successful intestinal colonization by *V. cholerae*.

The *Vibrio cholerae* pathogenicity island (VPI)

The genetic region that includes the operon of genes encoding TCP structure and biogenesis components is a 39.5-kb horizontally acquired element termed the *Vibrio cholerae* pathogenicity island (VPI) (Fig. 5.1B and Table 5.1), alternatively termed the TCP pathogenicity island. This genetic region was first associated with *V. cholerae* virulence when the coding sequence for *tcpA* was identified (Taylor *et al.*, 1987). A screen designed to identify factors regulated in a similar manner as *tcpA* revealed a number of genes located in the region surrounding *tcpA* (Peterson and Mekalanos, 1988). In addition to TCP biogenesis components, these open reading frames within this chromosomal region encode factors contributing to virulence, regulation of virulence factors, and regulation of bacterial motility. These genes are annotated as accessory colonization factors (*acfA, acfB, acfC, acfD*), ToxR-activated genes (*tagA*), as well as an aldehyde dehydrogenase (*aldA*) (Peterson and Mekalanos, 1988; Parsot and Mekalanos, 1991; Parsot *et al.*, 1991). *tcpI*, originally named *tagB*, is located upstream of the *tcp* operon and is transcribed in the reverse direction of the *tcp* operon (Harkey *et al.*, 1994). TcpI has significant homology to other membrane bound methyl-accepting chemotaxis proteins (MCPs) that contribute to bacterial motility. A study utilizing a TcpI–PhoA hybrid protein revealed that inactivation of *tcpI* results in increased TcpA production (Taylor, 1989). A gene that encodes an essential regulator of virulence genes, *toxT*, is located within the TCP structural and biogenesis operon (DiRita *et al.*, 1991). The identification of ToxT, the later discoveries of TcpP and TcpH (Carroll *et al.*, 1997, Häse and Mekalanos, 1998) and additional regulatory factors encoded by the VPI, along with characterization of the entire *V. cholerae* virulence regulon they control, uncovered a complex and environmentally responsive regulatory cascade that will be covered fully later in this chapter.

Several properties associated with the VPI demonstrate that it is a horizontally acquired genetic element. This genetic region encodes an integrase at its distal 3′ end, and a transposase at the proximal 5′ end, with *att* sites surrounding each of these gene sequences (Kovach *et al.*, 1996; Karaolis *et al.*, 1998). This, in conjunction with the low G-C content (35%) of the VPI compared with surrounding chromosomal sequence, indicates that this region encoding virulence factors was horizontally acquired. Along with its precise response to environmental cues, the regulation of these acquired virulence factors is controlled in a complex fashion via island encoded, as well as ancestral genome encoded factors. The VPI is an essential element for virulence of pandemic *V. cholerae* strains.

Cholera toxin (CT)

A second major factor required for the pathogenesis of cholera is the production and extracellular secretion of cholera toxin (CT). Cholera toxin is a member of the AB_5 family of toxins, composed of a single 28kDa A subunit and five 11kDa B subunits (Finkelstein *et al.*, 1974). The holotoxin, containing all six subunits, is extracellularly secreted from the bacterial cell via the type two secretion apparatus, termed the extracellular protein secretion system (EPS) (Sandkvist *et al.*, 1997). Following secretion, CT utilizes the glycolipid transport pathway of the host epithelial cell to enter into the cell by the following sequence of events. CT binds to the glycolipid ganglioside receptor GM1 on the outer leaflet of the plasma membrane. This complex, consisting of GM1 and CT, is hypothesized to enhance the formation of GM1 lipid rafts in the membrane (Chinnapen *et al.*, 2007). The GM1/CT complex can then enter the cell by one of three endocytic mechanisms – (1) caveolar-mediated, (2) clathrin-mediated or (3) non-caveolar, non-clathrin-mediated (CLIC) – depending on the cell type being entered, (Chinnapen *et al.*, 2007). Once in the cell, CT transits from the cell membrane, through the Golgi network, and arrives at the endoplasmic reticulum. In order for CT to access the cell cytoplasm and deliver its toxic activity, the A subunit is reduced, separating into A1 and A2 segments. The A1 segment temporarily unfolds, and utilizes the cellular quality control retrograde transport pathway to gain access to the cytoplasm. In the cytoplasm, the A1 segment quickly refolds thus avoiding proteosomal degradation (Lencer and Tsai, 2003). Within the cytosol, CT A1 subunit causes the cleavage of nicotinamide adenine dinucleotide (NAD) into byproducts ADP-ribose and nicotinamide. This results in the ADP-ribosylation of a G protein, and this action prevents GTPase function, causing a constitutive upregulation of adenylyl cyclase (Finkelstein, 1992). The consequent overproduction of cyclic AMP causes the efflux of chloride ions and water producing the rice watery stool that is characteristic of cholera infection. The structural genes for the cholera toxin, *ctxAB*, are encoded by the filamentous CTXϕ, which persists in the bacterial cells as two forms, a lysogenic form in which the phage genome has integrated into the chromosome and/or a replicating plasmid (Waldor and Mekalanos, 1996). The CTXϕ is also able to exit the bacterial cell in the same way mature CT is secreted, via the EPS secretory apparatus (Davis *et al.*, 2000). For *V. cholerae* cells to be susceptible for CTXϕ entry, they must express TCP, which is the CTXϕ receptor. Consequently, cells that do not express TCP and cannot infect humans are resistant to infection by the CTXϕ (Waldor and Mekalanos, 1996). This association between the two major *V. cholerae* virulence factors provides a mechanism for CTXϕ to preferentially lysogenize bacteria that already express factors required to colonize humans, and facilitate efficient toxin delivery.

The *Vibrio cholerae* virulence regulon

The regulation of genes participating in the virulence cascade in *V. cholerae* is controlled by a series of regulatory proteins responding to a number of environmental input signals. The first regulator of *V. cholerae* virulence gene expression identified was ToxR. This factor was initially characterized for its role in activating the expression of genes encoding cholera toxin (*ctxAB*) (Miller and Mekalanos, 1984). The *toxR* gene encodes a 32.5-kDa transmembrane, DNA-binding protein. Its transcriptional activation domain is located in the cytoplasmic N-terminal portion of the protein (Miller *et al.*, 1987). The protein forms homodimers in the membrane and activates the transcription of *ctxAB* via binding directly to the promoter region of *ctxAB* (Miller *et al.*, 1987; Pfau and Taylor, 1996). The *toxR* sequence is present on the ancestral chromosome and a BLAST search (Altschul *et al.*, 1990) using the NCBI database reveals a high conservation of the protein among a number of *V. cholerae* isolates, and highly related proteins encoded by a variety of *Vibrio* species. This implies that ToxR is involved in the regulation of genes on the ancestral chromosome, not just those associated with virulence regulation. This was substantiated with the identification of two genes, located on the ancestral chromosome, encoding outer membrane proteins, OmpU and OmpT, that are regulated by ToxR. Expression of *ompU* is activated and *ompT* is repressed by ToxR in response to signals in the extracellular environment (Miller and Mekalanos, 1988; Chakrabarti *et al.*, 1996; Sperandio *et al.*, 1996). Further studies linked ToxR to the regulatory cascade activating expression of genes encoding virulence factors in addition to CT, including TCP, and a number of other factors encoded on the VPI (Taylor *et al.*, 1987; Peterson and Mekalanos, 1988).

Examination of the genetic sequence downstream of *toxR* revealed a second gene within the operon, and it was named *toxS* (Miller *et al.*, 1989). The *toxS* gene encodes a 19-kDa transmembrane protein that interacts with ToxR, via the ToxR periplasmic domain (Fig. 5.2) (Miller *et al.*, 1989). To determine if ToxR was sufficient for the activation of expression of all the virulence regulon associated genes, ToxR activity was evaluated using a heterologous system, *E. coli*. The presence of ToxR was found not to be sufficient for activation of VPI encoded genes, *tagA* or *aldA* (Parsot and Mekalanos, 1991). This represented the initial evidence that suggested the existence of an additional regulatory factor participating in virulence gene regulation, forming a regulatory cascade. This factor was identified and termed ToxT (DiRita *et al.*, 1991). The *toxT* coding sequence is located approximately 200 base pairs downstream of *tcpF* (Fig. 5.1C) (Higgins *et al.*, 1992). The ToxT protein is approximately 32 kDa in size and is a member of the AraC family of transcriptional regulators. The protein has two distinct domains, a C-terminal DNA binding domain (Prouty *et al.*, 2005) and a N-terminal domain critical for protein–protein interactions, as has been shown for several other AraC family members (Childers *et al.*, 2007, Shakhnovich *et al.*, 2007). ToxR participates in the activation of the expression of *toxT* (DiRita *et al.*, 1991). Additionally, *toxT* is autoregulated via transcriptional read-through from the *tcp* operon promoter (Brown and Taylor, 1995). To identify the requirements for ToxT-dependent activation, promoter activation of virulence genes was again examined in *E. coli*. ToxT alone was sufficient to activate the *ctx* promoter, in addition to its ability to activate the promoters for *tcpA*, *tcpI*, *aldA* and *tagA*, none of which are directly activated by ToxR (DiRita *et al.*, 1991). These findings, in addition to the evidence that the over-expression of ToxT can suppress a *toxR* mutant phenotype, comprised the initial indications that ToxT is the additional regulatory factor, in addition to ToxR,

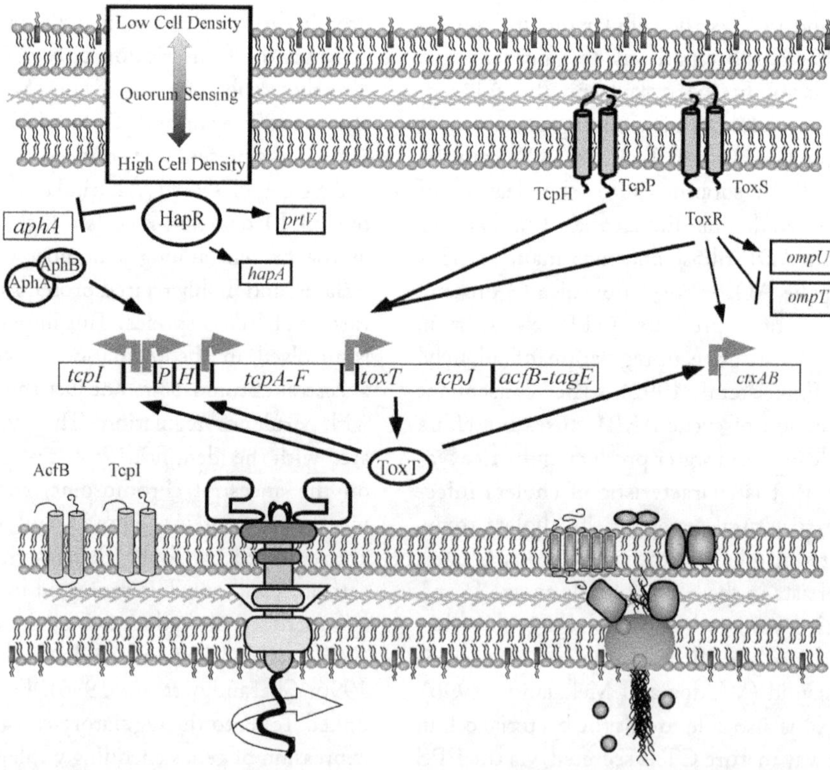

Figure 5.2 Model for virulence gene regulation circuitry in *V. cholerae*. Regulatory proteins ToxT and TcpP/H are encoded by genes within the VPI, while others depicted are encoded within the ancestral genome. Gene regulation is in response to environmental factors, including temperature, pH, and cell density. See text for details.

that participates in the induction of genes encoding virulence factors in *V. cholerae* (DiRita, *et al.*, 1991).

Recent studies have identified promoter requirements for ToxT binding, and have identified core ToxT binding elements as well as a toxbox binding motif (Yu and DiRita, 2002; Withey and DiRita, 2006). ToxT preferentially binds to DNA sequences rich in consecutive runs of T or A nucleotides (Hulbert and Taylor, 2002; Yu and DiRita, 2002). ToxT was also demonstrated to bind to two different classes of promoters, suggesting that ToxT binds as either a monomer, in the case of *acfA*, *acfD*, and *aldA*, or a homodimer, in the case of *ctx* and *tcp* (Withey and DiRita, 2005a,b; Shakhnovich *et al.*, 2007). An exhaustive alanine scanning mutagenesis of the coding region for the N-terminal domain of ToxT identified residues critical for ToxT dimerization and activation (Childers *et al.*, 2007). Additionally, ToxT dimerization and function can be specifically blocked with the small molecule inhibitor virstatin, resulting in a loss of virulence factor activation. This finding provides a basis for potential therapeutics for cholera using inhibitors of ToxT activity (Shakhnovich *et al.*, 2007).

Interestingly, although it was demonstrated that ToxR directly binds to the promoter region of *toxT* and is essential for activation of *toxT* expression in *V. cholerae*, it is not sufficient for activation, and in isolation it is unable to successfully activate *toxT* expression (Higgins and DiRita, 1994). This finding led to the identification of an additional level in the virulence regulatory cascade. A pair of integral membrane proteins, TcpP and TcpH, are encoded within an operon on the VPI, upstream of the *tcp* operon (Figs. 5.1B and C and 5.3 and Table 5.1) (Carroll *et al.*, 1997, Häse and Mekalanos, 1998). TcpP and TcpH comprise a pair similar to ToxR and ToxS, with TcpP as the DNA binding protein and TcpH required for its stabilization. TcpH was initially identified as being required for virulence gene expression (Carroll *et al.*, 1997). Later investigations revealed that a specific region of the coding sequence of *tcpH* has a high frequency of a frameshift mutation, that results in the loss of the expression of the *tcp* biogenesis operon (Carroll *et al.*, 1997). Cells harbouring this frameshift mutation are enriched for by repeated cycles of *in vitro* growth. TcpH is also required for maximal *toxT* activation, and it has been demonstrated that *tcpH* expression is environmentally regulated by such inputs as temperature and pH (Thomas *et al.*, 1995, Carroll *et al.*, 1997). Recent work has characterized the role of TcpH in protecting TcpP from a two step proteolytic degradation by an unknown periplasmic protease and the membrane associated protease, YaeL (Matson and DiRita, 2005). TcpP was identified shortly after TcpH and found to cooperate with ToxR to activate transcription at the *toxT* promoter (Häse and Mekalanos, 1998). This activation was found to occur both in *E. coli* and in a *V. cholerae toxR tcpP* double mutant, for which *tcp* expression was partially restored by the overproduction of TcpPH on a plasmid. This restoration was enhanced when *toxRS* was additionally induced in the same strain (Krukonis *et al.*, 2000).

Investigation into what controls the regulation of *tcpPH* expression revealed a link between the VPI, and the ancestral chromosome encoded regulators, AphA and AphB. This additional level in the regulation of the *V. cholerae* virulence cascade was identified by the discovery that activation of *tcpPH* requires the binding of the novel transcriptional activator AphA, in conjunction with AphB, a LysR family member (Kovacikova and Skorupski, 1999; Skorupski and Taylor, 1999). Analysis of the interaction of AphA and AphB with DNA in the *tcpPH* promoter region revealed that they bind overlapping sites upstream of the sigma binding site, with the AphB site located in the promoter proximal position (Kovacikova and Skorupski, 2001). The crystal structure of AphA revealed that it is a novel member within the winged helix transcription factor superfamily (De Silva *et al.*, 2005). These studies also provided a model for potential AphA–AphB protein–protein interaction (De Silva *et al.*, 2005). While investigating the role of AphB in inducing expression of *tcpPH*, analysis of both the El Tor and classical biotype sequences revealed a base pair difference at the AphB binding site in the *tcpPH* promoter. This change accounts for the difference in virulence gene expression levels between the two biotypes, with the classical biotype exhibiting higher expression levels than El Tor (Kovacikova and Skorupski,

2000). This insight, coupled with *tcpPH* binding domain chimera experiments, also identified that it is the AphB response to environmental cues, and its resulting binding to the promoter of *tcpPH*, that differentially induces the virulence regulon between the two biotypes (Kovacikova and Skorupski, 2000). Taken together, these results have demonstrated that the regulation of *tcpPH* expression is the key checkpoint in the virulence regulatory cascade.

An example of the expression of virulence factors responding to changes in the surrounding bacterial population is the repression of expression of *aphA* by the cell density-dependent quorum-sensing system (Kovacikova and Skorupski, 2002). Under conditions of high cell density, a cascade of events that include the dephosphorylation of LuxO, the resulting lack of induction of small RNA molecules, and the increased stabilization of *hapR* mRNA results in an abundance of HapR in the cell (Lenz *et al.*, 2004). HapR binds to a site in the promoter region of *aphA*, and this results in the repression of *aphA* expression, which results in the down-regulation of the virulence operon (Kovacikova and Skorupski, 2002). This binding is via an essential N-terminal helix–turn–helix DNA-binding domain (De Silva *et al.*, 2007). AphA was also found to be auto-repressive, binding to its own promoter region in a regulatory fashion (Lin *et al.*, 2007). Activation of *aphA* has recently been determined to be controlled by the regulators Lrp (leucine responsive regulatory protein) and VpsR, which controls biofilm formation under appropriate conditions (Lin *et al.*, 2007).

Expression of additional genes, located on the ancestral chromosome, was demonstrated to be regulated by AphA under conditions of low cell density, including repression of those genes in the acetoin biosynthetic operon, *alsDSO* (Kovacikova *et al.*, 2005). Acetoin is an intermediate produced in the conversion of pyruvate to 2,3-butanediol, and is critical for preventing the overacidification that results from the metabolism of carbon sources such as glucose (Kovacikova *et al.*, 2005). The repression of this operon also results in the down-regulation of the *acgAB* genes, which are located in the same operon, downstream of *alsDSO*. AcgA and AcgB contain EAL and GGDEF domains, respectively, which often contribute to regulation of cyclic-di-GMP levels in the cell in a manner that leads to altered regulation of specific gene expression. As is typical for proteins that contain these domains, over-expression of AcgA resulted in a loss in biofilm formation and increase in motility, while over-expression of AcgB resulted in increased biofilm formation and a slight reduction in motility (Kovacikova *et al.*, 2005). These results suggest that these proteins may contribute to the regulation of the levels of second messenger cyclic-di-GMP (c-di-GMP) in the cell (Kovacikova *et al.*, 2005). AphA was also found to activate *vps* gene expression (Kovacikova *et al.*, 2005), the products of which are instrumental in the production of exopolysaccharide, a cellular component essential to biofilm formation. It is not yet fully understood why a regulator of metabolic pathways was co-opted for use as a virulence regulator (Kovacikova *et al.*, 2005). It is hypothesized that this may be due to the precise regulatory control quorum sensing has on *aphA* expression.

Additional factors affecting *Vibrio cholerae* pilus expression

Histone-like nucleoid structuring (H-NS) protein

A common regulator found in Gram-negative α-, β- and γ-proteobacteria, H-NS (Tendeng and Bertin, 2003), has a significant role in the repression of the *V. cholerae* virulence cascade (Nye *et al.*, 2000; Tendeng *et al.*, 2000). Also termed VicH, *V. cholerae* H-NS is a small, 135 amino acid, 15-kDa protein, that is presumed to be abundant in *V. cholerae*, similar to what has been found in *E. coli*, which harbours about 20,000 copies per bacterial cell (Tendeng and Bertin, 2003; Dorman, 2004). As a global bacterial repressor, the H-NS protein has been demonstrated to function by binding to AT rich, curved regions of DNA. The repressive effect of H-NS on the expression of *V. cholerae* genes has been identified at various levels of the virulence regulon. For example, H-NS has been demonstrated to repress gene expression from the AT rich *toxT*, *tcp* and *ctx* promoters (Nye *et al.*, 2000). Furthermore, the activator ToxT and H-NS have been demonstrated to have an antagonistic relationship at the *tcp* and *ctx* promoters, occupying overlapping sites (Nye *et al.*, 2000; Yu and DiRita, 2002).

Pathogenicity islands and horizontally acquired genetic elements are often AT rich, and the GC percentage of these elements is typically distinct from the GC content of the rest of the chromosomal DNA (Lucchini *et al.*, 2006; Navarre *et al.*, 2006). This has led to a recent hypothesis that H-NS, as a global bacterial repressor, can effectively silence the expression of recently acquired genetic elements (Lucchini *et al.*, 2006; Navarre *et al.*, 2006). This silencing theoretically prevents these acquired factors from reducing the overall fitness of the cell prior to establishing appropriate regulation within its new genomic context. This allows for successful bacterial evolution by the retention of potentially beneficial genetic elements. H-NS is hypothesized to serve as the 'genome sentinel', providing time for factors encoded within the ancestral genome to be recruited for a role in regulating expression of genes located on these acquired elements (Dorman, 2007). This trend is consistent with the makeup of the *Vibrio cholerae* pathogenicity island (VPI), having a GC content of 35.5%, differing from the remainder of the *V. cholerae* genome, which has a GC content of 47% (Heidelberg *et al.*, 2000, Karaolis *et al.*, 2001).

Cyclic-di-GMP signalling

A recent addition to the bacterial repertoire of mechanisms to sense and respond to changing environmental conditions is the signalling cascade that transmits changes in cellular concentration of the second messenger, cyclic di-GMP (c-di-GMP) (Jenal and Malone, 2006). The c-di-GMP molecule is synthesized from two GTP molecules by the enzyme diguanylate cyclase (DGC) and hydrolysed into pGpG by phosphodiesterase (PDE). These enzymes have distinctive domains that identify their opposing activities, with DGC containing the amino acid motif of GGDEF, while PDE proteins typically possess an EAL domain. Analysis of the *V. cholerae* proteome for these domains reveals 31 potential DGC proteins, 12 PDE proteins and 10 proteins that contain both domains (Galperin *et al.*, 2001). High and low intracellular levels of c-di-GMP are often associated with opposing bacterial behaviours. In *V. cholerae*, high c-di-GMP levels coincide with reduced transcription of *toxT* and associated virulence genes, as well as an induction of genes associated with polysaccharide production and biofilm formation (Tischler and Camilli, 2004, Tischler and Camilli, 2005). Levels of c-di-GMP also play a role in the regulatory effect on genes encoding flagellar related proteins. High c-di-GMP levels correspond with repression of genes encoding flagellins and basal body proteins, while low c-di-GMP (and high PDE activity) results in an increase in motility (Beyhan *et al.*, 2006b).

Several of the 31 DGC and PDE proteins have been specifically characterized for their role in *V. cholerae* gene regulation. Screens designed to identify factors contributing to the intestinal colonization of *V. cholerae* identified a 3-component signal transduction system, *vieSAB* (Lee *et al.*, 1998). Sequence analyses revealed genes encoding proteins with homology to a sensor kinase (*vieS*) and two response regulators (*vieAB*), organized within an operon. Levels of VieS are constitutive during *in vitro* growth, while *vieA* expression is autoregulated (Lee *et al.*, 1998). Expression of *vieA* leads to repression of biofilm formation, as well as induction of *ctxAB* and *toxT* expression. This switch in phenotype is a function of its EAL domain, with functional phosphodiesterase activity reducing levels of intercellular c-di-GMP (Tischler and Camilli, 2004, 2005; Tamayo *et al.*, 2005). The transcription of *vieB* is specifically induced during conditions of intestinal colonization (Camilli and Mekalanos, 1995; Lee *et al.*, 1998). This correlates with low levels of c-di-GMP during *in vivo* growth, due to VieA activity, and results in successful intestinal colonization (Tischler and Camilli, 2005). In addition, a *vieA* mutant displays a reduction in flagellar motility and chemotaxis, which may indicate an additional contribution towards effective intestinal colonization (Beyhan *et al.*, 2006a).

Several proteins that contain both GGDEF and EAL domains have been characterized for their roles in biofilm formation and biofilm structural maintenance, including RocS, MbaA, and CdgC (Bomchil *et al.*, 2003; Rashid *et al.*, 2003; Lim *et al.*, 2007). Although they contain the sequences for both domains, in the case of RocS and CdgC, they have been characterized as having only a single enzymatic activity each, with RocS behaving as a DGC, and CdgC possessing PDE activity (Rashid *et al.*, 2003, Lim *et al.*, 2007). Recently, a c-di-GMP binding domain, PilZ, has been identified (Amikam and Galperin, 2006). Five *V. cholerae* proteins of unknown function harbour this domain, and genes encoding these proteins have been annotated as *plzA*, *plzB*, *plzC*, *plzD*, and *plzE* (Pratt *et al.*, 2007). PlzC and PlzD have been shown to bind c-di-GMP, and several conserved residues within the PilZ domain are essential for this binding (Pratt *et al.*, 2007). Phenotypes of strains unable to produce individual Plz proteins revealed that some of these proteins are involved in bacterial cell motility, biofilm formation and successful intestinal colonization of infant mice (Pratt *et al.*, 2007). These results further establish the regulatory role that the c-di-GMP second messenger plays in cellular functions, and begins to explain how this molecule transmits its signal within the *V. cholerae* regulatory cascades.

Bile

Bile is a heterogeneous complex mixture of sodium salts secreted from the human gallbladder. The presence and concentration of bile within the small intestine has been demonstrated to be an environmental signalling factor that contributes to *V. cholerae* virulence and colonization. Although bile is considered to have generalized antimicrobial effects as a result of its detergent properties, enteric bacteria, including *V. cholerae*, have developed strategies to regulate the expression of genes encoding efflux pumps and porins to actively protect against bile sensitivity (McMurry *et al.*, 1983; Levy, 1992; Bina and Mekalanos, 2001; Bina *et al.*, 2006). ToxR, in a ToxT independent manner, has been demonstrated to regulate the expression of genes encoding outer membrane proteins OmpU and OmpT (Miller and Mekalanos, 1988; Provenzano and Klose, 2000). Improper regulation of the expression of these genes can increase bile sensitivity. Bile, present at variable concentrations throughout the small intestine, has been demonstrated to directly affect regulation of flagellar motility as well as affect the function of regulators of the *V. cholerae* virulence cascade. The transcriptional activity of the regulator ToxT is reduced in the presence of bile in the extracellular environment. A domain in the N-terminus of ToxT appears to mediate at least part of this bile response (Schuhmacher and Klose, 1999; Prouty *et al.*, 2005). In addition, recent studies have identified a potential mechanism whereby unsaturated fatty acids present in bile enhance repression of virulence gene expression in a H-NS dependent manner (Chatterjee *et al.*, 2007). Bile has also been implicated in the induction of biofilm associated gene expression, and the increased bile resistant phenotype exhibited by biofilm associated cells (Hung *et al.*, 2006). The mechanism for bile resistance and gene responsiveness in *V. cholerae* is currently an active area of investigation.

Quorum sensing

Within communities of *V. cholerae*, the ability to respond to signals from the extracellular environment allows for coordination of behaviour, including production of factors required for biofilm formation, secretion of proteases, and regulation of colonization. Quorum sensing is the ability of interspecies and intraspecies bacterial communities to communicate via chemical signals (Waters and Bassler, 2005). Bacteria sense and respond to other organisms within their environment via secretion and recognition of small molecules called autoinducers. This response circuitry allows bacterial communities to behave in a manner akin to a multicellular organism, and synchronize behaviours for the benefit of the community. *V. cholerae* has a complex quorum sensing system that incorporates three parallel signalling inputs that converge on a central regulator, HapR (Jobling and Holmes, 1997; Miller *et al.*, 2002; Lenz *et al.*, 2005). This system is responsible for sensing and responding to surrounding cell density and repressing the downstream virulence cascade, as well as repressing genes encoding exopolysaccharide production and biofilm formation. Additionally, quorum sensing is responsible for inducing expression of the genes encoding secreted proteases HapA and PrtV (Jobling and Holmes, 1997; Kovacikova and Skorupski, 2002; Zhu *et al.*, 2002; Zhu and Mekalanos, 2003). It is hypothesized that this response provides the bacteria with the ability to form biofilms while persisting within the aquatic environment and during early stages of intestinal colonization, recognized by the bacteria as low cell density conditions (Watnick and Kolter, 1999; Watnick *et al.*, 2001; Zhu and Mekalanos, 2003). At these time points, a lack of autoinducer molecules results in the absence of signalling through the three parallel quorum sensing inputs. This sets up a signalling cascade that induces the production of small RNA molecules that destabilize the *hapR* mRNA (Lenz *et al.*, 2004). This leads to appropriate inducing conditions for *aphA* activation, and initiation of the *V. cholerae* virulence cascade (Kovacikova *et al.*, 2004; Kovacikova *et al.*, 2005). During conditions of high bacterial concentrations, the lack of small RNA production leads to stabilization of the *hapR* message, and consequent protein production and increased protein levels. During this phase, HapR serves as a repressor, halting the production of virulence factors. It is simultaneously an activator of genes encoding proteases, potentially expediting an exit from the site of intestinal colonization prior to the death of the host (Jobling and Holmes, 1997; Skorupski and Taylor, 1999; Zhu and Mekalanos, 2003) (Jude, Skorupski, and Taylor, unpublished).

Surface attachment by *Vibrio cholerae*

Genetic organization and function of the MSHA pilus. A factor that participates specifically in *V. cholerae* colonization in the aquatic environment is the mannose-sensitive haemagglutinin pilus (MSHA). D-Mannose-dependent haemagglutination, a phenotype observed in El Tor and O139 *V. cholerae* strains, but not O1 classical strains, has been attributed to the MSHA pilus (Jonson *et al.*, 1991). This pilus is distinct in structure and function from TCP (Strom and Lory, 1993). MSHA is classified as a type 4a pilus, based on physical characteristics of the pilin subunit MshA, namely a typical short leader peptide sequence and a N-methylated phenylalanine residue in the first position of the mature protein (Strom and Lory, 1993, Jonson *et al.*, 1994). The *mshA* gene has been mapped within a genomic region 16.7 kb in size, consisting of 16 *msh* genes (Häse *et al.*, 1994; Marsh *et al.*, 1996; Marsh and Taylor, 1999). The genes present in this region include, in order: *mshIJKLMNEGFBACDOPQ* (Marsh and Taylor, 1999). These genes are transcriptionally organized into two operons with distinct promoters, and include the transcriptional unit *mshIJKLMNEG*, encoding proteins responsible for pilus biogenesis and *mshFBACDOPQ*, which includes the gene encoding the major pilin, MshA (Marsh and Taylor, 1999).

The MSHA pilus functions during aquatic surface colonization, participating in both attachment to biotic and abiotic surfaces as well as in biofilm formation (Watnick *et al.*, 1999; Chiavelli *et al.*, 2001). MSHA mediates adherence of *V. cholerae* O139 to chitinous zooplankton, and to a lesser extent, that of *V. cholerae* O1 El Tor biotype strains (Chiavelli *et al.*, 2001). As expected, adherence of *V. cholerae* O1 classical biotype was not altered in a *mshA* mutant derivative (Chiavelli *et al.*, 2001). The role of MSHA in the ordered process of biofilm formation on diverse aquatic surfaces was also examined. Biofilm formation occurs in a stepwise fashion, and includes initial bacterial cell surface attachment which is accelerated by flagellar based motility to access the surface. The attachment process is followed by cell–cell interactions, and the production of extracellular matrix components such as exopolysaccharide, ultimately producing a three-dimensional surface structure (Watnick and Kolter, 1999). It was demonstrated that expression of the genes that encode MSHA components is induced in response to an initial surface attachment, while flagellar gene expression is repressed, providing the appropriate cues leading to mature biofilm formation (Moorthy and Watnick, 2004). Although critical in aquatic colonization steps, MSHA does not participate in intestinal colonization, which has been examined in both the infant mouse model and humans (Thelin and Taylor, 1996; Tacket *et al.*, 1998). Furthermore, recent studies have indicated that *msh* gene expression is specifically repressed under conditions of *in vivo* intestinal colonization, and that this repression is mediated by the virulence regulator ToxT (Hsiao *et al.*, 2006). This gene modulation is proposed to be a mechanism for bacterial avoidance of binding by S-IgA (Hsiao *et al.*, 2006). S-IgA is the most abundant isotype of antibody found in mucosal secretions (Macpherson *et al.*, 2001), such as colostrum and breast milk. S-IgA provides a first line of defence against pathogens (Woof and Kerr, 2006). The surface of S-IgA is modified by the addition of mannose residues, a result of N- and O-linked glycosylation (Royle *et al.*, 2003). These modifications provide S-IgA with the ability to bind to a larger range of substrates, including bacterial lectins, via associations that are distinct from antigen specific immune responses (Moshier *et al.*, 1996). When MSHA is present on bacterial cells while in the intestine, it is bound by S-IgA glycosylated with mannose residues, resulting in bacterial-antibody complexes (Hsiao *et al.*, 2006). These complexes were observed to be unable to migrate though a mucin column *in vitro*, suggestive that a similar function may occur *in vivo*, with S-IgA-mediated agglutination blocking the ability of bacterial cells to access the epithelial cell surface (Hsiao *et al.*, 2006). Furthermore, a strain constitutively expressing MshA is able to colonize IgA$^-$ suckling mice, at a level equivalent to WT bacteria (Hsiao *et al.*, 2006). These results demonstrate that IgA is an essential component in the host defence against *V. cholerae* colonization of the intestine (Hsiao *et al.*, 2006). This study also determined that during infection, ToxT represses the expression of genes encoding the MSHA pilus components, providing a mechanism to avoid the binding of S-IgA to the bacterial surface (Hsiao *et al.*, 2006).

The MSHA structure was further characterized as the receptor for multiple filamentous bacteriophages including fs1, fs2, VGJ, KSF-1ϕ, and 493, all of which were purified from either O139, or non-O1 non-O139 isolates (Ehara *et al.*, 1997; Jouravleva *et al.*, 1998; Campos *et al.*, 2003; Faruque *et al.*, 2005). These phages are able to mediate horizontal gene transfer (Faruque *et al.*, 2005). The high expression levels of MSHA during periods of aquatic survival provide ample opportunity for MSHA mediated phage transduction to occur between strains (Faruque *et al.*, 2005). These results, coupled with a careful examination of the timeline for the emergence and decline of El Tor and O139 outbreaks, may suggest a general molecular mechanism for horizontal gene transfer and evolution of these *V. cholerae* strains over time (Jouravleva *et al.*, 1998; Faruque *et al.*, 2005).

GbpA-mediated aquatic and intestinal surface attachment

For *V. cholerae* to switch from the aquatic environment to the epithelium of the small intestine, bacterial cells must be able to transition from binding to available aquatic surfaces to successful attachment to the intestinal surface. The secreted protein GbpA has recently been identified as an attachment factor with the bifunctional ability to bind to structures present within both of these distinct environments (Kirn *et al.*, 2005). The binding specificity of GbpA is for the sugar GlcNAc, the building block of the most abundant aquatic carbon source, chitin (Meibom *et al.*, 2004; Kirn *et al.*, 2005). GbpA is required both to bind to the chitinous surfaces of copepods of organisms such as *Daphnia* spp., as well as to promote effective intestinal colonization in the infant mouse model of cholera (Kirn *et al.*, 2005). Recent evidence has identified a link between the presence and levels of GbpA and the *V. cholerae* quorum sensing response. The level of GbpA is greatest during times of low cell density. When cell density is high, as observed in late stages of intestinal colonization, GbpA is degraded by proteases HapA and PrtV (Jude, Skorupski and Taylor, unpublished). Genes encoding HapA and PrtV are activated by HapR (Jobling and Holmes, 1997; Vaitkevicius *et al.*, 2006). This provides a mechanism to regulate GbpA mediated attachment to, and detachment from, surfaces in response to environmental cues. This response provides experimental evidence for the hypothesis that HapA protease functions as a detachase (Finkelstein *et al.*, 1992), aiding in the release of *V. cholerae* from the intestine when bacterial cell density is high.

Mechanism of *Vibrio cholerae* motility

The ability of *V. cholerae* to propel itself in a directional manner within its various environments is powered by its single polar flagellum. This flagellum, aside from being localized to a single bacterial pole, is sheathed with a LPS decorated covering, a surface continuous with the bacterial outer membrane (Sjoblad *et al.*, 1983). Unlike other related *Vibrio* species including *V. parahaemolyticus*, *V. cholerae* expresses only a polar flagellum, and does not express lateral flagella (McCarter, 2001). In contrast to the proton-motive force that is required to drive lateral flagella, polar flagella of *Vibrio sp.* are powered by a sodium-driven motor, which is responsible for movement of *Vibrio* spp. at speeds up to 60 µm/s (Atsumi *et al.*, 1992; McCarter, 2001). This motor is able to rotate extremely rapidly providing this high rate of swimming speed (Magariyama *et al.*, 1994; McCarter, 2001). Much research has been devoted towards establishing whether or not the expression and function of polar flagella is a requirement for bacterial colonization of a host (Ottemann and Miller, 1997; Josenhans and Suerbaum, 2002). Flagella have the potential to function in colonization via two potential mechanisms: rotation of flagella to transit to surfaces, or use of flagella structures to directly attach to surfaces. The distinction between the flagella functioning for motility and the flagella serving as an attachment factor was determined by comparing colonization of a *fla* mutant to a strain with a non-motile flagellum (*mot*). These results determined that flagella are not utilized as an attachment factor in the *in vivo* mouse model, and that any deficiency in colonization is due to changes in motility (Lee *et al.*, 2001). Various reports examining both classical and El Tor biotypes have attempted to determine if non-motile mutants, with and without the presence of flagella, are functional for intestinal colonization (Guentzel and Berry, 1975; Richardson, 1991; Gardel and Mekalanos, 1996). In the El Tor biotype strains, the lack of motility results in a drop in colonization, but in the classical biotype results were contradictory amongst studies. Some studies demonstrated a role for motility in colonization, whereas other studies concluded that a lack of motility causes no colonization defect (Guentzel and Berry, 1975; Freter *et al.*, 1981; Richardson, 1991; Gardel and Mekalanos, 1996; Lee *et al.*, 2001). Murine studies were unable to identify a clear requirement for motility/presence of flagella in intestinal colonization. When non-motile strains were tested in volunteers, these strains also effectively colonized and elicited immune responses, but were found to be less reactogenic, causing only mild forms of symptoms usually associated with cholera infection (Mekalanos and Sadoff, 1994; Taylor *et al.*, 1994; Coster *et al.*, 1995; Kenner *et al.*, 1995). These results suggest that flagella are potentially critical to the process of human intestinal colonization by *V. cholerae*. To gain a better understanding of these interactions and the requirements of motility, as well as the physical presence of flagella for virulence of organisms such as *V. cholerae*, some time must be devoted to the structural and regulatory components required for flagellar assembly. The correlation between motility defects with a potential inverse relationship between bacterial motility and virulence gene expression will be examined in more depth later in this chapter.

Flagellar synthesis

Flagellar structural components

A large number of proteins are responsible for the assembly and proper function of the *V. cholerae* polar flagellum (Fig. 5.2B). Analysis of the annotated *V. cholerae* genome reveals five sequences encoding core flagellins of high sequence similarity (*flaA, flaB, flaC, flaD, flaE*) (Heidelberg *et al.*, 2000). Deletion of the genes encoding the various flagellins individually, as well as in combination, revealed that FlaA is the only flagellin protein essential for motility (Klose and Mekalanos, 1998a). To secure the flagellin filament, composed of FlaA subunits, to the bacterial membrane, the bacteria must produce basal body and hook structures (Fig. 5.3 and Table 5.2A). The basal body encompasses a series of ring and rod structures located within the bacterial inner and outer membranes. These structures provide a framework for anchoring the flagellum to the pole of the bacterial cell. The ring structures include the L ring in the outer membrane (encoded by *flgH*), the P ring within the peptidoglycan matrix (encoded by *flgI*), and the MS and C rings of the cytoplasmic membrane (encoded by *fliF* and *fliG*, respectively). Recently, an additional ring structure has been identified and characterized thus far only within sequenced *Vibrio* species, of which all possess a sodium driven motor that drives the flagellar rotation (Terashima *et al.*, 2006). Referred to as the T ring, and visible by transmission electron microscopy (TEM), it is predicted to be composed of the MotX and MotY proteins. MotX and MotY are stator proteins that are required to power

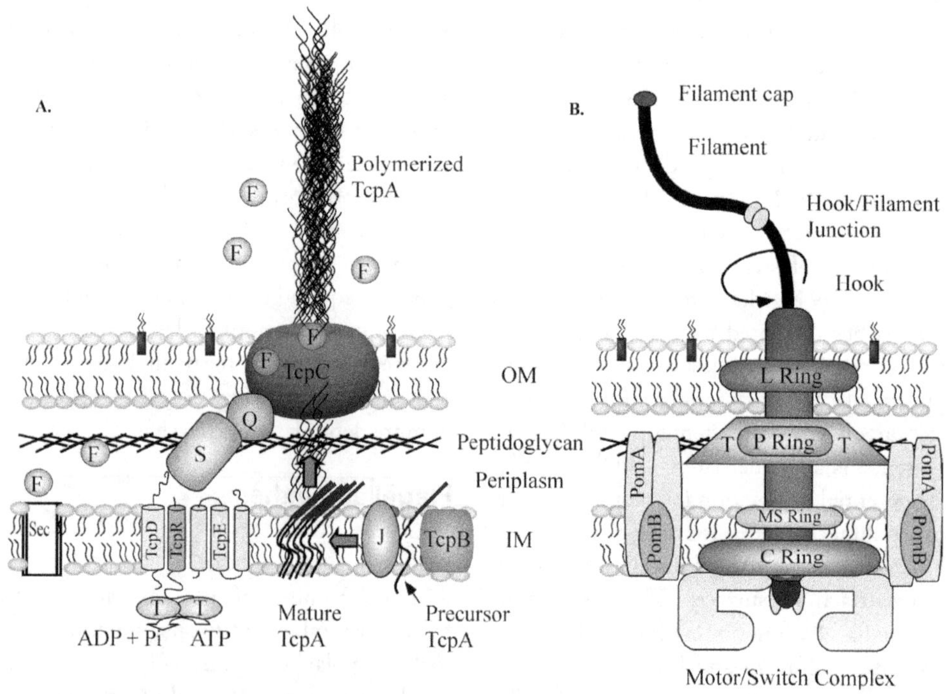

Figure 5.3 Models of *V. cholerae* TCP and polar flagellum. (A) Model depicting the arrangement of the proteins involved in TCP pilus biogenesis apparatus. (B) Model depicting the protein structures involved in the formation and rotation of the *V. cholerae* flagellum.

the motor, and consequently cause the *V. cholerae* polar flagellum to rotate. This ring is located between the P and MS rings (Terashima *et al.*, 2006). The rod structural proteins are encoded by the following genes (from proximal to distal): *flgC, flgB, fliE, flgF, flgG*. The hook (encoded by *flgE*) and hook–filament junction (encoded by *flgL* and *flgK*) connect the flagellin filament to the basal body complex. The filament structure is capped by the protein encoded by *fliD*.

Vibrio sodium-driven flagellar motor

The flagella are powered by the flagellar motor. In the case of *Vibrio* species, as well as other halophilic species, this is a sodium driven motor (Imae and Atsumi, 1989). This is in contrast to the flagellar motors powered by protons, present in bacterial species such as *Bacillus subtilis* and *E. coli*, and the *Vibrio* species that express lateral flagella (Imae and Atsumi, 1989; Atsumi *et al.*, 1992). The sodium-driven motor, powered by a sodium gradient and potently blocked by amiloride inhibitors, is highly active under low-viscosity swimming conditions encountered in the aquatic environment (Atsumi *et al.*, 1992, Kojima *et al.*, 1999; Yorimitsu and Homma, 2001). Observations made of the speed of flagellar rotation calculate the *Vibrio* motor as one the fastest observed in nature, with rotation rates ranging from 1100–1700 rotations per second (Magariyama *et al.*, 1994). Many of the details known about the function and subunits of the sodium driven motor have been elucidated in other species of *Vibrio*, and are presumed to function in a similar manner in *V. cholerae*. There are two distinct classes of genes encoding motor proteins, which include a general class of genes that is required for the proper function of all flagellar motors, and a second class of genes that is unique to those powered by sodium motive force. All motors consist of a motor switch complex encoded by *fliG*, *fliM* and *fliN* (Fig. 5.2) (Macnab, 1995). Additionally, all motors require an ion-conducting channel to power the motor rotation. This channel has ion specificity for either Na^+ or H^+, which serves to classify the motor as sodium driven or proton driven. In *V. cholerae*, the polar flagellar motor channel is specific for Na^+, and proteins composing this channel are encoded by *pomA* and *pomB* (Asai *et al.*, 1997). Loss of either *pomA* or *pomB* expression results in non-motile, but flagellated strains

Table 5.2A Genes that encode proteins that constitute the flagellar apparatus of *V. cholerae*

VC no.	Gene designation	Role/function, if known
VC2187	*flaC*	Flagellin
VC2188	*flaA*	Flagellin*
VC2189	–	Hypothetical protein
VC2190	*flgL*	Hook–filament junction
VC2191	*flgK*	Hook–filament junction
VC2192	*flgJ*	Flagellar protein
VC2193	*flgI*	P ring
VC2194	*flgH*	L ring
VC2195	*flgG*	Distal rod
VC2196	*flgF*	Proximal rod
VC2197	*flgE*	Flagellar hook protein
VC2198	*flgD*	Basal body rod modification protein
VC2199	*flgC*	Proximal rod
VC2200	*flgB*	Proximal rod
VC2203	–	Putative flagellar protein
VC2204	*flgM*	Negative regulator of flagellar synthesis
VC2601	*motX*	Stator, sodium-type flagellar protein T-ring
VC1008	*motY*	Stator, sodium-type flagellar protein T-ring
VC0892	*pomA*	Na channel
VC0893	*pomB*	Na channel
VC2120	*flhB*	Motor switch
VC2121	*fliR*	Flagellar biosynthetic protein
VC2122	*fliQ*	Flagellar biosynthetic protein
VC2123	*fliP*	Flagellar biosynthetic protein
VC2124	*fliO*	Flagellar biosynthetic protein
VC2125	*fliN*	Motor switch protein
VC2126	*fliM*	Motor switch protein
VC2127	*fliL*	Putative flagellar protein
VC2128	*fliK*	Putative flagellar hook length control protein
VC2129	*fliJ*	Putative flagellar protein
VC2130	*fliI*	Flagellar-specific ATP synthase
VC2131	*fliH*	Putative flagellar assembly protein
VC2132	*fliG*	C ring
VC2133	*fliF*	MS ring
VC2134	*fliE*	Proximal rod
VC2138	*fliS*	Flagellar protein
VC2139	*flaI*	Flagellar rod protein
VC2140	*fliD*	Fliament cap
VC2141	*flaG*	Export?
VC2142	*flaB*	Flagellin
VC2143	*flaD*	Flagellin
VC2144	*flaE*	Flagellin
VC2069	*flhA*	Export

Table 5.2B Genes that encode proteins that regulate flagellar gene expression in *V. cholerae*

VC no.	Gene designation	Role/function, if known
VC2135	*flrC*	Flagellar regulatory protein
VC2136	*flrB*	Flagellar regulatory protein
VC2137	*flrA*	σ^{54}-dependent activator
VC2066	*fliA*	σ^{28}
VC2067	*flhG*	Flagellar regulatory protein
VC2068	*flhF*	Flagellar regulatory protein

(Asai *et al.*, 1997; Kojima *et al.*, 1999). Further work has determined that PomA and PomB physically interact with each other, with PomA serving to stabilize PomB (Yorimitsu *et al.*, 1999; Sato and Homma, 2000). PomA and PomB form a complex in 2:1 stoichiometry, consisting of four PomA subunits, and two PomB subunits, and together provide the mechanism for torque generation for the sodium driven motor (Sato and Homma, 2000; Yakushi *et al.*, 2004; Yorimitsu *et al.*, 2004). MotX and MotY, forming the previously described T ring, are two other essential proteins unique to the sodium driven motor (Terashima *et al.*, 2006). TEM images of flagellar basal body complexes suggest that this T ring is responsible for interaction and possibly stabilization of the motor stator complex, consisting of PomA and PomB, to the bacterial cell pole (Terashima *et al.*, 2006). The loss of MotX or MotY results in a flagellar structure that lacks the T ring basal body component and is non-functional (Terashima *et al.*, 2006). Sodium driven motors also require Na^+ translocating NADH ubiquinone oxidoreductase (NQR), an enzyme complex involved in energy production, which drives the sodium motive force required to power the motor (Häse and Mekalanos, 1998; Häse and Barquera, 2001). Other unique requirements for the proper function of the sodium-driven motor, including mechanisms for stabilizing and maintaining high rotational speed compared with H^+-driven motors, require further analysis.

Regulation of flagellar gene expression

To produce the complex molecular machine of the flagellar structure, assembly of proteins must be carried out in a precise order. Therefore, the genes encoding the various proteins are sequentially expressed (McCarter, 2001). Analysis of the annotated *V. cholerae* genome reveals six chromosomal regions encoding flagellar specific genes (Heidelberg *et al.*, 2000; Prouty *et al.*, 2001). Examination of the transcriptional hierarchy of the flagellar apparatus provided the basis for the division of the flagellar associated genes into four classes, I, II, III and IV, based on the order of transcription, as well as the various gene promoter requirements for two alternative sigma factors, σ^{54} and σ^{28} (Prouty *et al.*, 2001). Initiation of flagellar synthesis is dependent on the transcription of *flrA*, which encodes a σ^{54}-dependent activator of downstream flagellar gene expression. *flrA* is the only gene within class I, and its own regulatory requirements remain unclear (Table 5.2B) (Prouty *et al.*, 2001). FlrA, and σ^{54} associated RNA polymerase (RNAP) holoenzyme proceed to induce expression of class II genes, which include *flrBC* (Table 5.2B), the *fli* operon (*fliEFGHIJKLMNOPQR*), *flhB*, *flhAFG*, *cheABWYZA* (Table 5.2A), and *fliA*, encoding a homologue of σ^{28}, which is required for class IV gene expression (Table 5.2B) (Klose and Mekalanos, 1998b; Prouty *et al.*, 2001). The induction of class III gene expression is dependent on FlrC, FlrA, and σ^{54}. These genes include *motX*, *flgBCDEFGHIJ*, *flgKL*, *flaA*, *flaG*, *flaI*, *fliD* and *fliS* (Table 5.2A) (Correa *et al.*, 2000; Prouty *et al.*, 2001). Class IV genes, dependent on σ^{28} associated RNAP holoenzyme for activation, include *flaB*, *flaC*, *flaD*, *flaE*, *flgMN*, *pomAB*, and *motY* (Table 5.2A), but to achieve high levels of induction of *flaB*, *flaC*, *flaD* and *flaE*, FlrC, FlrA, and σ^{54} are also required (Klose and Mekalanos, 1998a; Prouty *et al.*, 2001). The hierarchy imparted by this regulatory cascade is essential for the proper production of a flagellar apparatus with a sodium-driven motor.

Chemotaxis

In order for the flagellum to respond appropriately to the surrounding environment, it requires a mechanism to respond to signals from the environment and transmit those signals to alter the behaviour of the flagellum motor. *V. cholerae*, like other motile bacterial species, utilizes chemotaxis for this purpose. Bacterial chemotaxis modulates the speed and direction of flagellar based movement in response to extracellular signals, changing speed of rotation as well as the rotational direction between clockwise (CW) to counterclockwise (CCW). The directionality of the single *V. cholerae* polar flagellum results in two distinct swimming behaviours (Butler and Camilli, 2005). A consistent CCW rotation results in a steady, forward swimming pattern known as a run, while a switch to a CW rotation results in the bacteria exhibiting random tumbling motions. Bacterial tumbling provides a mechanism for the bacterium to rapidly change its swimming direction.

Chemotaxis proteins interact in a signal transduction sensing cascade that is critical for the proper flagellar rotation within changing environmental conditions. This cascade links the recognition of chemoeffectors by membrane bound methyl-accepting chemotaxis proteins (MCPs) to a physical interaction with the flagellar motor complex. This interaction is mediated by the appropriate phosphorylation of several specific response regulators. The *V. cholerae* genome includes coding sequences for an estimated 67 chemotaxis genes, a large number in comparison to the 10 chemotaxis genes found within the *E. coli* genome. These are organized into 3 operons, with the exception of the majority of those that encode MCPs, which are located throughout the genome (Table 5.3) (Heidelberg *et al.*, 2000). Only operon 2, containing *cheA-2*, which encodes one of the autohistidine kinases, has been demonstrated to be essential for *V. cholerae* chemotaxis (Lee *et al.*, 2001; Gosink *et al.*, 2002). Yet, the fact that there is such a high level of genomic variability between *V. cholerae* and *E. coli* chemotaxis sets of genes leads to the hypothesis that the *V. cholerae* chemotaxis cascade is much more complex than that of *E. coli*, possibly due to its requirement/ability to transition between various aquatic and intestinal environments.

The initial understanding of the *V. cholerae* chemotaxis cascade was derived from studies of the *E. coli* system. This cascade initiates with the interaction with one of a number of MCPs, which respond to the presence of chemoeffectors. Following this interaction, MCPs undergo a change in methylation state, mediated by CheR, a methyltransferase and CheB, a methylesterase, as well as a response regulator (Springer and Koshland, 1977; Yonekawa *et al.*, 1983). The CheW coupling protein recognizes the changed MCP methylation state and transfers a signal to CheA, which responds via autophosphorylation (Bourret *et al.*, 1993). CheA activation and subsequent transfer of phosphate from CheA-P to either CheY or CheB, two competing soluble response regulators, results in the determination of the directionality of the flagellar rotation (Li *et al.*, 1995). Phosphorylated CheY (CheY-P) interaction with the flagellar switch complex, comprised of FliM protein, is responsible for the switch to CW rotation (Bren and Eisenbach, 1998; Bren and Eisenbach, 2001). The ratio of phosphorylated and non-phosphorylated forms of CheY is also critical for the direction of flagellar rotation (Bren and Eisenbach, 2001). CheZ, a phosphatase, is responsible for a rapid dephosphorylation of CheA-P, and this keeps periods of CW random tumbling to a minimum (Hess *et al.*, 1988). Studies of chemotactic null mutants in *V. cholerae* have resulted in classification of *cheA*, *cheY*, *cheW* or *cheR* mutants as CCW biased, while the loss of *cheZ* or *cheB* leads to CW-biased rotation (Butler and Camilli, 2005).

The potential effect that appropriate chemotaxis has during intestinal colonization by *V. cholerae* is complex. Examination of specific mutants rendered non-chemotactic revealed an enhanced colonization of up to 70 times that observed in wild type strains (Lee *et al.*, 2001). This competitive advantage was limited to mutants that were non-chemotactic due to a CCW biased flagellar rotation (Butler and Camilli, 2004). In contrast, CW biased non-chemotactic mutants exhibited a log-fold attenuation in colonization compared to wild type (Butler and Camilli, 2004). The hypercolonizing mutants display aberrant coloniza-

Table 5.3 Genes that encode factors that participate in *V. cholerae* chemotaxis

VC no.	Gene designation	Role/function, if known
Cluster 1		
VC1394	*mcp*	Methyl-accepting protein
VC1395	*cheY-1*	Soluble response regulator
VC1396	–	Hypothetical protein
VC1397	*cheA-1*	Auto-histidine kinase
VC1398	*cheY-2*	Soluble repsonse regulator
VC1399	*cheR-1*	Methyltransferase
VC1401	*cheB-1*	Soluble repsonse regulator
VC1402	*cheW*	Coupling protein
VC1403	*mcp*	Methyl-accepting protein
VC1404	–	Hypothetical protein
VC1405	*mcp*	Methyl-accepting protein
VC1406	*mcp*	Methyl-accepting protein
Cluster 2		
VC2059	*cheW*	Coupling protein
VC2060	–	Hypothetical protein – *cheW* domain
VC2061	*parA*	Cell division
VC2062	*cheB-2*	Soluble response regulator/methylesterase
VC2063	*cheA-2*	Auto-histidine kinase
VC2064	*cheZ*	Phosphatase
VC2065	*cheY-3*	Soluble response regulator
Cluster 3		
VCA1088	*mcp*	Methyl accepting protein
VCA1089	*cheB-3*	Soluble response regulator
VCA1090	*cheD*	Deaminase
VCA1091	*cheR-3*	Methyltransferase
VCA1092	*mcp*	Methyl-accepting protein
VCA1093	*cheW-2*	Coupling protein
VCA1094	*cheW-3*	Coupling protein
VCA1095	*cheA-3*	Histidine kinase
VCA1096	*cheY-4*	Response regulator
Orphan genes		
VC0098	–	Methyl-accepting chemotaxis protein
VC0216	–	Methyl-accepting chemotaxis protein
VC0282	–	Methyl-accepting chemotaxis protein
VC0449	–	Methyl-accepting chemotaxis protein
VC0512	–	Methyl-accepting chemotaxis protein
VC0514	–	Methyl-accepting chemotaxis protein
VC1248	–	Methyl-accepting chemotaxis protein
VC1289	–	Methyl-accepting chemotaxis protein
VC1298	–	Methyl-accepting chemotaxis protein

Table 5.3 *continued*

VC no.	Gene designation	Role/function, if known
VC1313	–	Methyl-accepting chemotaxis protein
VC1316	*cheY*	Response regulator
VC1394	–	Methyl-accepting chemotaxis protein
VC1403	–	Methyl-accepting chemotaxis protein
VC1405	–	Methyl-accepting chemotaxis protein
VC1406	–	Methyl-accepting chemotaxis protein
VC1413	–	Methyl-accepting chemotaxis protein
VC1535	–	Methyl-accepting chemotaxis protein
VC1602	*cheV-1*	Coupling and adaptational
VC1643	–	Methyl-accepting chemotaxis protein
VC1859	–	Methyl-accepting chemotaxis protein
VC1868	–	Methyl-accepting chemotaxis protein
VC1898	–	Methyl-accepting chemotaxis protein
VC1967	–	Methyl-accepting chemotaxis protein
VC2006	*cheV-2*	Coupling and adaptational
VC2161	–	Methyl-accepting chemotaxis protein
VC2201	*cheR-2*	Methyltransferase
VC2202	*cheV-3*	Coupling and adaptational
VC2439	–	Methyl-accepting chemotaxis protein
VCA0008	–	Methyl-accepting chemotaxis protein
VCA0031	–	Methyl-accepting chemotaxis protein
VCA0068	–	Methyl-accepting chemotaxis protein
VCA0176	–	Methyl-accepting chemotaxis protein
VCA0268	–	Methyl-accepting chemotaxis protein
VCA0658	–	Methyl-accepting chemotaxis protein
VCA0663	–	Methyl-accepting chemotaxis protein
VCA0773	–	Methyl-accepting chemotaxis protein
VCA0864	–	Methyl-accepting chemotaxis protein
VCA0895	–	Chemotaxis transducer-related protein
VCA0906	–	Methyl-accepting chemotaxis protein
VCA0923	–	Methyl-accepting chemotaxis protein
VCA0954	*put cheV*	Coupling and adaptational
VCA0974	–	Methyl-accepting chemotaxis protein
VCA0979	–	Methyl-accepting chemotaxis protein
VCA0988	–	Methyl-accepting chemotaxis protein
VCA1031	–	Methyl-accepting chemotaxis protein, authentic frameshift
VCA1034	–	Methyl-accepting chemotaxis protein
VCA1056	–	Methyl-accepting chemotaxis protein
VCA1069	–	Methyl-accepting chemotaxis protein
VCA1088	–	Methyl-accepting chemotaxis protein
VCA1092	–	Methyl-accepting chemotaxis protein

tion patterns of the infant mouse compared with WT, which include colonization throughout the length of the small intestine, and accumulation on the surface mucus and lumen rather than within the intestinal crypts (Freter and O'Brien, 1981; Freter *et al.*, 1981a; Freter *et al.*, 1981b, Lee *et al.*, 2001). These phenotypes lead to the hypothesis that chemotactic gradients direct WT bacteria to avoid colonization of the upper small intestine, possibly by chemorepellent directed movement away from sites or chemoattractant directed movement towards the distal intestine (Butler and Camilli, 2005). A secondary chemoeffector gradient directs bacteria to the epithelial crypts, presumably, to optimal sites of colonization, away from potential peristaltic flushing (Butler and Camilli, 2005). Following passage through the human intestine, *V. cholerae* recovered from stool displayed a CCW biased flagellar rotation and had a competitive advantage when organisms were immediately introduced into the infant mouse model (Butler and Camilli, 2004). A recent study has linked this non-chemotactic CCW biased phenotype to a transient loss in expression of *cheW-1*, a coupling protein (Butler *et al.*, 2006). This is the first report of bacteria altering their ability to chemotax in response to human infection. Although these human passaged bacteria have not yet been tested for increased infectious capacity when re-introduced to humans, this provides a potential explanation for the clonality of environmental and clinical isolates presumed to be highly infectious, obtained during cholera outbreaks (Faruque *et al.*, 2005).

Vibrio species, and specifically *V. cholerae*, are responsive to a large variety of chemoattractant compounds, some of which are present in both the aquatic and intestinal sites of colonization. *V. cholerae* exhibits chemotaxis towards mucin components present on the mucilaginous surface of Anabaena blue green algae (Islam *et al.*, 2006). Mucin is similarly present on the intestinal surface, in addition to other chemoattractants, including amino acids, carbohydrates, and bile (O'Toole *et al.*, 1999). Another aquatic *Vibrio* species, *V. furnissii*, has been studied extensively for its mechanisms involved in chemotaxis towards chitin, oligosaccharides and sugars, which it later utilizes to degrade and transport via a specific permease (Bassler *et al.*, 1991a; Bassler *et al.*, 1991b; Yu *et al.*, 1993; Bouma and Roseman, 1996). A study comparing the *V. furnissii* and *V. cholerae* genomes revealed a high similarity within genes encoding factors involved in the chitinolytic cascade. *V. cholerae* has been shown to chemotax towards chitin oligosaccharides, specifically $(GlcNAc)_2$ (Li and Roseman, 2004). A GlcNAc induced MCP was also identified in *V. cholerae*, and was further characterized as participating in chemotaxis towards GlcNAc monomer and disaccharide forms, which are present in both the aquatic and intestinal environments (Meibom *et al.*, 2004). *Vibrio* chemorepellents have been less well characterized. Chemotaxis, and its role in the cyclic transition between the aquatic and intestinal environments remains an active area of investigation.

Relationship between motility and virulence factor expression

The ability of *V. cholerae* to cyclically transition between the aquatic environment and the human small intestine requires the precise regulation and interaction of its virulence factors. There exists a complex relationship between cell motility and the induction of the virulence regulon (Fig. 5.4). In the aquatic environment, the bacteria exist in heterogeneous populations of free swimming, planktonic cells, as well as bacteria attached to surfaces, often associated with other organisms in a biofilm (Watnick and Kolter, 1999; Watnick *et al.*, 1999; Zhu and Mekalanos, 2003; Moorthy and Watnick, 2004; Reguera and Kolter, 2005). The bacteria that colonize the intestine can begin the process either in a planktonic state or as ingested microcolonies. Similarly, bacteria shed from the intestine are a mixture of these two forms, planktonic cells and bacterial aggregates (Faruque *et al.*, 2006). The gene regulation and behavioural responses of *V. cholerae* while transitioning and persisting within these two environments will be examined in detail.

The aquatic environment in which *V. cholerae* persists is often characterized by a high salt concentration, as well as an abundance of readily available chitinous carbon sources, such as algal sheaths and zooplankton exoskeletons (Lipp *et al.*, 2002). Although *V. cholerae* can be found as free-swimming planktonic cells within

Persistence in the aquatic environment
- Bacteria bound to surfaces (zooplankton/phytoplankton)
  - •mediated by MSHA/ GbpA
- Form biofilms (EPS)
- Less motile
- High c-di-GMP levels
- Protection from predator grazing

Transition to the intestinal environment
- Ingestion of:
  - •planktonic bacteria
  - •aquatic biofilms (acid tolerant)
  - •clusters of bacterial aggregates/microcolonies
  - •hyperinfective CCW biased non-chemotactic bacteria
- Cell density low
- GbpA present
- Bile concentration high
  - •increased motility
  - •inhibition of ToxT activation of virulence factors
- Low c-di-GMP
- Loss of MSHA

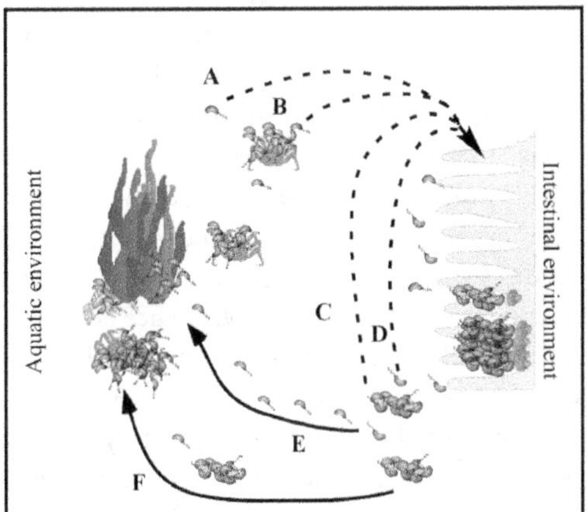

Persistence in the intestinal environment
- Colonize intestinal crypts
  - •direct epithelial binding (GbpA, others)
  - •bile concentration lowered
  - •mucus present
  - •motility reduced
  - •ToxT available for virulence gene activation
  - •microcolony formation (TCP)
- Some shedding of planktonic bacteria and bacterial aggregates in stool

Transition to the aquatic environment
- Cell density high
- Quorum sensing system activated
  - •transcription of virulence factors down
  - •activation/Secretion of proteases (HapA/PrtV)
  - •GbpA degraded by HapA/PrtV
- RpoS mediated mucosal escape response
  - •induction quorum sensing system
  - •induction motility/chemotaxis
- Release of hyperinfective, non-chemotactic cells

Figure 5.4 Response of virulence and motility during environmental shift. Persistence within the aquatic and intestinal environments, and the transition between these two environments result in a series of responses that include modulation of the virulence cascade and flagellar motility. Dashed arrows indicate bacterial forms that progress to colonize the intestine including (A) planktonic bacteria, (B) biofilms on aquatic surfaces, (C) clusters of bacterial aggregates from stool and (D) hyperinfective non-chemotactic bacteria shed in stool. Bacteria shed in stool can also colonize the aquatic environment as (annotated by continuous black arrows), (E) planktonic, free swimming bacteria, or (F) by forming biofilms on aquatic surfaces.

the aquatic environment, they are susceptible to protozoan grazing (Matz *et al.*, 2005). Following attachment to surfaces, the ability of the bacteria to form biofilms on aquatic surfaces provides further protection for the bacteria from increased temperature, acid stress, and grazing (Nalin *et al.*, 1979; Huq *et al.*, 1984; Zhu and Mekalanos, 2003, Castro-Rosas and Escartin, 2005). Association within a biofilm also potentially concentrates the number of bacteria in a small volume, allowing adequate numbers of organisms to achieve an infectious dose (Sack *et al.*, 1998; Colwell *et al.*, 2003; Faruque *et al.*, 2006). When living in a biofilm, the bacteria have an additional mechanism to defend against grazing, through secretion of the protease, PrtV (Vaitkevicius *et al.*, 2006). Expression of *prtV* is induced by the quorum-sensing regulator HapR under conditions of localized high cell density (Zhu *et al.*, 2002). Survival and persistence within an aquatic biofilm provides the bacteria the defences required for transitioning to colonize the human intestine.

The transition from the aquatic to the intestinal environment exposes bacteria to a vast range of bile concentrations. Bile is secreted into the duodenum from the gall bladder at an estimated concentration range of 0.2–2% for individual bile salts (Hofmann, 1998). Bile aids in digestion, but has also been demonstrated to cause increased bacterial motility, decreased expression of virulence factors, and altered regulation of outer membrane porin expression (Gupta and Chowdhury, 1997; Schuhmacher and Klose, 1999; Bina *et al.*, 2006; Chatterjee *et al.*, 2007). Bile, added at a concentration of 0.4%

to potentially mimic the environment which may be encountered within the intestinal lumen, induces motility of the bacteria, as well as reduces ToxT mediated activation of genes encoding virulence factors, including CT and TCP (Gupta and Chowdhury, 1997; Schuhmacher and Klose, 1999). This phenotype is potentially mediated by blocking dimerization of ToxT, which is necessary for optimal induction of *ctxAB* and *tcpA* expression, as occurs with the ToxT inhibitor, virstatin (Shakhnovich *et al.*, 2007). Flagellar based motility is also specifically increased in the presence of unsaturated fatty acids in bile, via increased expression of *flrA* (Chatterjee *et al.*, 2007). Increased motility in this environment is hypothesized to be beneficial in manoeuvring to the epithelial surface, the site of attachment and colonization. At the epithelial surface, bile concentrations are predicted to be lower than those observed within the lumen due to the mucus barrier, and once in these sites of colonization, the lowered bile concentration could reduce motility and coincidentally induce the expression of virulence factors (Schuhmacher and Klose, 1999). Additionally, the expression of genes encoding outer membrane proteins OmpU and OmpT has been investigated for its role in responding to changes in environmental conditions. Expression of *ompU* is activated by ToxR, while *ompT* expression is repressed (Crawford *et al.*, 1998, Li *et al.*, 2000). Under conditions of high bile concentrations, *ompU* transcription is induced (Provenzano *et al.*, 2000). OmpU, when expressed ectopically in the absence of OmpT, causes decreased bile sensitivity (Provenzano and Klose, 2000). In contrast, when OmpT is expressed ectopically in the absence of OmpU, bile sensitivity is increased, and reduction in *ctxAB* transcription and loss of TCP production is observed (Provenzano and Klose, 2000). These results demonstrate that the ToxR dependent induction of *ompU* expression affects bile resistance, and links the appropriate regulation of the genes encoding these porins by ToxR with the appropriate expression of virulence factors (Provenzano and Klose, 2000).

Once ingested, the bacteria must transition from their attachment on aquatic surfaces, to the surfaces of the intestinal epithelium. Quorum sensing sufficient biofilms were demonstrated to be better able to colonize, leading to the hypothesis that bacteria ingested as part of an aggregate, in addition to providing a large infectious dose, can better survive the acid barrier of the stomach, and are able to exit from the aggregate and transit to sites of intestinal colonization (Zhu and Mekalanos, 2003; Faruque *et al.*, 2006; Liu *et al.*, 2007).

Following *in vivo* attachment to the site of colonization, mediated by a cohort of factors including GbpA, microcolony formation can occur (Kirn *et al.*, 2005). Results from two recent studies have implied a complex relationship between flagellar motility powered by the sodium driven motor, expression of genes encoding exopolysaccharide (EPS) critical for biofilm formation, and expression of genes encoding virulence factors by *V. cholerae* O139. In the first study, O139 strains lacking a polar flagellar structure (*flaA* mutant) were found to show the rugose phenotype when cultured on solid medium, correlating with a corresponding increase in EPS production (Watnick *et al.*, 2001). Both studies found that a *flaA* mutant of *V. cholerae* O139 was defective for colonization of the infant mouse (Watnick *et al.*, 2001; Lauriano *et al.*, 2004). In the second study, the *flaA* mutant also exhibited a reduction in levels of TcpA and CT produced under *in vitro* growth conditions (Lauriano *et al.*, 2004). To address the relative contribution of each of these two phenotypes, the lack of flagella and increase in EPS production, to the reduced colonization, a *flaAvpsF* mutant was constructed. The *flaAvpsF* strain, while still lacking a flagellar structure, no longer produced EPS. In the *flaAvpsF* mutant, *in vitro* production of TcpA and CT (Lauriano *et al.*, 2004), as well as *in vivo* colonization ability of the infant mouse (Watnick *et al.*, 2001, Lauriano *et al.*, 2004) were restored to wild type levels. The authors hypothesize that it is the improper production of EPS *in vivo*, rather than the absence of motility due to lack of a flagellum, that is the primary cause for the reduction in colonization ability of the *flaA* mutant strain (Watnick *et al.*, 2001; Lauriano *et al.*, 2004).

The second study also demonstrated that EPS production in *flaA* mutants could be reduced by blocking the function of the sodium driven motor, either genetically, by deletion of *motX*, or chemically, by the addition of phenamil

(Lauriano *et al.*, 2004). When a gene encoding a component of the sodium driven motor was deleted in a *flaA* background (*flaAmotX*), the loss of aberrant EPS production did not restore *in vivo* colonization to wild type levels, as it had in the *flaAvpsF* mutant. In fact, the *flaAmotX* mutant was more defective for colonization than either of the single mutants, and the low levels of TcpA and CT produced by this strain were similar to those observed in a single *flaA* mutant strain (Lauriano *et al.*, 2004). These results may suggest other roles that the sodium driven motor has in relationship with the regulation of genes encoding virulence factors. These studies have revealed a potential complex interaction between the function of the sodium driven flagellar motor, the flagellar structure, and expression of genes producing factors essential for colonization *in vivo*.

Additional factors on the VPI participate in the modulation between motility and virulence factor production. The *tcpI* gene encodes an integral inner membrane protein with homology to MCPs, specifically ones involved in switching the directionality of flagellar rotation (Harkey *et al.*, 1994). Fusing TcpI to PhoA results in an increase in motility, and it is hypothesized that TcpI reduces chemotaxis directed motility while allowing maximal expression of *tcpA* (Harkey *et al.*, 1994). The gene encoding AcfB is also located on the VPI, activated by ToxT, and it too has features similar to TcpI, predicted to be an inner membrane protein with sequence similarity to MCPs (Everiss *et al.*, 1994). Both of these proteins are implicated in regulating bacterial motility through sensing of chemotaxis factors in the periplasm (Everiss *et al.*, 1994, Harkey *et al.*, 1994). Under *tcp* inducing conditions, motility is repressed, correlating with the increased TcpI and AcfB protein levels. This provides an additional link for the reduction in motility under virulence inducing conditions.

Transition from the intestine back to the aquatic environment requires the ability to coordinate the repression of genes encoding virulence factors with induction of protease expression and motility factors. A recent study examined the role of the stationary phase alternative sigma factor RpoS in this environmental transition (Nielsen *et al.*, 2006). It was determined that RpoS participates in what was termed the 'mucosal escape response'. It is presumed that conditions of localized high cell density within the small intestine, along with other stress including amino acid starvation and osmotic stress, combine to foster RpoS accumulation. This high cell density is also responsible for the induction of the quorum sensing system, and production of increased levels of HapR protein. As previously mentioned, this regulator is responsible for repression of genes encoding virulence gene activators and activation of genes encoding secreted proteases (Jobling and Holmes, 1997; Skorupski and Taylor, 1999). RpoS was demonstrated at this transitional stage to contribute to a reduction in virulence gene expression, possibly as a consequence of the role of RpoS in *hapR* gene expression (Nielsen *et al.*, 2006). Additionally, induction of genes encoding factors responsible for chemotaxis, motility, and exopolysaccharide are RpoS dependent (Nielsen *et al.*, 2006). Another study examined the shedding of aggregated bacterial clusters in the stools of cholera patients (Faruque *et al.*, 2006). Contained in the rice water stools, these aggregated clusters exhibited an enhanced infectivity compared to whole stool or planktonic cells, when introduced in the infant mouse model of cholera infection (Faruque *et al.*, 2006). These studies highlight the various mechanisms by which infectious *V. cholerae* can exit the intestinal site of colonization and be introduced to additional colonization sites.

Identification of factors involved in colonization based on *in vivo* expression

In order to identify novel factors required for growth *in vivo* and/or virulence, and to establish connections between previously characterized regulatory cascades, screens have been designed to examine gene expression patterns (both activated and repressed) during growth *in vivo*. Several techniques have been successfully adapted for *V. cholerae* to monitor its gene expression patterns in the infant mouse and in humans. Results from the use of these techniques have confirmed expression of previously characterized virulence factors as well as led to the identification of novel factors essential to the process of intestinal colonization and survival. These include variations on *in vivo* expression technology (IVET), including RIVET and IVIAT, signature tagged mutagenesis (STM), and a modification of differential fluorescence induction (Camilli *et al.*, 1994; Chiang and Mekalanos, 1998; Lee *et al.*, 1998; Lee *et al.*, 2001; Merrell *et al.*, 2002; Hang *et al.*, 2003; Hsiao *et al.*, 2006).

IVET/RIVET

In vivo expression technology (IVET) is utilized to identify genes that are specifically expressed during growth *in vivo*. This technique has been adapted for use in many bacterial species, including *Salmonella typhimurium*, *Pseudomonas aeruginosa*, and *Staphylococcus aureus* (Mahan *et al.*, 1993; Wang *et al.*, 1996; Lowe *et al.*, 1998). This technology was applied to the study of *V. cholerae* in 1994, as a variation of the original pIVET1 system, pioneered to utilize a *purA* expression selection system in *S. typhimurium* (Mahan *et al.*, 1993; Slauch *et al.*, 1994). The system used in *V. cholerae*, termed recombinase-based *in vivo* expression technology (RIVET), involves cloning a library of small DNA fragments to create random transcriptional fusions to the *lacZY* and *tnpR* resolvase genes harboured on pIVET5. Pools of pale blue colonies from XGal agar are inoculated into infant mice. If the promoter is induced *in vivo* (*ivi* promoter), *tnpR* is expressed and the resolvase catalyses the heritable loss of a *tet* gene encoding tetracycline resistance that is flanked by *res* sites in the *V. cholerae* chromosomal *lacZ* locus. Bacteria harvested from the intestine are then screened for sensitivity to tetracycline in order to identify *ivi* genes. Although the technique has the limitation that the *ivi* genes must exhibit a very low level of expression *in vitro*, it was used successfully to identify genes such as the *vieSAB* locus involved in c-di-GMP signalling cascades (Camilli and Mekalanos, 1995; Lee *et al.*, 1998). Further modifications to the RIVET technique allowed for examination of the temporal expression patterns of genes regulated by the ToxR regulon during intestinal growth (Lee *et al.*, 1999), as well as identification of regulators of *ctx* and *toxT*, which included genes encoding factors involved in metabolism, chemotaxis and motility, biosynthesis and transport of extracellular factors, and EPS production (Lee *et al.*, 2001).

IVIAT

In vivo-induced antigen technology (IVIAT) was developed to identify antigens produced by pathogenic bacteria during human infection (Handfield *et al.*, 2000). IVIAT was initially developed to identify antigens produced *in vivo* by *Actinobacillus actinomycetemcomitans* during human oral infection, and was subsequently adapted for use in *V. cholerae* (Handfield *et al.*, 2000; Hang *et al.*, 2003). This technology incorporates the use of patient sera obtained post-bacterial infection. Purification of this sera by adsorption against bacteria grown *in vitro* results in a pool of antibodies that specifically recognize the products of *in vivo* induced bacterial antigens (IVIs). Colonies from an *E. coli* expression library constructed from *V. cholerae* genomic DNA fragments are partially lysed, transferred to nitrocellulose, and immunoblotted with the adsorbed sera. Reactive clones are confirmed, and sequenced to identify the *V. cholerae* genomic insertion. In this initial study, screening of 30,000 colonies led to the identification of 38 potential IVI antigens. This technique identified the presence of antibodies in the human sera that recognized proteins of multiple functional classes, including pilus components and related secretins, TcpF, CT, chemotaxis and motility specific factors, as well as the quorum sensing protein, LuxP (Hang *et al.*, 2003). Interestingly, this technique identified certain antigens present during human infection that had been previously determined to be not required in the colonization of the infant mouse. Antibodies recognizing PilA and PilQ, two proteins involved in a putative type 4 pilus, were present in the human sera, suggesting their presence during human infection (Hang *et al.*, 2003). Previous *in vivo* studies utilizing the infant mouse model determined that PilA is not required for mouse colonization, although the presence of anti-PilA antibodies in mouse sera following colonization was not determined (Fullner and Mekalanos, 1999). This result suggests a potential distinction between the requirements for the colonization of the mouse and human intestine, but needs to be further characterized. In the future, a saturating screen of the *V. cholerae* genome utilizing IVIAT may identify additional *in vivo* specific antigens that participate in intestinal colonization and survival.

STM

An alternative technique utilized for the identification of factors required for survival *in vivo* is signature-tagged mutagenesis (STM). Initially used in *S. typhimurium*, STM was developed to allow screening of pools of bacteria carrying random transposon insertions for the ability to colonize a host. Each mutant carries a unique DNA sequence tag incorporated into the transposon insertion (signature tagged transposon) that provides a mechanism to identify input clones that are missing after infection (Hensel *et al.*, 1995). This technique proved to be useful for the high throughput screening of *V. cholerae* mutants (Chiang and Mekalanos, 1998; Merrell *et al.*, 2002). The STM mutant libraries in *V. cholerae* were initially constructed by randomly mutagenizing a *V. cholerae* strain with a pool of signature-tagged mini Tn5 transposons (mTn5) (Chiang and Mekalanos, 1998). An original group of 1700 Amp^S transconjugates were screened to identify tags able to be amplified by PCR, as well as to eliminate cross-hybridization of tags. The final library of 1100 mutants was grouped into pools of 48 strains each, and inoculated at a dose of 1×10^6 to 1×10^7 cfu per mouse, with each pool administered to two mice to eliminate the possibility of a random loss of a clone. Following colonization, intestines were harvested, and pooled genomic DNA was utilized as a template for the generation of labelled PCR products of the STs present in the colonizing bacteria. Labeled tags were hybridized to a dot blot array of STs present in the input pool of mutants. Signature tags from the input pool that did not hybridize with a tag, represented mutants that were unable to colonize the small intestine. Mapping of these transposon insertions identified genes encoding potential factors involved in survival *in vivo*. The advantage of this technique is that pools of mutants can be inoculated into the same animal, allowing large numbers of mutants to be tested *in vivo* at the same time. One consideration when using this technique is that the number of clones that can be successfully screened in a given pool is limited by the number that can simultaneously infect a given animal, sometimes referred to as a bottleneck.

In a second STM screen, using a *V. cholerae* strain with enhanced colonization capacity, Merrell *et al.* were able to double the number of strains inoculated in a single animal, and relieve this bottleneck, to some extent (Merrell *et al.*, 2002). This was done by mutagenizing this *V. cholerae* strain separately with 96 mTn5s that each express a unique tag. These tags were designed to all be readily amplified by PCR and not cross hybridize with any other tag, eliminating the need to prescreen the pool of mutants for PCR amplification and cross-hybridization prior to inoculation of the mouse (Mei *et al.*, 1997). 100 sets of 96 mutant clones were distributed into 96 well plates, creating 100 groups of 96 differentially tagged mutant strains. In this screen, transposons expressing 96 unique tags generated 9600 random transposon mutants. Due to the fact that each pool contained a mutant expressing one of each of the 96 STs, a master filter spotted with input tags could be constructed, allowing for hybridization by tags amplified from successfully colonized strains. This STM screen identified 251 mutant strains that were unable to colonize the small intestine, 2.5% of the total number of strains tested, which correlates with the 2% rate of colonization deficient strains identified in the previous STM screen by Chiang *et al.* This screen identified genes encoding factors involved in the LPS and purine biosynthetic pathways, regulatory factors, acid tolerance response components, and the NQR/electron transport system, in addition to genes already known to encode factors important in colonization, such as the *tcp* operon genes (Chiang and Mekalanos, 1998; Merrell *et al.*, 2002).

Differential fluorescence induction

The screens described above are designed to identify factors that are essential for *in vivo* survival. Screens have also been developed to identify genes for which expression is repressed when organisms transition from *in vitro* to *in vivo* environments. A modification of a differential fluorescence induction screen was utilized for this purpose in *V. cholerae* (Hsiao *et al.*, 2006). In this screen the parental *V. cholerae* strain was engineered to include a *gfp* gene driven by a TetR repressible promoter. Mutagenesis of this strain was accomplished with insertion of a *mariner* transposon harbouring a promoterless *tetR* gene. Strains in which the transposon inserts within an induced gene are GFP- due to TetR production. A library of transposon mutants was grown *in vitro* and sorted by flow cytometry to generate a pool of GFP- bacteria (transposon insertion into promoters active during *in vitro* growth conditions). This pool was subsequently inoculated into infant mice. Bacteria harvested from the intestines of the mice following colonization were again sorted by flow cytometry, this time selecting for the population of GFP+ bacteria. These bacteria harbour a transposon insertion within a gene that is expressed *in vitro*, but no longer expressed *in vivo*. These genes can then be identified by sequencing the transposon insertion. One finding from this screen was that the genes encoding the MSHA pilus are repressed during *in vivo* colonization of the infant mouse (Hsiao *et al.*, 2006). As discussed earlier in this chapter, this *in vivo* repression is postulated to allow the bacteria to evade binding of the mannose glycosylated S-IgA to the MSHA on the bacterial surface, which otherwise results in a bacterial–antibody complex that is unable to access the sites of colonization on intestinal epithelial cells (Hsiao *et al.*, 2006). This approach also identified genes encoding factors involved in flagellar biosynthesis and chemotaxis as repressed during colonization of the infant mouse (Hsiao *et al.*, 2006). Genes encoding chemotaxis factors CheA-2, CheY-1, and CheZ (Table 5.3), as well as flagellar biosynthesis and regulatory factors FliA, FlhF and FlhA (Table 5.2A and B) were identified using this screen. Although these findings have not yet been confirmed by RT-PCR, this preliminary data further establishes a relationship between the reduction in motility and induction of virulence factor expression during intestinal colonization (Harkey *et al.*, 1994; Gardel and Mekalanos, 1996; Schuhmacher and Klose, 1999; Krukonis and DiRita, 2003), and provides *in vivo* evidence of the repression of expression of genes encoding factors involved in motility during intestinal colonization (Hsiao *et al.*, 2006).

The complex regulation of *V. cholerae* virulence factors is required due to its ability to transition between its two natural environments. The precise control and timing of gene expression and protein degradation is still in the process of being fully characterized. The close association between bacterial motility and expression of virulence factors is also crucial for optimal bacterial survival within its environments.

References

Altschul, S.F., Gish, W., Miller, W., Myers, E.W. and Lipman, D.J. (1990). Basic local alignment search tool. J. Mol. Biol. *215*, 403–410.

Amikam, D. and Galperin, M.Y. (2006). PilZ domain is part of the bacterial c-di-GMP binding protein. Bioinformatics. *22*, 3–6.

Asai, Y., Kojima, S., Kato, H., Nishioka, N., Kawagishi, I. and Homma, M. (1997). Putative channel components for the fast-rotating sodium-driven flagellar motor of a marine bacterium. J. Bacteriol. *179*, 5104–5110.

Atsumi, T., McCarter, L. and Imae, Y. (1992). Polar and lateral flagellar motors of marine *Vibrio* are driven by different ion-motive forces. Nature. *355*, 182–184.

Bassler, B.L., Gibbons, P.J., Yu, C. and Roseman, S. (1991a). Chitin utilization by marine bacteria. Chemotaxis to chitin oligosaccharides by *Vibrio furnissii*. J. Biol. Chem. *266*, 24268–24275.

Bassler, B.L., Yu, C., Lee, Y.C. and Roseman, S. (1991b). Chitin utilization by marine bacteria. Degradation and catabolism of chitin oligosaccharides by *Vibrio furnissii*. J. Biol. Chem. *266*, 24276–24286.

Beyhan, S., Tischler, A.D., Camilli, A. and Yildiz, F.H. (2006a). Differences in gene expression between the classical and El Tor biotypes of *Vibrio cholerae* O1. Infect. Immun. *74*, 3633–3642.

Beyhan, S., Tischler, A.D., Camilli, A. and Yildiz, F.H. (2006b). Transcriptome and phenotypic responses of *Vibrio cholerae* to increased cyclic di-GMP level. J. Bacteriol. *188*, 3600–3613.

Bina, J.E. and Mekalanos, J.J. (2001). *Vibrio cholerae tolC* is required for bile resistance and colonization. Infect. Immun. *69*, 4681–4685.

Bina, J.E., Provenzano, D., Wang, C., Bina, X.R. and Mekalanos, J.J. (2006). Characterization of the *Vibrio cholerae vexAB* and *vexCD* efflux systems. Arch. Microbiol. *186*, 171–181.

Bomchil, N., Watnick, P. and Kolter, R. (2003). Identification and characterization of a *Vibrio cholerae* gene, *mbaA*, involved in maintenance of biofilm architecture. J. Bacteriol. *185*, 1384–1390.

Bose, N. and Taylor, R.K. (2005). Identification of a TcpC-TcpQ outer membrane complex involved in the biogenesis of the toxin-coregulated pilus of *Vibrio cholerae*. J. Bacteriol. *187*, 2225–2232.

Bouma, C.L. and Roseman, S. (1996). Sugar transport by the marine chitinolytic bacterium *Vibrio furnissii*. Molecular cloning and analysis of the mannose/glucose permease. J. Biol. Chem. *271*, 33468–33475.

Bourret, R.B., Davagnino, J. and Simon, M.I. (1993). The carboxy-terminal portion of the CheA kinase mediates regulation of autophosphorylation by transducer and CheW. J. Bacteriol. *175*, 2097–2101.

Bren, A. and Eisenbach, M. (1998). The N terminus of the flagellar switch protein, FliM, is the binding domain for the chemotactic response regulator, CheY. J. Mol. Biol. *278*, 507–514.

Bren, A. and Eisenbach, M. (2001). Changing the direction of flagellar rotation in bacteria by modulating the ratio between the rotational states of the switch protein FliM. J. Mol. Biol. *312*, 699–709.

Brown, R.C. and Taylor, R.K. (1995). Organization of *tcp*, *acf*, and *toxT* genes within a ToxT-dependent operon. Mol. Microbiol. *16*, 425–439.

Butler, S.M. and Camilli, A. (2004). Both chemotaxis and net motility greatly influence the infectivity of *Vibrio cholerae*. Proc. Natl. Acad. Sci. USA *101*, 5018–5023.

Butler, S.M. and Camilli, A. (2005). Going against the grain: chemotaxis and infection in *Vibrio cholerae*. Nat. Rev. Microbiol. 3, 611–620.

Butler, S.M., Nelson, E.J., Chowdhury, N., Faruque, S.M., Calderwood, S.B. and Camilli, A. (2006). Cholera stool bacteria repress chemotaxis to increase infectivity. Mol. Microbiol. *60*, 417–426.

Camilli, A., Beattie, D.T. and Mekalanos, J.J. (1994). Use of genetic recombination as a reporter of gene expression. Proc. Natl. Acad. Sci. USA *91*, 2634–2638.

Camilli, A. and Mekalanos, J.J. (1995). Use of recombinase gene fusions to identify *Vibrio cholerae* genes induced during infection. Mol. Microbiol. *18*, 671–683.

Campos, J., Martinez, E., Marrero, K., Silva, Y., Rodriguez, B.L., Suzarte, E., Ledon, T. and Fando, R. (2003). Novel type of specialized transduction for CTXϕ or its satellite phage RS1 mediated by filamentous phage VGJϕ in *Vibrio cholerae*. J. Bacteriol. *185*, 7231–7240.

Carroll, P.A., Tashima, K.T., Rogers, M.B., DiRita, V.J. and Calderwood, S.B. (1997). Phase variation in *tcpH* modulates expression of the ToxR regulon in *Vibrio cholerae*. Mol. Microbiol. *25*, 1099–1111.

Castro-Rosas, J. and Escartin, E.F. (2005). Increased tolerance of *Vibrio cholerae* O1 to temperature, pH, or drying associated with colonization of shrimp carapaces. Int. J. Food Microbiol. *102*, 195–201.

Chakrabarti, S.R., Chaudhuri, K., Sen, K. and Das, J. (1996). Porins of *Vibrio cholerae*: purification and characterization of OmpU. J. Bacteriol. *178*, 524–530.

Chatterjee, A., Dutta, P.K. and Chowdhury, R. (2007). Effect of fatty acids and cholesterol present in bile on expression of virulence factors and motility of *Vibrio cholerae*. Infect. Immun. *75*, 1946–1953.

Chiang, S.L. and Mekalanos, J.J. (1998). Use of signature-tagged transposon mutagenesis to identify *Vibrio cholerae* genes critical for colonization. Mol. Microbiol. *27*, 797–805.

Chiavelli, D.A., Marsh, J.W. and Taylor, R.K. (2001). The mannose-sensitive hemagglutinin of *Vibrio cholerae* promotes adherence to zooplankton. Appl. Environ. Microbiol. *67*, 3220–3225.

Childers, B.M., Weber, G.G., Prouty, M.G., Castaneda, M.M., Peng, F. and Klose, K.E. (2007). Identification of residues critical for the function of the *Vibrio cholerae* virulence regulator ToxT by scanning alanine mutagenesis. J. Mol. Biol. *367*, 1413–1430.

Chinnapen, D.J., Chinnapen, H., Saslowsky, D. and Lencer, W.I. (2007). Rafting with cholera toxin: endocytosis and trafficking from plasma membrane to ER. FEMS Microbiol. Lett. *266*, 129–137.

Colwell, R.R., Huq, A., Islam, M.S., Aziz, K.M., Yunus, M., Khan, N.H., Mahmud, A., Sack, R.B., Nair, G.B., Chakraborty, J., Sack, D.A. and Russek-Cohen, E. (2003). Reduction of cholera in Bangladeshi villages by simple filtration. Proc. Natl. Acad. Sci. USA *100*, 1051–1055.

Correa, N.E., Lauriano, C.M., McGee, R. and Klose, K.E. (2000). Phosphorylation of the flagellar regulatory protein FlrC is necessary for *Vibrio cholerae* motility and enhanced colonization. Mol. Microbiol. *35*, 743–755.

Coster, T.S., Killeen, K.P., Waldor, M.K., Beattie, D.T., Spriggs, D.R., Kenner, J.R., Trofa, A., Sadoff, J.C., Mekalanos, J.J. and Taylor, D.N. (1995). Safety, immunogenicity, and efficacy of live attenuated *Vibrio cholerae* O139 vaccine prototype. Lancet *345*, 949–952.

Craig, L., Taylor, R.K., Pique, M.E., Adair, B.D., Arvai, A.S., Singh, M., Lloyd, S.J., Shin, D.S., Getzoff, E.D., Yeager, M., Forest, K.T. and Tainer, J.A. (2003). Type IV pilin structure and assembly: X-ray and EM analyses of *Vibrio cholerae* toxin-coregulated pilus and *Pseudomonas aeruginosa* PAK pilin. Mol. Cell. *11*, 1139–1150.

Crawford, J.A., Kaper, J.B. and DiRita, V.J. (1998). Analysis of ToxR-dependent transcription activation of *ompU*, the gene encoding a major envelope protein in *Vibrio cholerae*. Mol. Microbiol. *29*, 235–246.

Davis, B.M., Lawson, E.H., Sandkvist, M., Ali, A., Sozhamannan, S. and Waldor, M.K. (2000). Convergence of the secretory pathways for cholera toxin and the filamentous phage, CTXϕ. Science *288*, 333–335.

De Silva, R.S., Kovacikova, G., Lin, W., Taylor, R.K., Skorupski, K. and Kull, F.J. (2005). Crystal structure of the virulence gene activator AphA from *Vibrio cholerae* reveals it is a novel member of the winged helix transcription factor superfamily. J. Biol. Chem. *280*, 13779–13783.

De Silva, R.S., Kovacikova, G., Lin, W., Taylor, R.K., Skorupski, K. and Kull, F.J. (2007). Crystal structure of the *Vibrio cholerae* quorum sensing regulatory protein HapR. J. Bacteriol. *189*, 5683–5691.

DiRita, V.J., Parsot, C., Jander, G. and Mekalanos, J.J. (1991). Regulatory cascade controls virulence in *Vibrio cholerae*. Proc. Natl. Acad. Sci. USA *88*, 5403–5407.

Dorman, C.J. (2004). H-NS: a universal regulator for a dynamic genome. Nat. Rev. Microbiol. *2*, 391–400.

Dorman, C.J. (2007). H-NS, the genome sentinel. Nat. Rev. Microbiol. *5*, 157–161.

Ehara, M., Shimodori, S., Kojima, F., Ichinose, Y., Hirayama, T., Albert, M.J., Supawat, K., Honma, Y., Iwanaga, M. and Amako, K. (1997). Characterization of filamentous phages of *Vibrio cholerae* O139 and O1. FEMS Microbiol. Lett. *154*, 293–301.

Eveiss, K.D., Hughes, K.J., Kovach, M.E. and Peterson, K.M. (1994). The *Vibrio cholerae acfB* colonization determinant encodes an inner membrane protein that is related to a family of signal-transducing proteins. Infect. Immun. *62*, 3289–3298.

Faruque, S.M., Bin Naser, I., Fujihara, K., Diraphat, P., Chowdhury, N., Kamruzzaman, M., Qadri, F., Yamasaki, S., Ghosh, A.N. and Mekalanos, J.J. (2005).

Genomic sequence and receptor for the *Vibrio cholerae* phage KSF-1ϕ: evolutionary divergence among filamentous vibriophages mediating lateral gene transfer. J. Bacteriol. *187*, 4095–4103.

Faruque, S.M., Biswas, K., Udden, S.M., Ahmad, Q.S., Sack, D.A., Nair, G.B. and Mekalanos, J.J. (2006). Transmissibility of cholera: *in vivo*-formed biofilms and their relationship to infectivity and persistence in the environment. Proc. Natl. Acad. Sci. USA *103*, 6350–6355.

Faruque, S.M., Naser, I.B., Islam, M.J., Faruque, A.S., Ghosh, A.N., Nair, G.B., Sack, D.A. and Mekalanos, J.J. (2005). Seasonal epidemics of cholera inversely correlate with the prevalence of environmental cholera phages. Proc. Natl. Acad. Sci. USA *102*, 1702–1707.

Finkelstein, R.A. (1992). Cholera Enterotoxin (Choleragen) A Historical Perspective In p. 155–187. Barua, D. and Greenough, W. B. (ed.), Current Topics in Infectious Diseases, Plenum Publishing Company, New York

Finkelstein, R.A., Boesman-Finkelstein, M., Chang, Y. and Häse, C.C. (1992). *Vibrio cholerae* hemagglutinin/protease, colonial variation, virulence, and detachment. Infect. Immun. *60*, 472–478.

Finkelstein, R.A., Boesman, M., Neoh, S.H., LaRue, M.K. and Delaney, R. (1974). Dissociation and recombination of the subunits of the cholera enterotoxin (choleragen). J. Immunol. *113*, 145–150.

Freter, R., Allweiss, B., O'Brien, P.C., Halstead, S.A. and Macsai, M.S. (1981). Role of chemotaxis in the association of motile bacteria with intestinal mucosa: in vitro studies. Infect. Immun. *34*, 241–249.

Freter, R. and O'Brien, P.C. (1981). Role of chemotaxis in the association of motile bacteria with intestinal mucosa: fitness and virulence of nonchemotactic *Vibrio cholerae* mutants in infant mice. Infect. Immun. *34*, 222–233.

Freter, R., O'Brien, P.C. and Macsai, M.S. (1981). Role of chemotaxis in the association of motile bacteria with intestinal mucosa: *in vivo* studies. Infect. Immun. *34*, 234–240.

Fullner, K.J. and Mekalanos, J.J. (1999). Genetic characterization of a new type IV-A pilus gene cluster found in both classical and El Tor biotypes of *Vibrio cholerae*. Infect. Immun. *67*, 1393–1404.

Galperin, M.Y., Nikolskaya, A.N. and Koonin, E.V. (2001). Novel domains of the prokaryotic two-component signal transduction systems. FEMS Microbiol. Lett. *203*, 11–21.

Gardel, C.L. and Mekalanos, J.J. (1996). Alterations in *Vibrio cholerae* motility phenotypes correlate with changes in virulence factor expression. Infect. Immun. *64*, 2246–2255.

Gosink, K.K., Kobayashi, R., Kawagishi, I. and Häse, C.C. (2002). Analyses of the roles of the three *cheA* homologs in chemotaxis of *Vibrio cholerae*. J. Bacteriol. *184*, 1767–1771.

Guentzel, M.N. and Berry, L.J. (1975). Motility as a virulence factor for *Vibrio cholerae*. Infect. Immun. *11*, 890–897.

Gupta, S. and Chowdhury, R. (1997). Bile affects production of virulence factors and motility of *Vibrio cholerae*. Infect. Immun. *65*, 1131–1134.

Handfield, M., Brady, L.J., Progulske-Fox, A. and Hillman, J.D. (2000). IVIAT: a novel method to identify microbial genes expressed specifically during human infections. Trends Microbiol. *8*, 336–339.

Hang, L., John, M., Asaduzzaman, M., Bridges, E.A., Vanderspurt, C., Kirn, T.J., Taylor, R.K., Hillman, J.D., Progulske-Fox, A., Handfield, M., Ryan, E. T. and Calderwood, S.B. (2003). Use of *in vivo*-induced antigen technology (IVIAT) to identify genes uniquely expressed during human infection with *Vibrio cholerae*. Proc. Natl. Acad. Sci. USA *100*, 8508–8513.

Harkey, C.W., Everiss, K.D. and Peterson, K.M. (1994). The *Vibrio cholerae* toxin-coregulated-pilus gene *tcpI* encodes a homolog of methyl-accepting chemotaxis proteins. Infect. Immun. *62*, 2669–2678.

Häse, C.C. and Barquera, B. (2001). Role of sodium bioenergetics in *Vibrio cholerae*. Biochim. Biophys. Acta. *1505:* 169–178.

Häse, C.C., Bauer, M.E. and Finkelstein, R.A. (1994). Genetic characterization of mannose-sensitive hemagglutinin (MSHA)-negative mutants of *Vibrio cholerae* derived by Tn5 mutagenesis. Gene *150*, 17–25.

Häse, C.C. and Mekalanos, J.J. (1998). TcpP protein is a positive regulator of virulence gene expression in *Vibrio cholerae*. Proc. Natl. Acad. Sci. USA *95*, 730–734.

Heidelberg, J.F., Eisen, J.A., Nelson, W.C., Clayton, R.A., Gwinn, M.L., Dodson, R.J., Haft, D.H., Hickey, E.K., Peterson, J.D., Umayam, L., Gill, S.R., Nelson, K.E., Read, T.D., Tettelin, H., Richardson, D., Ermolaeva, M.D., Vamathevan, J., Bass, S., Qin, H., Dragoi, I., Sellers, P., McDonald, L., Utterback, T., Fleishmann, R.D., Nierman, W.C., White, O., Salzberg, S.L., Smith, H.O., Colwell, R.R., Mekalanos, J.J., Venter, J.C. and Fraser, C.M. (2000). DNA sequence of both chromosomes of the cholera pathogen *Vibrio cholerae*. Nature *406*, 477–483.

Hensel, M., Shea, J.E., Gleeson, C., Jones, M.D., Dalton, E. and Holden, D.W. (1995). Simultaneous identification of bacterial virulence genes by negative selection. Science. *269*, 400–403.

Herrington, D.A., Hall, R.H., Losonsky, G., Mekalanos, J.J., Taylor, R.K. and Levine, M.M. (1988). Toxin, toxin-coregulated pili, and the *toxR* regulon are essential for *Vibrio cholerae* pathogenesis in humans. J. Exp. Med. *168*, 1487–1492.

Hess, J.F., Oosawa, K., Kaplan, N. and Simon, M.I. (1988). Phosphorylation of three proteins in the signaling pathway of bacterial chemotaxis. Cell. *53*, 79–87.

Higgins, D.E. and DiRita, V.J. (1994). Transcriptional control of *toxT*, a regulatory gene in the ToxR regulon of *Vibrio cholerae*. Mol. Microbiol. *14*, 17–29.

Higgins, D.E., Nazareno, E. and DiRita, V.J. (1992). The virulence gene activator ToxT from *Vibrio cholerae* is a member of the AraC family of transcriptional activators. J. Bacteriol. *174*, 6974–6980.

Hofmann, A.F. (1998). Bile secretion and the enterohepatic circulation of bile acids In p. 937–948. Feldman, M., Friedman, L. S. and Sleisenger, M. H. (ed.). Sleisenger and Fordtran's gastrointestinal and liver disease, 7th edn. W.B. Saunders, Philadelphia

Hsiao, A., Liu, Z., Joelsson, A. and Zhu, J. (2006). *Vibrio cholerae* virulence regulator-coordinated evasion of host immunity. Proc. Natl. Acad. Sci. USA *103*, 14542–14547.

Hulbert, R.R. and Taylor, R.K. (2002). Mechanism of ToxT-dependent transcriptional activation at the *Vibrio cholerae tcpA* promoter. J. Bacteriol. *184*, 5533–5544.

Hung, D.T., Zhu, J., Sturtevant, D. and Mekalanos, J.J. (2006). Bile acids stimulate biofilm formation in *Vibrio cholerae*. Mol. Microbiol. *59*, 193–201.

Huq, A., West, P.A., Small, E.B., Huq, M.I. and Colwell, R.R. (1984). Influence of water temperature, salinity, and pH on survival and growth of toxigenic *Vibrio cholerae* serovar O1 associated with live copepods in laboratory microcosms. Appl. Environ. Microbiol. *48*, 420–424.

Imae, Y. and Atsumi, T. (1989). Na^+-driven bacterial flagellar motors. J. Bioenerg. Biomembr. *21*, 705–716.

Islam, M.S., Goldar, M.M., Morshed, M.G., Bakht, H.B., Islam, M.S. and Sack, D.A. (2006). Chemotaxis between *Vibrio cholerae* O1 and a blue-green alga, *Anabaena* sp. Epidemiol. Infect. *134*, 645–648.

Jenal, U. and Malone, J. (2006). Mechanisms of cyclic-di-GMP signaling in bacteria. Annu. Rev. Genet. *40*, 385–407.

Jobling, M.G. and Holmes, R.K. (1997). Characterization of *hapR*, a positive regulator of the *Vibrio cholerae* HA/protease gene *hap*, and its identification as a functional homologue of the *Vibrio harveyi luxR* gene. Mol. Microbiol. *26*, 1023–1034.

Jonson, G., Holmgren, J. and Svennerholm, A.M. (1991). Identification of a mannose-binding pilus on *Vibrio cholerae* El Tor. Microb. Pathog. *11*, 433–441.

Jonson, G., Lebens, M. and Holmgren, J. (1994). Cloning and sequencing of *Vibrio cholerae* mannose-sensitive haemagglutinin pilin gene: localization of *mshA* within a cluster of type 4 pilin genes. Mol. Microbiol. *13*, 109–118.

Josenhans, C. and Suerbaum, S. (2002). The role of motility as a virulence factor in bacteria. Int. J. Med. Microbiol. *291*, 605–614.

Jouravleva, E.A., McDonald, G.A., Marsh, J.W., Taylor, R.K., Boesman-Finkelstein, M. and Finkelstein, R.A. (1998). The *Vibrio cholerae* mannose-sensitive hemagglutinin is the receptor for a filamentous bacteriophage from *V. cholerae* O139. Infect. Immun. *66*, 2535–2539.

Karaolis, D.K., Johnson, J.A., Bailey, C.C., Boedeker, E.C., Kaper, J.B. and Reeves, P.R. (1998). A *Vibrio cholerae* pathogenicity island associated with epidemic and pandemic strains. Proc. Natl. Acad. Sci. USA *95*, 3134–3139.

Karaolis, D.K., Lan, R., Kaper, J.B. and Reeves, P.R. (2001). Comparison of *Vibrio cholerae* pathogenicity islands in sixth and seventh pandemic strains. Infect. Immun. *69*, 1947–1952.

Kaufman, M.R., Seyer, J.M. and Taylor, R.K. (1991). Processing of TCP pilin by TcpJ typifies a common step intrinsic to a newly recognized pathway of extracellular protein secretion by gram-negative bacteria. Genes Dev. *5*, 1834–1846.

Kenner, J.R., Coster, T.S., Taylor, D.N., Trofa, A.F., Barrera-Oro, M., Hyman, T., Adams, J.M., Beattie, D.T., Killeen, K.P., Spriggs, D.R., *et al.* (1995). Peru-15, an improved live attenuated oral vaccine candidate for *Vibrio cholerae* O1. J. Infect. Dis. *172*, 1126–1129.

Kirn, T.J., Bose, N. and Taylor, R.K. (2003). Secretion of a soluble colonization factor by the TCP type 4 pilus biogenesis pathway in *Vibrio cholerae*. Mol. Microbiol. *49*, 81–92.

Kirn, T.J., Jude, B.A. and Taylor, R.K. (2005). A colonization factor links *Vibrio cholerae* environmental survival and human infection. Nature. *438*, 863–866.

Kirn, T.J., Lafferty, M.J., Sandoe, C.M. and Taylor, R.K. (2000). Delineation of pilin domains required for bacterial association into microcolonies and intestinal colonization by *Vibrio cholerae*. Mol. Microbiol. *35*, 896–910.

Kirn, T.J. and Taylor, R.K. (2005). TcpF is a soluble colonization factor and protective antigen secreted by El Tor and classical O1 and O139 *Vibrio cholerae* serogroups. Infect. Immun. *73*, 4461–4470.

Klose, K.E. and Mekalanos, J.J. (1998a). Differential regulation of multiple flagellins in *Vibrio cholerae*. J. Bacteriol. *180*, 303–316.

Klose, K.E. and Mekalanos, J.J. (1998b). Distinct roles of an alternative sigma factor during both free-swimming and colonizing phases of the *Vibrio cholerae* pathogenic cycle. Mol. Microbiol. *28*, 501–520.

Kojima, S., Asai, Y., Atsumi, T., Kawagishi, I. and Homma, M. (1999). Na^+-driven flagellar motor resistant to phenamil, an amiloride analog, caused by mutations in putative channel components. J. Mol. Biol. *285*, 1537–1547.

Kovach, M.E., Shaffer, M.D. and Peterson, K.M. (1996). A putative integrase gene defines the distal end of a large cluster of ToxR-regulated colonization genes in *Vibrio cholerae*. Microbiology *142*, 2165–2174.

Kovacikova, G., Lin, W. and Skorupski, K. (2004). *Vibrio cholerae* AphA uses a novel mechanism for virulence gene activation that involves interaction with the LysR-type regulator AphB at the *tcpPH* promoter. Mol. Microbiol. *53*, 129–142.

Kovacikova, G., Lin, W. and Skorupski, K. (2005). Dual regulation of genes involved in acetoin biosynthesis and motility/biofilm formation by the virulence activator AphA and the acetate-responsive LysR-type regulator AlsR in *Vibrio cholerae*. Mol. Microbiol. *57*, 420–433.

Kovacikova, G. and Skorupski, K. (1999). A *Vibrio cholerae* LysR homolog, AphB, cooperates with AphA at the *tcpPH* promoter to activate expression of the ToxR virulence cascade. J. Bacteriol. *181*, 4250–4256.

Kovacikova, G. and Skorupski, K. (2000). Differential activation of the *tcpPH* promoter by AphB determines biotype specificity of virulence gene expression in *Vibrio cholerae*. J. Bacteriol. *182*, 3228–3238.

Kovacikova, G. and Skorupski, K. (2001). Overlapping binding sites for the virulence gene regulators AphA, AphB and cAMP-CRP at the *Vibrio cholerae tcpPH* promoter. Mol. Microbiol. *41*, 393–407.

Kovacikova, G. and Skorupski, K. (2002). Regulation of virulence gene expression in *Vibrio cholerae* by quorum sensing: HapR functions at the *aphA* promoter. Mol. Microbiol. *46*, 1135–1147.

Krukonis, E.S. and DiRita, V.J. (2003). From motility to virulence: Sensing and responding to environmental signals in *Vibrio cholerae*. Curr. Opin. Microbiol. *6*, 186–190.

Krukonis, E.S., Yu, R.R. and DiRita, V.J. (2000). The *Vibrio cholerae* ToxR/TcpP/ToxT virulence cascade: distinct roles for two membrane-localized transcriptional activators on a single promoter. Mol. Microbiol. *38*, 67–84.

LaPointe, C.F. and Taylor, R.K. (2000). The type 4 prepilin peptidases comprise a novel family of aspartic acid proteases. J. Biol. Chem. *275*, 1502–1510.

Lauriano, C.M., Ghosh, C., Correa, N.E. and Klose, K.E. (2004). The sodium-driven flagellar motor controls exopolysaccharide expression in *Vibrio cholerae*. J. Bacteriol. *186*, 4864–4874.

Lee, S.H., Angelichio, M.J., Mekalanos, J.J. and Camilli, A. (1998). Nucleotide sequence and spatiotemporal expression of the *Vibrio cholerae vieSAB* genes during infection. J. Bacteriol. *180*, 2298–2305.

Lee, S.H., Butler, S.M. and Camilli, A. (2001). Selection for *in vivo* regulators of bacterial virulence. Proc. Natl. Acad. Sci. USA *98*, 6889–6894.

Lee, S.H., Hava, D.L., Waldor, M.K. and Camilli, A. (1999). Regulation and temporal expression patterns of *Vibrio cholerae* virulence genes during infection. Cell. *99*, 625–634.

Lencer, W.I. and Tsai, B. (2003). The intracellular voyage of cholera toxin: going retro. Trends Biochem. Sci. *28*, 639–645.

Lenz, D.H., Miller, M.B., Zhu, J., Kulkarni, R.V. and Bassler, B.L. (2005). CsrA and three redundant small RNAs regulate quorum sensing in *Vibrio cholerae*. Mol. Microbiol. *58*, 1186–1202.

Lenz, D.H., Mok, K.C., Lilley, B.N., Kulkarni, R.V., Wingreen, N.S. and Bassler, B.L. (2004). The small RNA chaperone Hfq and multiple small RNAs control quorum sensing in *Vibrio harveyi* and *Vibrio cholerae*. Cell. *118*, 69–82.

Levy, S.B. (1992). Active efflux mechanisms for antimicrobial resistance. Antimicrob. Agents Chemother. *36*, 695–703.

Li, C.C., Crawford, J.A., DiRita, V.J. and Kaper, J.B. (2000). Molecular cloning and transcriptional regulation of *ompT*, a ToxR-repressed gene in *Vibrio cholerae*. Mol. Microbiol. *35*, 189–203.

Li, J., Swanson, R.V., Simon, M.I. and Weis, R.M. (1995). The response regulators CheB and CheY exhibit competitive binding to the kinase CheA. Biochemistry *34*, 14626–14636.

Li, X. and Roseman, S. (2004). The chitinolytic cascade in Vibrios is regulated by chitin oligosaccharides and a two-component chitin catabolic sensor/kinase. Proc. Natl. Acad. Sci. USA *101*, 627–631.

Lim, B., Beyhan, S. and Yildiz, F.H. (2007). Regulation of *Vibrio* polysaccharide synthesis and virulence factor production by CdgC, a GGDEF-EAL domain protein, in *Vibrio cholerae*. J. Bacteriol. *189*, 717–729.

Lin, W., Kovacikova, G. and Skorupski, K. (2007). The quorum sensing regulator HapR downregulates the expression of the virulence gene transcription factor AphA in *Vibrio cholerae* by antagonizing Lrp- and VpsR-mediated activation. Mol. Microbiol. *64*, 953–967.

Lipp, E.K., Huq, A. and Colwell, R.R. (2002). Effects of global climate on infectious disease: the cholera model. Clin. Microbiol. Rev. *15*, 757–770.

Liu, Z., Stirling, F.R. and Zhu, J. (2007). Temporal quorum-sensing induction regulates *Vibrio cholerae* biofilm architecture. Infect. Immun. *75*, 122–126.

Lowe, A.M., Beattie, D.T. and Deresiewicz, R.L. (1998). Identification of novel staphylococcal virulence genes by *in vivo* expression technology. Mol. Microbiol. *27*, 967–976.

Lucchini, S., Rowley, G., Goldberg, M.D., Hurd, D., Harrison, M. and Hinton, J.C. (2006). H-NS mediates the silencing of laterally acquired genes in bacteria. PLoS Pathog. *2*, e81.

Macnab, R.M. (1995). Flagellar Switch p. 181–199. In Hoch, J. A. and Silhavy, T. J. (ed.). Two-component signal transduction, ASM Press, Washington, D.C

Macpherson, A.J., Hunziker, L., McCoy, K. and Lamarre, A. (2001). IgA responses in the intestinal mucosa against pathogenic and non-pathogenic microorganisms. Microbes Infect. *3*, 1021–1035.

Magariyama, Y., Sugiyama, S., Muramoto, K., Maekawa, Y., Kawagishi, I., Imae, Y. and Kudo, S. (1994). Very fast flagellar rotation. Nature. *371*, 752.

Mahan, M.J., Slauch, J.M. and Mekalanos, J.J. (1993). Selection of bacterial virulence genes that are specifically induced in host tissues. Science. *259*, 686–688.

Marsh, J.W., Sun, D. and Taylor, R.K. (1996). Physical linkage of the *Vibrio cholerae* mannose-sensitive hemagglutinin secretory and structural subunit gene loci: identification of the *mshG* coding sequence. Infect. Immun. *64*, 460–465.

Marsh, J.W. and Taylor, R.K. (1999). Genetic and transcriptional analyses of the *Vibrio cholerae* mannose-sensitive hemagglutinin type 4 pilus gene locus. J. Bacteriol. *181*, 1110–1117.

Matson, J.S. and DiRita, V.J. (2005). Degradation of the membrane-localized virulence activator TcpP by the YaeL protease in *Vibrio cholerae*. Proc. Natl. Acad. Sci. USA *102*, 16403–16408.

Matz, C., McDougald, D., Moreno, A.M., Yung, P.Y., Yildiz, F.H. and Kjelleberg, S. (2005). Biofilm formation and phenotypic variation enhance predation-driven persistence of *Vibrio cholerae*. Proc. Natl. Acad. Sci. USA *102*, 16819–16824.

McCarter, L.L. (2001). Polar flagellar motility of the Vibrionaceae. Microbiol. Mol. Biol. Rev. *65*, 445–462.

McMurry, L.M., Aronson, D.A. and Levy, S.B. (1983). Susceptible *Escherichia coli* cells can actively excrete tetracyclines. Antimicrob. Agents Chemother. *24*, 544–551.

Mei, J.M., Nourbakhsh, F., Ford, C.W. and Holden, D.W. (1997). Identification of Staphylococcus aureus virulence genes in a murine model of bacteraemia using signature-tagged mutagenesis. Mol. Microbiol. *26*, 399–407.

Meibom, K.L., Li, X.B., Nielsen, A.T., Wu, C.Y., Roseman, S. and Schoolnik, G.K. (2004). The *Vibrio cholerae* chitin utilization program. Proc. Natl. Acad. Sci. USA *101*, 2524–2529.

Mekalanos, J.J. and Sadoff, J.C. (1994). Cholera vaccines: fighting an ancient scourge. Science. *265*, 1387–1389.

Merrell, D.S., Hava, D.L. and Camilli, A. (2002). Identification of novel factors involved in colonization and acid tolerance of *Vibrio cholerae*. Mol. Microbiol. *43*, 1471–1491.

Miller, M.B., Skorupski, K., Lenz, D.H., Taylor, R.K. and Bassler, B.L. (2002). Parallel quorum sensing systems converge to regulate virulence in *Vibrio cholerae*. Cell. *110*, 303–314.

Miller, V.L., DiRita, V.J. and Mekalanos, J.J. (1989). Identification of *toxS*, a regulatory gene whose product enhances ToxR-mediated activation of the cholera toxin promoter. J. Bacteriol. *171*, 1288–1293.

Miller, V.L. and Mekalanos, J.J. (1984). Synthesis of cholera toxin is positively regulated at the transcriptional level by ToxR. Proc. Natl. Acad. Sci. USA *81*, 3471–3475.

Miller, V.L. and Mekalanos, J.J. (1988). A novel suicide vector and its use in construction of insertion mutations: osmoregulation of outer membrane proteins and virulence determinants in *Vibrio cholerae* requires *toxR*. J. Bacteriol. *170*, 2575–2583.

Miller, V.L., Taylor, R.K. and Mekalanos, J.J. (1987). Cholera toxin transcriptional activator ToxR is a transmembrane DNA binding protein. Cell. *48*, 271–279.

Moorthy, S. and Watnick, P.I. (2004). Genetic evidence that the *Vibrio cholerae* monolayer is a distinct stage in biofilm development. Mol. Microbiol. *52*, 573–587.

Moshier, A., Reddy, M.S. and Scannapieco, F.A. (1996). Role of type 1 fimbriae in the adhesion of *Escherichia coli* to salivary mucin and secretory immunoglobulin A. Curr. Microbiol. *33*, 200–208.

Nalin, D.R., Daya, V., Reid, A., Levine, M.M. and Cisneros, L. (1979). Adsorption and growth of *Vibrio cholerae* on chitin. Infect. Immun. *25*, 768–770.

Navarre, W.W., Porwollik, S., Wang, Y., McClelland, M., Rosen, H., Libby, S.J. and Fang, F.C. (2006). Selective silencing of foreign DNA with low GC content by the H-NS protein in *Salmonella*. Science. *313*, 236–238.

Nielsen, A.T., Dolganov, N.A., Otto, G., Miller, M.C., Wu, C.Y. and Schoolnik, G.K. (2006). RpoS controls the *Vibrio cholerae* mucosal escape response. PLoS Pathog. *2*, e109.

Nye, M.B., Pfau, J.D., Skorupski, K. and Taylor, R.K. (2000). *Vibrio cholerae* H-NS silences virulence gene expression at multiple steps in the ToxR regulatory cascade. J. Bacteriol. *182*, 4295–4303.

O'Toole, R., Lundberg, S., Fredriksson, S.A., Jansson, A., Nilsson, B. and Wolf-Watz, H. (1999). The chemotactic response of *Vibrio anguillarum* to fish intestinal mucus is mediated by a combination of multiple mucus components. J. Bacteriol. *181*, 4308–4317.

Ottemann, K.M. and Miller, J.F. (1997). Roles for motility in bacterial–host interactions. Mol. Microbiol. *24*, 1109–1117.

Parsot, C. and Mekalanos, J.J. (1991). Expression of the *Vibrio cholerae* gene encoding aldehyde dehydrogenase is under control of ToxR, the cholera toxin transcriptional activator. J. Bacteriol. *173*, 2842–2851.

Parsot, C., Taxman, E. and Mekalanos, J.J. (1991). ToxR regulates the production of lipoproteins and the expression of serum resistance in *Vibrio cholerae*. Proc. Natl. Acad. Sci. USA *88*, 1641–1645.

Peterson, K.M. and Mekalanos, J.J. (1988). Characterization of the *Vibrio cholerae* ToxR regulon: identification of novel genes involved in intestinal colonization. Infect. Immun. *56*, 2822–2829.

Pfau, J.D. and Taylor, R.K. (1996). Genetic footprint on the ToxR-binding site in the promoter for cholera toxin. Mol. Microbiol. *20*, 213–222.

Pfau, J.D. and Taylor, R.K. (1998). Mutations in *toxR* and *toxS* that separate transcriptional activation from DNA binding at the cholera toxin gene promoter. J. Bacteriol. *180*, 4724–4733.

Pratt, J.T., Tamayo, R., Tischler, A.D. and Camilli, A. (2007). PilZ domain proteins bind cyclic diguanylate and regulate diverse processes in *Vibrio cholerae*. J. Biol. Chem. *282*, 12860–12870.

Prouty, M.G., Correa, N.E. and Klose, K.E. (2001). The novel σ^{54}- and σ^{28}- dependent flagellar gene transcription hierarchy of *Vibrio cholerae*. Mol. Microbiol. *39*, 1595–1609.

Prouty, M.G., Osorio, C.R. and Klose, K.E. (2005). Characterization of functional domains of the *Vibrio cholerae* virulence regulator ToxT. Mol. Microbiol. *58*, 1143–1156.

Provenzano, D. and Klose, K.E. (2000). Altered expression of the ToxR-regulated porins OmpU and OmpT diminishes *Vibrio cholerae* bile resistance, virulence factor expression, and intestinal colonization. Proc. Natl. Acad. Sci. USA 97, 10220–10224.

Provenzano, D., Schuhmacher, D.A., Barker, J.L. and Klose, K.E. (2000). The virulence regulatory protein ToxR mediates enhanced bile resistance in *Vibrio cholerae* and other pathogenic *Vibrio* species. Infect. Immun. *68*, 1491–1497.

Rashid, M.H., Rajanna, C., Ali, A. and Karaolis, D.K. (2003). Identification of genes involved in the switch between the smooth and rugose phenotypes of *Vibrio cholerae*. FEMS Microbiol. Lett. *227*, 113–119.

Reguera, G. and Kolter, R. (2005). Virulence and the environment: a novel role for *Vibrio cholerae* toxin-coregulated pili in biofilm formation on chitin. J. Bacteriol. *187*, 3551–3555.

Richardson, K. (1991). Roles of motility and flagellar structure in pathogenicity of *Vibrio cholerae*: analysis of motility mutants in three animal models. Infect. Immun. *59*, 2727–2736.

Royle, L., Roos, A., Harvey, D.J., Wormald, M.R., van Gijlswijk-Janssen, D., Redwan, e.-R.M., Wilson, I.A., Daha, M.R., Dwek, R.A. and Rudd, P.M. (2003). Secretory IgA N- and O-glycans provide a link between the innate and adaptive immune systems. J. Biol. Chem. *278*, 20140–20153.

Sack, D.A., Tacket, C.O., Cohen, M.B., Sack, R.B., Losonsky, G.A., Shimko, J., Nataro, J.P., Edelman, R., Levine, M.M., Giannella, R.A., Schiff, G. and Lang, D. (1998). Validation of a volunteer model of cholera with frozen bacteria as the challenge. Infect. Immun. *66*, 1968–1972.

Sandkvist, M., Michel, L.O., Hough, L.P., Morales, V.M., Bagdasarian, M., Koomey, M., DiRita, V.J. and Bagdasarian, M. (1997). General secretion pathway (eps) genes required for toxin secretion and outer membrane biogenesis in *Vibrio cholerae*. J. Bacteriol. *179*, 6994–7003.

Sato, K. and Homma, M. (2000). Multimeric structure of PomA, a component of the Na^+-driven polar flagellar motor of *Vibrio alginolyticus*. J. Biol. Chem. *275*, 20223–20228.

Schuhmacher, D.A. and Klose, K.E. (1999). Environmental signals modulate ToxT-dependent virulence factor expression in *Vibrio cholerae*. J. Bacteriol. *181*, 1508–1514.

Shakhnovich, E.A., Hung, D.T., Pierson, E., Lee, K. and Mekalanos, J.J. (2007). Virstatin inhibits dimerization of the transcriptional activator ToxT. Proc. Natl. Acad. Sci. USA *104*, 2372–2377.

Sjoblad, R.D., Emala, C.W. and Doetsch, R.N. (1983). Invited review: bacterial flagellar sheaths: structures in search of a function. Cell Motil. *3*, 93–103.

Skorupski, K. and Taylor, R.K. (1999). A new level in the *Vibrio cholerae* ToxR virulence cascade: AphA is required for transcriptional activation of the *tcpPH* operon. Mol. Microbiol. *31*, 763–771.

Slauch, J.M., Mahan, M.J. and Mekalanos, J.J. (1994). *In vivo* expression technology for selection of bacterial genes specifically induced in host tissues. Methods Enzymol. *235*, 481–492.

Sperandio, V., Bailey, C., Giron, J.A., DiRita, V.J., Silveira, W.D., Vettore, A.L. and Kaper, J.B. (1996). Cloning and characterization of the gene encoding the OmpU outer membrane protein of *Vibrio cholerae*. Infect. Immun. *64*, 5406–5409.

Springer, W.R. and Koshland, D.E.J. (1977). Identification of a protein methyltransferase as the *cheR* gene product in the bacterial sensing system. Proc. Natl. Acad. Sci. USA *74*, 533–537.

Strom, M.S. and Lory, S. (1993). Structure-function and biogenesis of the type IV pili. Annu. Rev. Microbiol. *47*, 565–596.

Tacket, C.O., Taylor, R.K., Losonsky, G., Lim, Y., Nataro, J.P., Kaper, J.B. and Levine, M.M. (1998). Investigation of the roles of toxin-coregulated pili and mannose-sensitive hemagglutinin pili in the pathogenesis of *Vibrio cholerae* O139 infection. Infect. Immun. *66*, 692–695.

Tamayo, R., Tischler, A.D. and Camilli, A. (2005). The EAL domain protein VieA is a cyclic diguanylate phosphodiesterase. J. Biol. Chem. *280*, 33324–33330.

Taylor, D.N., Killeen, K.P., Hack, D.C., Kenner, J.R., Coster, T.S., Beattie, D.T., Ezzell, J., Hyman, T., Trofa, A., Sjogren, M.H., *et al.* (1994). Development of a live, oral, attenuated vaccine against El Tor cholera. J. Infect. Dis. *170*, 1518–1523.

Taylor, R.K. (1989). Genetic studies of enterotoxin and other potential virulence factors of *Vibrio cholerae* In Hopwood, D. A. and Chater, K. F. (ed.) Genetics of bacterial diversity, Academic Press Ltd, London, pp. 309–329.

Taylor, R.K., Kirn, T.J., Bose, N., Stonehouse, E., Tripathi, S.A., Kovac, P. and Wade, W.F. (2004). Progress towards development of a cholera subunit vaccine. Chem. Biodivers. *1*, 1036–1057.

Taylor, R.K., Miller, V.L., Furlong, D.B. and Mekalanos, J.J. (1987). Use of *phoA* gene fusions to identify a pilus colonization factor coordinately regulated with cholera toxin. Proc. Natl. Acad. Sci. USA *84*, 2833–2837.

Tendeng, C., Badaut, C., Krin, E., Gounon, P., Ngo, S., Danchin, A., Rimsky, S. and Bertin, P. (2000). Isolation and characterization of *vicH*, encoding a new pleiotropic regulator in *Vibrio cholerae*. J. Bacteriol. *182*, 2026–2032.

Tendeng, C. and Bertin, P.N. (2003). H-NS in Gram-negative bacteria: a family of multifaceted proteins. Trends Microbiol. *11*, 511–518.

Terashima, H., Fukuoka, H., Yakushi, T., Kojima, S. and Homma, M. (2006). The *Vibrio* motor proteins, MotX and MotY, are associated with the basal body of Na-driven flagella and required for stator formation. Mol. Microbiol. *62*, 1170–1180.

Thelin, K.H. and Taylor, R.K. (1996). Toxin-coregulated pilus, but not mannose-sensitive hemagglutinin, is required for colonization by *Vibrio cholerae* O1 El Tor biotype and O139 strains. Infect. Immun. *64*, 2853–2856.

Thomas, S., Williams, S.G. and Manning, P.A. (1995). Regulation of *tcp* genes in classical and El Tor strains of *Vibrio cholerae* O1. Gene. *166*, 43–48.

Tischler, A.D. and Camilli, A. (2004). Cyclic diguanylate (c-di-GMP) regulates *Vibrio cholerae* biofilm formation. Mol. Microbiol. *53*, 857–869.

Tischler, A.D. and Camilli, A. (2005). Cyclic diguanylate regulates *Vibrio cholerae* virulence gene expression. Infect. Immun. *73*, 5873–5882.

Tripathi, S.A. and Taylor, R.K. (2007). Membrane association and multimerization of TcpT, the cognate ATPase ortholog of the *Vibrio cholerae* TCP biogenesis apparatus. J. Bacteriol. *189*, 4401–4409.

Vaitkevicius, K., Lindmark, B., Ou, G., Song, T., Toma, C., Iwanaga, M., Zhu, J., Andersson, A., Hammarstrom, M.L., Tuck, S. and Wai, S.N. (2006). A *Vibrio cholerae* protease needed for killing of *Caenorhabditis elegans* has a role in protection from natural predator grazing. Proc. Natl. Acad. Sci. USA *103*, 9280–9285.

Waldor, M.K. and Mekalanos, J.J. (1996). Lysogenic conversion by a filamentous phage encoding cholera toxin. Science. *272*, 1910–1914.

Wang, J., Mushegian, A., Lory, S. and Jin, S. (1996). Large-scale isolation of candidate virulence genes of *Pseudomonas aeruginosa* by *in vivo* selection. Proc. Natl. Acad. Sci. USA *93*, 10434–10439.

Waters, C.M. and Bassler, B.L. (2005). Quorum sensing: cell-to-cell communication in bacteria. Annu. Rev. Cell Dev. Biol. *21*, 319–346.

Watnick, P.I., Fullner, K.J. and Kolter, R. (1999). A role for the mannose-sensitive hemagglutinin in biofilm formation by *Vibrio cholerae* El Tor. J. Bacteriol. *181*, 3606–3609.

Watnick, P.I. and Kolter, R. (1999). Steps in the development of a *Vibrio cholerae* El Tor biofilm. Mol. Microbiol. *34*, 586–595.

Watnick, P.I., Lauriano, C.M., Klose, K.E., Croal, L. and Kolter, R. (2001). The absence of a flagellum leads to altered colony morphology, biofilm development and virulence in *Vibrio cholerae* O139. Mol. Microbiol. *39*, 223–235.

Withey, J.H. and DiRita, V.J. (2005a). *Vibrio cholerae* ToxT independently activates the divergently transcribed *aldA* and *tagA* genes. J. Bacteriol. *187*, 7890–7900.

Withey, J.H. and DiRita, V.J. (2005b). Activation of both *acfA* and *acfD* transcription by *Vibrio cholerae* ToxT requires binding to two centrally located DNA sites in an inverted repeat conformation. Mol. Microbiol. *56*, 1062–1077.

Withey, J.H. and DiRita, V.J. (2006). The toxbox: specific DNA sequence requirements for activation of *Vibrio cholerae* virulence genes by ToxT. Mol. Microbiol. *59*, 1779–1789.

Woof, J.M. and Kerr, M.A. (2006). The function of immunoglobulin A in immunity. J. Pathol. *208*, 270–282.

Yakushi, T., Kojima, M. and Homma, M. (2004). Isolation of *Vibrio alginolyticus* sodium-driven flagellar motor complex composed of PomA and PomB solubilized by sucrose monocaprate. Microbiology. *150*, 911–920.

Yonekawa, H., Hayashi, H. and Parkinson, J.S. (1983). Requirement of the *cheB* function for sensory adaptation in *Escherichia coli*. J. Bacteriol. 156: 1228–1235.

Yorimitsu, T. and Homma, M. (2001). Na^+-driven flagellar motor of *Vibrio*. Biochim. Biophys. Acta *1505*, 82–93.

Yorimitsu, T., Kojima, M., Yakushi, T. and Homma, M. (2004). Multimeric structure of the PomA/PomB channel complex in the Na^+-driven flagellar motor of *Vibrio alginolyticus*. J. Biochem. (Tokyo) *135*, 43–51.

Yorimitsu, T., Sato, K., Asai, Y., Kawagishi, I. and Homma, M. (1999). Functional interaction between PomA and PomB, the Na^+-driven flagellar motor components of *Vibrio alginolyticus*. J. Bacteriol. *181*, 5103–5106.

Yu, C., Bassler, B.L. and Roseman, S. (1993). Chemotaxis of the marine bacterium *Vibrio furnissii* to sugars. A potential mechanism for initiating the chitin catabolic cascade. J. Biol. Chem. *268*, 9405–9409.

Yu, R.R. and DiRita, V.J. (2002). Regulation of gene expression in *Vibrio cholerae* by ToxT involves both antirepression and RNA polymerase stimulation. Mol. Microbiol. *43*, 119–134.

Zhu, J. and Mekalanos, J.J. (2003). Quorum sensing-dependent biofilms enhance colonization in *Vibrio cholerae*. Dev. Cell. *5*, 647–656.

Zhu, J., Miller, M.B., Vance, R.E., Dziejman, M., Bassler, B.L. and Mekalanos, J.J. (2002). Quorum-sensing regulators control virulence gene expression in *Vibrio cholerae*. Proc. Natl. Acad. Sci. USA *99*, 3129–3134.

Genetics of O-Antigens, Capsules and the Rugose Variant of *Vibrio cholerae*

6

Afsar Ali

Abstract

The human pathogen *Vibrio cholerae* produces three major cell-surface associated polysaccharides, including (i) lipopolysaccharide (LPS), (ii) capsule and (iii) rugose exopolysaccharide. While LPS and capsule primarily help the bacterium to evade host defence mechanisms, the rugose exopolysaccharide may aid the bacterium in persisting in the nutrient-deficient aquatic environments. This chapter describes underlying genetic elements and mechanism(s) involved in the biosynthesis and regulation of LPS, capsule and rugose exopolysaccharide elicited by *V. cholerae*.

Genetics of O-antigen biosynthesis in *Vibrio cholerae*

As is true for many gram-negative bacteria, the lipopolysaccharide (LPS) is an important structural and functional component of *V. cholerae*. There are three major constituents of LPS, including (i) the lipid A, (ii) core-polysaccharide (core-PS), and (iii) O-antigen polysaccharide (O-PS). Like other gram-negative bacteria, the lipid A of *V. cholerae* is highly conserved and is required for the outer membrane integrity in the bacterium. Although much is known about the biosynthesis of lipid A in many gram-negative bacteria, very little information is available on the characteristics of the genes involved in the biosynthesis of lipid A in *V. cholerae*. Available information on genetics of biosynthesis of *V. cholerae* lipopolysaccharide is presented in the following sections.

Core-PS

The core-PS of the LPS of gram-negative bacteria including *V. cholerae* is also known to have a role in maintaining outer membrane stability (Heinriches *et al.*, 1998). In *V. cholerae*, there have been questions as to whether the core-PS also plays a part in virulence of the organism. Not much information has been available on the genetics of the core-PS biosynthesis, while structural studies have revealed that *V. cholerae* O1 and O139 core-PS structures are largely similar (Chatterjee and Chaudhuri, 2003). Within the past few years, there have been studies of the core-PS biosynthesis gene clusters of *V. cholerae* of different serogroups (Nesper *et al.*, 2002). Following the new nomenclature introduced by Reeves *et al.* (1996), the core-PS biosynthesis locus was designated as the *wav* gene cluster. The *wav* gene cluster located in chromosome 1 comprises ORFs VC0223 to VC0240 based on the genomic sequence of *V. cholerae* El Tor strain N16961 (Heidelberg *et al.*, 2000). The genetic organization of the putative *wav* gene cluster of *V. cholerae* has been described by Nesper *et al.*, (2002). Subsequent investigations of 38 different *V. cholerae* strains by Southern blotting, PCR and sequence analyses revealed the presence of five different putative *wav* gene cluster types. The O1 El Tor *wav* gene cluster type (Type 1) was prevalent among clinical isolates of different serogroups associated with cholera and environmental O1 strains. On the other hand, 19 unrelated non-O1, non-O139 environmental and human isolates, not associated with cholera, contained four new *wav* gene cluster types that differed from each other in distinct gene loci. These data provided evidence for horizontal transfer of *wav* genes and for limited structural diversity of the core-PS among *V. cholerae* isolates. That the type 1 *wav* gene locus was predominant in strains associated with clinical cholera suggests that a specific core-PS structure contributes to the virulence of *V. cholerae* strains. Some other interesting observations derived from the sequence of the genome of O1 El Tor strain N16961 (Heidelberg *et al.*, 2000) include (i) the localization of the O1-antigen biosynthesis gene cluster (*wbe*) upstream of VC0240 and indication that in *V. cholerae* most of the LPS biosynthetic genes are clustered, (ii) the VC0237 encoding the O-antigen ligase Waa L and (iii) the ORF VC0233 (*waaA*) encoded the Kdo transferase that ligates the lipid-A to Kdo. The occurrence of 10 *wav* genes (*wavA*, *wavB*, *waaC*, *wavC*, *waaA*, *wavI*, *waaF*, *waaL*, *wavL* and *gmhD*) in the genomes of all 38 different *V. cholerae* strains indicates the presence of a common core-PS backbone structure for all the strains investigated.

Although the gene products, including heptosyltransferase (WaaC) and the glycosyltransferasees (WaaF) are involved in the biosynthesis of the inner core of LPS, further studies are required to fully characterize the products of the different genes in the core-PS genetic organization of *V. cholerae*.

O-antigen polysaccharide (O-PS)

The genetic regions required for the biosynthesis of the *V. cholerae* O1 O-antigen of both Inaba and Ogawa serotypes were cloned and expressed in *E. coli* K-12 (Manning *et al.*, 1986). These genes could be localized to a region of about 16–19 kb within a 20-kb *Sac*I fragment by extensive restriction and heteroduplex analyses of independent cosmid clones. This region was shown to contain all the information necessary for synthesis of the O-antigen in a heterologous host (Manning *et al.*, 1986; Ward *et al.*, 1987) and referred to as the *wbe* region (Reeves *et al.*, 1996). Subsequent transposon mutagenesis of this region in the *V. cholerae* O1 chromosome confirmed that this region was required for O-antigen biosynthesis (Ward *et al.*, 1989). Subsequently, three additional genes, designated *wbeU*, *wbeV* and *wbeW*, were described and were found essential for O-antigen biosynthesis in *V. cholerae* O1 since mutations in these genes also led to the loss of O-antigen biosynthesis (Fallarino *et al.*, 1997). The organization of O-PS gene clusters of *V. cholerae* O1 is shown in Fig. 6.1a. These O-antigen biosynthesis genes of *V. cholerae* O1 are organized in a cluster located on chromosome 1 in strain N16961, between the open reading frames (ORFs) VC0240 (*gmhD*) and VC0264 (*rjg*) (Heidelberg *et al.*, 2000). The gene *gmhD* encoding D-glycero-D-mannoheptose-1-phosphate guanosyltransferase, involved in LPS core biosynthesis is located in the left junction, while the *wbe* gene cluster is flanked on the right junction by *rjg* encoding a conserved hypothetical protein with similarities to mRNA 3′ end processing factor.

The *wbe* gene cluster of *V. cholerae* O1 consists of five major regions: (1) perosamine biosynthesis region comprising the genes *manC*, *manB*, *gmd* and *wbeE* (Stroeher *et al.*, 1995); (2) O-antigen transport involving *wbeG*, *wzm* and *wzt* genes; (Manning *et al.*, 1994; 1995) (3) tetronate biosynthesis region comprising *wbeK*, *wbeL*, *wbeM*, *wbeN* and *wbeO* genes (Stroeher *et al.*, 1998); (4) O-antigen modification involving *wbeT* gene (Hisatsune *et al.*, 1993; Stroeher *et al.*, 1992); and (5) additional genes required including *wbeU*, *wbeV* and *wbeW* (Fallarino *et al.*, 1997). A putative defective insertion sequence, IS1358dI, is present between *wbeP* and *wbeT* of the O-antigen gene cluster and the JUMP start sequence described by Hobbs and Reeves (1994), which contains important regulatory information (Bailey *et al.*, 1997), is located at the start of the *wbe* operon immediately downstream of the left junction gene, *gmhD*.

Stroeher and colleagues described several gene products required for perosamine biosynthesis (Stroeher *et al.*, 1998). These authors noted that the gene involved in the synthesis of GDP perosamine from the substrate GDP-4-keto-6-deoxymannose remains to be identified. The gene *wbeG* encodes a glycosyl transferase (Heidelberg *et al.*, 2000). Its function was not previously known, although it was speculated to encode O-antigen polymerase or perosamine transferase (Stroeher *et al.*, 1998). The genes *wzm* and *wzt* appear to be involved in polysac-

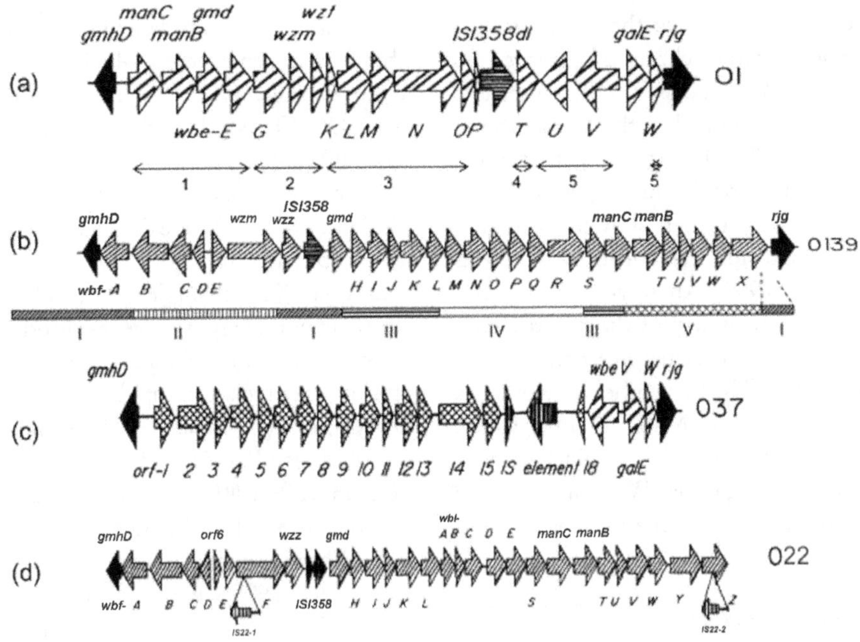

Figure 6.1 Organization of the O-PS biosynthetic genes (wb* region) of *V. cholerae* of four different serogroups. The directions of transcription of the different ORFs are shown by the arrows. The left (gmhD) and the right (rjg) junction genes are indicated by solid arrows. (a) The wbe region (manC through wbeW) of serogroup O1 is 22,000 bp in length. The five major regions of the O1 gene cluster are shown by the horizontal lines 1 to 5. (b) The wbf region (wbfA through wbfX) of the serogroup O139 is 35,807 bp in length. The bars with different patterns indicate the various classes, I to V, of genes in the O139 O-PS gene cluster (Chatterjee and Chaudhuri, 2004). It may be noted, however, that the ORF corresponding to wbfX was designated as wbfY (belonging to class V) and another previously unidentified ORF, wbfZ, was identified as belonging to class I, at the 3′end of wbfY by Yamasaki *et al.* (1999). (c) The wb* region (ORF-1 through wbeN) of serogroup O37 is 27,552 bp in length. The left junction is at gmhD gene, but the right junctions of O37 and O139 are different. On the other hand, the O1 right junction genes, wbeV through rjg, are conserved in the O37 wb* region; (d) the wb* region of serogroup O22 is 35,900 bp in length. The arrows below wbfF and wbfZ show genes found in insertion sequences, encoding proteins homologous to transposases. All genes within the O22 O-PS gene cluster, with the exception of wbfA and wblA to wblE, have very high homology to corresponding genes found in O139 serogroup. (Figure is reprinted from the Biochem. Biophys. Acta (Chatterjee and Chaudhuri, 2004) with permission of the publisher.)

charide export. The Wzm is predicted to be an integral membrane protein containing six α-helical transmembrane spanning domains and is thought to act as a carrier or pore. Wzt is homologous to export proteins and contains ATP-binding domains. It is likely that Wzm and Wzt are present as homodimers in the export complex (Manning *et al.*, 1995).

In tetronate biosynthesis, the identities and actions of different enzymes have been drawn mostly from homology studies. The similarity of WbeN to both LuxC and LuxE (*V. harveyi*) suggests that it has both enzymatic activities, i.e. fatty acid reductase and acyl protein synthetase. WbeM appears to be an iron-containing alcohol dehydrogenase, WbeL an acetyl-CoA synthetase and WbeO an acetyl-CoA transferase. Among the three additional genes identified to be associated with O-antigen biosynthesis of *V. cholerae* O1, the putative functions of WbeU and WbeW are mannosyl transferase and galactosyl transferase, respectively. WbeV is, however, a conserved hypothetical protein. The ORF between *wbeV* and *wbeW*, originally designated as ORF 35.7 (Fallarino, *et al.*, 1997), is homologous to *galE* of *E. coli*, coding for UDP-glucose-4-epimerase (Heidelberg *et al.*, 2000). Interestingly, the genes *wbeU* and *wbeV* are transcribed in the opposite direction to the rest of the *wbe* operon and appear to be translationally controlled (Fallarino, *et al.*, 1997). On the other hand, *galE* and *wbeW* are transcribed in the same direction as that of the *wbe* gene cluster. As mentioned earlier, chromosomal insertion mutations in *wbeV*, *wbeU* or *wbeW* caused the loss of O-antigen and also the resistance to the vibriophage CP-T1, suggesting that these genes are essential for O-antigen biosynthesis. Mutational studies further showed that *galE* was not required for O-antigen biosynthesis (Fallarino, *et al.*, 1997). These genes would not have been expressed in any of the constructs in which *V. cholerae* serotype-specific O-antigen was being made in *E. coli* K-12 or other heterologous hosts such as *Salmonella* (Manning *et al.*, 1986). It was possible that these hosts might have possessed *wbeU*, *wbeV* and *wbeW* homologues or the expression might be at the suboptimal level (Fallarino, *et al.*, 1997).

Vibrio cholerae O139

The O139 *wbf* region is about 35 kb in length and was collectively sequenced by several groups of workers (Bik *et al.*, 1996; Comstock *et al.*, 1996; Stroeher *et al.*, 1997a). The sequence of the left part of the *wbf* region, from *gmhD* to *gmd*, was reported to be 14,363 bp (Bik *et al.*, 1996), whereas that of the right part of the *wbf* region, from *wbfQ* to *wbfX*, was 12,938 bp (Comstock *et al.*, 1996). Emergence of *V. cholerae* O139 resulted from loss of the original *wbe* genes from *V. cholerae* O1 and acquisition of a new set of genes. Specifically, it was demonstrated that the serogroup O139 resulted from a 22-kb deletion of the *wbe* region of a *V. cholerae* O1 strain and replacement with a 35-kb *wbf* region encoding the O139 O-antigen (Comstock *et al.*, 1995).

The sequenced O1 and O139 *wb** regions contained only a 30-bp overlap at the right junction (Bik *et al.*, 1996; Fallarino *et al.*, 1997; Stroeher *et al.*, 1998). The gene, *gmhD*, involved in core biosynthesis forms the left junction, *rjg* forms the right junction and the insertion sequence IS1358 is situated at an intermediate position of the *wbf* region (Bik *et al.*, 1996; Fallarino *et al.*, 1997; Stroeher *et al.*, 1995a,b and c, 1998; Manning *et al.*, 1994). The region involved in the biosynthesis of the *V. cholerae* O139 surface polysaccharide is complex and composed of several independent transcriptional units (Stroeher *et al.*, 1998). Mutational analysis showed that there are some genes which are involved in both capsule and O-antigen biosynthesis, and some genes which are specific for either capsule or O-antigen biosynthesis (Bik *et al.*, 1996; Comstock *et al.*, 1995; Stroeher *et al.*, 1995a; Waldor *et al.*, 1994).

The two open reading frames (ORFs) (to the left of IS1358) were designated *wzm* and *wzz* whose putative functions are capsule transport and O-antigen chain length regulation, based on their homology with genes of other bacteria (Bik *et al.*, 1995). Mutations in both *wzm* and *wzz* appeared to affect capsule but not O-antigen biosynthesis (Bik *et al.*, 1996). The remaining region between IS1358 and *gmhD* was reported to contain *wbfE*, *wbfD*, *wbfC*, *wbfB* and *wbfA*, none of which showed any homology to polysaccharide biosynthesis genes and their roles were unknown (Bik *et al.*, 1995). It was further reported that an ORF downstream of IS1358 showed a high degree of homology to Gmd, the oxidoreductase from *V. cholerae* O1, indicating the presence of polysaccharide synthesis genes on both sides of IS1358 in *V. cholerae* O139. Comstock *et al.* identified a number of other ORFs, which showed homology to polysaccharide biosynthetic proteins, as well as defining the other end point of the novel O139 *V. cholerae* DNA (Comstock *et al.*, 1996).

Unlike the *V. cholerae* O1 *wbe* region, O139 *wbf* does not appear to be made up of clearly defined regions corresponding to specific biosynthetic pathways. The O139 polysaccharide biosynthetic cluster has a number of redundancies. One example is that there are at least two possible galactosyl transferases and a number of UDP-galactose-4-epimerases involved in the synthesis of UDP galactose. Fig. 6.1b presents the genetic organization of the *wbf* region of *V. cholerae* O139.

V. cholerae O139 was shown to have a *wzy* homologue (WbfQ/Wzy). The possible function of this homologue was uncertain as the Wzy protein was known to be involved in O-antigen polymerization, but in *V. cholerae* O139 there was only a single O-antigen subunit sugar substituted on the core. Subsequent mutation analysis of the *wbf*/capsule gene cluster of *V. cholerae* O139 clearly showed that the Wzy homologue was solely involved in capsule biosynthesis (Stroeher *et al.*, 1998). Another point of interest is that the *wbf* region of *V. cholerae* O139 contains only one partner of the two-component polysaccharide export system. The O139 contains only a gene (*wzt*) encoding the transmembrane protein but no energizing partner (Bik *et al.*, 1995). Further, *wzt* was shown to be an essential gene in *V. cholerae* O139 capsule biosynthesis (Bik *et al.*, 1996).

Yamasaki *et al.* (1999) reported independently the sequence of the O139 O-PS gene cluster, which in general confirmed the works published earlier by others (Bik *et al.*, 1996; Comstock *et al.*, 1996; Mooi and Bik, 1997; Stroeher *et al.*, 1997b), but also showed some differences. The size of the genes and homologous proteins encoded by the genes were different for *wbfK*, *wbfM*, *wbfQ*, *wbfS*, *manC*, *wbfX* and *wbfY* when compared with earlier reports. According to Comstock *et al.* (1996), the ORFs found next to ORF9 (*wbfW*) were ORF10 and ORF11, which were designated as *wbfX* and *wbfY*, respectively. On the other hand, Yamasaki *et al.*, (1999) reported a large ORF in the region corresponding to the *wbfX* and *wbfY*, and designated this large ORF as *wbfY*. It was observed that the separated ORFs, *wbfX* and *wbfY*, found previously, might have resulted from mutations. In addition, there was a potential gene between *wbfD* and the repeat unit and JUMPstart sequence upstream of *wbfE*, which was not previously identified, but the authors were not certain about its identity as the ORF was too short. The gene *wbfZ*, which was not previously documented, was identified at the 3′ end of *wbfY*. The sequence recorded by Yamasaki *et al.* (1999) contained in addition a 7.8-kb segment upstream of the O139-specific sequence and a 2.2-kb segment downstream of it.

Vibrio cholerae non-O1, non-O139 serogroups

The O-PS genetic organization of *V. cholerae* in serogroups other than O1 and O139 has not been extensively studied. Some relevant information on the serogroups O37 and O22 are available and discussed below. Some other members of the non-O1, non-O139 serogroups have been studied to find out homologies between their O-PS biosynthetic genes and those of O139, a brief account of which is also presented.

Vibrio cholerae O37

Among the different non-O1, non-O139 serogroups, the serogroup O37 deserves attention since the strains of this serogroup were implicated in localized cholera outbreaks in the past (Aldova *et al.*, 1968; Kamal, 1971), and thus, have clinical relevance. The O37 O-PS genetic region, shown in Fig. 6.1c, was sequenced and the entire O37 *wb** genetic region was found to be 27,552 bp long, while the O37-specific sequence was 23,388 bp long (Li *et al.*, 2002). The O37 *wb** cluster is unique since no significant DNA homology between the O37-specific sequences and the previously published *V. cholerae wb** cluster sequences was found except at the right junction. Twenty-three ORFs were identified in the O37 *wb** region, and many of the ORFs (ORF-1 to ORF-13 and ORF-18) encoded enzymes involved in polysaccharide biosynthesis; ORF-14 and ORF-15 encoded hypothetical proteins of unknown functions and ORF-14 had very weak homology to *yhf0* (which encoded a hypothetical protein) of *Bacillus subtilis.* A 1549-bp promoter region separates *gmhD* and ORF-1, and this region contains a putative promoter and ops elements found in known polysaccharide biosynthesis regions (Bailey *et al.*, 1997; Hobbs and Reeves, 1994). There is an IS element in the interval between the region that is unique to O37 and the right junction. The region downstream of the IS element has three ORFs almost identical to the O1 *wbe* cluster: *wbeV*, *galE* and *wbeW*, followed by *rjg*. The homology breakpoints (i.e. the DNA sequences where the common backbone sequence ends and the serogroup-specific sequence starts) in O37 strains are different from those of the O1 and O139 junctions. In the O37 serogroup, the left and right junctions are at *gmhD* and *wbeV*, respectively, instead of at *gmhD* and *rjg* as they are in O139.

Vibrio cholerae O22

The O-antigen cluster of O22 serogroup, as shown in Fig. 6.1d, is 35.9-kb in length (Yamasaki *et al.*, 1999). Insertion sequences designated as IS22-1 and IS22-2, which were flanked by repeat sequence, were found in the putative region of *wbfF* and *wbfZ*, respectively. If these IS were deleted, genes which correspond to *wbfF and wbfZ* would remain intact. Therefore, the IS sequence interrupts the translation of *wbfF* and *wbfZ*. The 35.9-kb region was found to contain 30 ORFs and the hypothetical *wbfF*, and the *wbfZ* found in O139 serogroup were also present at the 3′end of *wbfY* as shown in Fig. 6.1d. The transcriptional directions of *gmhD to wbfD* are all from right to left, while those of *wbfE* to *wbfZ* are all from left to right in agreement with those of genes in the O139 serogroup. The products IS22-1 and IS22-2 have 100% homology to each other (Yamasaki *et al.*, 1999). This IS encodes a protein homologous to *TnpA* of *Bacillus thuringiensis* (Baum, 1994) or hypothetical protein 2131 of spinach (Zhou *et al.*, 1988), suggesting that transposons might be involved in the horizontal transfer of O-antigen biosynthesis genes of the O22 serogroup.

The JUMPstart sequence, which is a conserved 39-bp sequence usually located in the non-coding upstream region of several bacterial gene clusters (Hobbs and Reeves, 1994), was also found in the region between *wbfD* and ORF6. In addition, just upstream of the JUMPstart sequence, seven tandemly repeated 7-bp units [G-(G/A)-(A/T)-(C/T)-C-T-A] were observed, whose function was not clear. Analysis of GC content in individual genes in O22 serogroup revealed that although the GC content of most genes investigated was in the range 41–54%, the GC content of *wbfA* and *wbfJ* to *wblE* was 28–35%, which is lower than that of the *V. cholerae* genome, indicating that *wbfA* and *wbfJ* to *wbfQ* may have originated from an organism other than *V. cholerae*.

Other non-O1 non-O139 *Vibrio cholerae*

Other non-O1 non-O139 strains have been investigated to find the extent of homology of their O-PS genes with those of O139, but detailed genetic analyses are generally not available. It was suggested that the evolution of the *wbf* region of *V. cholerae* O139 might have occurred via the acquisition of genes from a number of different ancestors. Several studies had shown that at least some of the DNA associated with O139 surface polysaccharide biosynthesis was also found in non-O1 *V. cholerae* serogroups (Bik *et al.*, 1995; 1996; Mooi and Bik, 1997). Two groups of workers (Bik *et al.*, 1996; Stroeher *et al.*, 1997) showed that an ORF in the region downstream of IS1358 of the O139 strain was approximately 80% identical to the *gmd* gene of *V. cholerae* O1. In addition, some of the regions downstream of IS1358 were found in non-O1 and non-cholera vibrios (Comstock *et al.*, 1995). These studies thus indicated that a number of the *V. cholerae* O139 surface polysaccharide genes had their origin within the *Vibrionaceae*. However, the exact mechanism by which these genes were transferred has remained elusive.

When DNA fragments derived from O139 O-antigen biosynthesis gene region were used as probes, the entire O139 O-antigen biosynthesis gene region could be divided into five classes, designated as I–V based on the reactivity pattern of the probes against reference strains of *V. cholerae* representing serogroups O1 to O193 (Yamasaki *et al.*, 1999). The genes in class I reacted with eight probes; the range of hybridization of these probes with other non-O1 non-O139 serogroups, as determined by the colony hybridization method, varied from 29 to 150 (Yamasaki *et al.*, 1999). Class II genes reacted with O22, O45, O53, O57, O61, O62, O75, O92, O105, O107, O112, O125, O132, O139, O141, O151 and O152 serogroups. Class III genes reacted with merely O22 and O139 serogroups. Class IV genes reacted with only the O139 serogroup. Class V reacted with three probes; the range of hybridization of these probes with other non-O1 non-O139 serogroups varied from 21 to 92. This study thus gave an idea of the similarity or dissimilarity of the O-PS genetic organizations of the 154 non-O1 non-O139 serogroups with that of O139. In another interesting study covering O139 and several non-O1 non-O139 serogroups of *V. cholerae*, Sozhamannan *et al.* (1999) sequenced the genes downstream of the O139 *wbfX* gene and analysed the genes flanking the *wb** gene cluster in other serogroups including 43 non-O1 non-O139 strains. The major results obtained included: (i) the *gmhD and rjg* genes were present in all of the *V. cholerae* strains examined and the IS1358 element was in 61% of the strains; (ii) the *gmhD* and *rjg* genes were probably arranged similarly flanking the O-antigen gene cluster; and (iii) *rjg* was not an essential gene for cell viability, and that *rjg* was a hot spot for genetic rearrangements. These results indicated a cassette-like organization of the *wb** region, with the conserved genes (*gmhD* and *rjg*) flanking the divergent, serogroup-specific *wb** genes and IS1358. It was suggested that homologous recombination via the junction genes could be the mechanism for the emergence of new pathogens.

The possibility of homologous recombination-mediated exchange of O-antigen biosynthetic gene clusters (*wb**) was tested by suitably designed experiments (DNA dot blotting, IS1004 fingerprinting and restriction fragment length polymorphism analysis) involving 300 *V. cholerae* strains of non-O1 non-O139 serogroups (Li *et al.*, 2002). Four non-O1 strains (serogroups O27, O37, O53 and O65) were found to have an O1 genetic backbone, suggesting an exchange of *wb** clusters. These strains were presumably derived from an epidemic strain by *wb** cluster exchange and subsequently diverged. DNA sequence analysis of the *wb** regions of one serogroup (O37) supported this conclusion. Although the virulence regions of these four strains were heterogeneous, they contained the entire VPI and a pre-CTXΦ (CTXΦ without the *ctx* genes). This study thus identified four non-O1 non-O139 serogroups that appear to have acquired pathogenic potential, the implication of which is discussed below.

Progenitor of *Vibrio cholerae* O139

Several studies have shown that *V. cholerae* O139 strain is phylogenetically and phenotypically very similar to O1 El Tor strains. The O139 Bengal strains, like El Tor strains, have tandemly repeated chromosomal cholera toxin genes and the complete CTX prophage (Das *et al.*, 1993; Johnson *et al.*, 1994), as well as genes for the TCP (Hall *et al.*, 1993; Morris Jr, 1994). Chromosomal DNAs from Bengal and El Tor strains give similar banding patterns by pulsed field gel electrophoresis and by Southern blot analysis with labelled rRNA probes (ribotyping) (Berche *et al.*, 1994; Popovic *et al.*, 1995; Morris Jr, 1994). They are also identical by multilocus enzyme electrophoresis analysis (Johnson *et al.*, 1994). However, at least two important differences exist between O1 El Tor and O139 Bengal strains. Manning *et al.* (1993) have demonstrated that *V. cholerae* O139 possesses a truncated LPS O-side chain which is un-reactive with O1-specific antiserum and that a large portion of DNA corresponding to the *wbe* region of O1 strains is absent from the O139 strain. In addition, the O139 strain expresses a polysaccharide capsule and the associated novel sugars (Faruque *et al.*, 2003; Johnson *et al.*, 1994; Manning *et al.*, 1993; Waldor *et al.*, 1994; Weintraub *et al.*, 1994; Chatterjee and Chaudhuri, 2003; Chatterjee and Chaudhuri, 2004) The mechanisms underlying the origination of O139 and the possible future emergence of other pathogenic *V. cholerae* strains have drawn the attention of recent investigators (Faruque *et al.*, 2003) and at least two probable hypotheses have been put forward. The first one proposed that a transposition event mediated by the element IS1358 resulted in the replacement of the O1 *wbe* genes with the O139 *wbf* genes (Mooi and Bik, 1997; Stroeher *et al.*, 1995, 1997; Stroeher and Manning, 1997). It was found that approximately 40% of all strains belonging to serogroups O1 to O155 contained this or a very closely related element (Stroeher *et al.*, 1998). *V. anguillarum*, a closely related species, also contained this element. Many of the *V. cholerae* and *V. anguillarum* serogroups that contained this element had multiple copies indicating the mobile nature of the elements. Among other important features of this element, IS1358, it was interesting to note its homology to the H-repeat of the *Rhs* element in *E. coli* (Hill *et al.*, 1994; Stroeher *et al.*, 1995; Zhao *et al.*, 1993). The H-repeat present in the *Rhs* element of *E. coli* had been proposed to play a role in large chromosomal rearrangement (Hill *et al.*, 1994; Zhao *et al.*, 1993). Similarly, an H-repeat element found associated with the *S. enterica wb** region had also been proposed to play a role in recombination between different *wb** operons (Xiang *et al.*, 1994). It was proposed that *V. cholerae* and *V. anguillarum* strains might exchange *wb** genes since the homologues of *bplA* and *vipA* of *V. anguillarum* O2 were found in a number of non-O1 *V. cholerae* serogroups (Stroeher *et al.*, 1998). The presence of IS1358 in many serogroups of both *V. anguillarum* and *V. cholerae* (Jedani *et al.*, 1998) is suggestive of such exchanges. However, the exact role of IS1358 in the evolution and rearrangements of *wb**/capsule regions was unknown and although IS1358 of *V. cholerae* O139 and a number of the *V. anguilarum* serogroups appeared to have all the features of a functional IS element, it had not been possible to demonstrate transposition.

The second hypothesis involved the role of junction genes in a homologous recombination event resulting in the replacement of the O1 *wbe* region by the O139 *wbf* region (Comstock *et al.*, 1996; Mooi and Bik, 1997; Stroeher *et al.*, 1998). With respect to the *wb** regions of O1 and O139 strains, the genes downstream of *wbfX* were not known and it was presumed that *wbfX* was the last gene of the *wbf* cluster. Sozhamannan *et al.* (1999) determined the sequences of 4 kb further downstream of the *wbfX* gene and, in addition to what has been stated earlier, their study revealed that the *rjg* gene downstream of *wbfX* was 1341 bp long, of which the first 17 bp were different in O1 and O139 strains. The preferential linkage of IS1358 to the *wb** regions of most *V. cholerae* strains analysed supports the idea that IS1358 was acquired by homologous recombination, rather than by transposition, as a random transposition event would have scattered the IS element across the chromosome (Stroeher *et al.*, 1997). While it has been suggested that O139 strains acquired their unique DNA (Bik *et al.*, 1995; 1996) by homologous recombination via the junction genes, the mechanism by which the genes in this DNA region were acquired from a donor strain remained largely unknown. It is possible that the DNA was transferred via conjugation, since many of the *V. cholerae* O139 Bengal isolates carried a large conjugative plasmid (Stroeher *et al.*, 1998). Alternatively, there may have been a phage-mediated transduction and incorporation of the relevant DNA.

Genesis of O139 O-PS gene cluster

Several earlier studies had already suggested possible non-pathogenic serogroups of *V. cholerae* as donors for the *wbf* cluster and involvement of generalized transducing phages or conjugative plasmids as vectors for the large DNA region in the recombination event (Mooi and Bik, 1997; Sozhamannan *et al.*, 1999). The idea that *V. cholerae* O22 might be a putative source of exogenous DNA resulting in the emergence of the *V. cholerae* O139 had been suggested by different workers on the basis of (i) the structural and chemical analysis of their LPS (Chatterjee *et al.*, 2003; Chatterjee and Chaudhuri, 2004), (ii) the O-antigenic relationship (Isshiki *et al.*, 1996), (iii) cloning and hybridization tests (Dumontier and Berche, 1998; Yamasaki *et al.*, 1999), (iv) the identical JUMP start sequences found in strains of the two serogroups (Yamasaki *et al.*, 1999), (v) the presence of seven tandemly repeated 7-bp units, G-(G/A)-(A/T)-(C/T)-C-T-A, in the two serogroups (Bik *et al.*, 1996) and (vi) the sequence analysis of a 4.5-kb fragment containing IS1358 and its adjacent genes in the two serogroups (Dumontier and Berche, 1998). Sequencing of the entire *wb** regions of the strains of these two serogroups was done and the similarities and differences between these two gene clusters have been shown (Fig. 6.1).

V. cholerae strains from serogroups O141 and O69 possessed some of these O139 antigen- and capsule-biosynthesis genes upstream of the IS1358 element (Bik *et al.*, 1996). However, O69 and O141 strains could not be exogenous DNA donors as genes *wbfF* and *wzz* in these strains differed substantially from those of *V. cholerae* O139 (Bik *et al.*, 1996). Further, the O139 antibodies did not cross-react with *V. cholerae* O69 and O141 antigens and, consequently, the O139, O69 and O141 did not produce cell wall polysaccharides with common antigenic determinants (Bik *et al.*, 1996). The O139 *wbf* region comprising *wbfA-wzz* genes is found in part in *V. cholerae* O69 and O141. Sequencing of these regions of O69 and O141 revealed that the O139 DNA had not been directly acquired from these serogroups. In contrast, it was reported that *V. cholerae* strains O22 and O155 possessed antigen factors in common with *V. cholerae* O139 (Bik *et al.*, 1996; Isshiki *et al.*, 1996). However, *V. cholerae* O155, in which several copies of IS1358 were found, did not possess any of the flanking genes (Dumontier and Berche, 1998).

All these findings taken together suggested that strains of *V. cholerae* O22 from the environment were the source or donor of the exogenous DNA resulting in the emergence of the new epidemic strain O139 (Dumontier and Berche, 1998; Faruque *et al.*, 2003) from a progenitor El Tor strain of *V. cholerae* O1.

Concluding remarks

Many new serogroups of *V. cholerae* are being discovered with the progress of time. There were around 155 known serogroups in 1998 (Stroeher *et al.*, 1998) and today this number has risen to more than 200 (Li *et al.*, 2002). The structure of the O-PS of *V. cholerae* of any serogroup has been found to be unique (Chatterjee and Chaudhuri, 2003; 2004). The genetic organization of the O-PS biosynthesis genes is thus quite susceptible to change, but the factors responsible for effecting such changes are still largely unknown. On the other hand, several non-O1 and non-O139 serogroups have been found to acquire pathogenic potential in epidemic genetic backgrounds (Li *et al.*, 2002) and may cause future epidemics. The existing *V. cholerae* vaccines would provide little or no protection against such newly emerged pathogenic strains. This situation emphasizes the importance of new and rapid vaccine development strategies.

It is expected that, like the *V. cholerae* O1 strains, all the strains belonging to the O139 serogroup will not be pathogenic. Molecular analysis of several O139 strains isolated in the recent past revealed the existence of genetically diverse strains and included both toxigenic and non-toxigenic variants. The new variants of *V. cholerae* O139 included strains with new ribotypes, CTX genotypes and altered antimicrobial resistance (Faruque *et al.*, 1997; Mitra *et al.*, 1996). The non-toxigenic O139 strains included *N*-[1-(thienyl)cyclohexyl] piperidine (TCP)-positive variants that were susceptible to toxigenic conversion by CTX phage (Faruque *et al.*, 2000; Waldor and Mekalanos, 1996) as well as TCP-negative strains. The *gmhD* locus is assumed to have served as a site of homologous recombination in the gene transfer event that led to the origination of O-PS gene cluster of *V. cholerae* O139. This locus when amplified by PCR from all toxigenic O139, El Tor and CTXΦ-susceptible non-toxigenic O139 strains was found to be similar in size, and was also similar to those of non-O1, non-O139 strains isolated from the environment (Faruque *et al.*, 2000). It was presumed that these environmental CT-negative O139 strains lacked the entire TCP pathogenicity island. From these studies, Faruque *et al.* (2000) concluded that these CT-negative O139 strains originated from non-O1 progenitors, and that the O139 antigen was present in different lineages and thus the O139 serogroup comprised both epidemic and non-epidemic strains derived separately from different progenitors. The identification of these progenitors and the mechanisms of gene transfer in such cases, however, remains to be the subject of further investigations.

Various changes in genetic and phenotypic properties of O139 strains have now been documented, which include (i) amplification or rearrangement of the CTX prophage, (ii) acquisition of new CTX prophage, (iii) restriction fragment length polymorphisms in conserved r-RNA genes (ribotypes), (iv) diversity in the multilocus enzyme electrophoresis patterns, and (v) change in the antimicrobial resistance pattern (Mukhopadhyay *et al.*, 1998). Factors that determine the emergence and domination of particular clones of toxigenic *V. cholerae* and displacement of existing ones have again remained largely unknown. Molecular studies have shown that *V. cholerae* strains possess a distinctive class of integrons, which are gene expression elements that may capture ORFs and convert them to functional genes (Mazel *et al.*, 1998). Until recently, all known integrons were associated with genes conferring antibiotic resistance (Recchia and Hall, 1995). However, the integrons in *V. cholerae* are suggested to have also played a role in the acquisition of pathogenic genes as well as genes for different biological functions (Mazel *et al.*, 1998). Further work on the role of integrons in *V. cholerae* is expected to give a better insight into the mechanism of acquisition of genes by this organism. It is believed that the continual emergence of new toxigenic strains is an essential component of the natural ecosystem, and this is likely to complicate further the development of an effective cholera vaccine (Li *et al.*, 2002).

Capsular polysaccharide (CPS) of *Vibrio cholerae*

Overview

V. cholerae O1 lacks a polysaccharide capsule and forms translucent colonies. However, most of the >200 non-O1 serotypes of *V. cholerae*, including the recently described epidemic O139 strains, are encapsulated and form opaque colonies (Johnson *et al.*, 1992; Kaper *et al.*, 1995). Although non-O1 *V. cholerae* strains cause sporadic cases of diarrhoea (Aldova *et al.*, 1968; Kamal, 1971; Morris *et al.*, 1981), they are primarily associated with extraintestinal diseases, including ear infections, wound infections and septicaemia (Blake *et al.*, 1980; Johnson *et al.*, 1992; Morris *et al.*, 1990). Encapsulated and un-encapsulated strains of non-O1 *V. cholerae* exhibit spontaneous and reversible phase variation similar to that of rugose and non-rugose strains (Johnson *et al.*, 1992). Despite the possession of CPS by a majority of non-O1 *V. cholerae* strains, the CPS's role as a virulence factor of *V. cholerae* was not fully appreciated until recently. Johnson *et al.* (1992) reported that a Tn*phoA*-induced, un-encapsulated mutant of a serotype O31 strain NRT-36S (Fig 6.2) was (i) significantly less virulent for mice and more sensitive to serum/complement-mediated lysis than was the isogenic, encapsulated wild-type strain, and (ii) unable to adhere to Caco-2 epithelial cells. Also, several other groups of investigators have found that the CPS of *V. cholerae* O139 is essential for its ability to colonize an infant mouse model and to resist complement-mediated lysis. Studies of the genes involved in CPS biosynthesis by these two non-O1 serotypes of *V. cholerae* are briefly discussed below.

Vibrio cholerae NRT-36S

NRT36S, a serotype O31, heavily encapsulated strain (Fig. 6.2), was isolated from an adult with traveller's diarrhoea at Narita Airport in Tokyo (Morris *et al.*, 1990). Using mini transposon (put Km-2) mutagenesis, a recent study (Chen *et al.*, 2007) reported that four genes required for CPS production by NRT36S are interspersed with those involved with LPS biosynthesis. Mutants defective in these genes were found to be complement sensitive and lack visible capsule as determined by staining with polycationic ferritin followed by electron microscopy. This study confirms the roles for these genes in the biosynthesis/export of capsular polysaccharide. Although the roles of these inactivated genes were not substantiated for the biosynthesis of LPS, these authors suggested that the gene products have dual functions; i.e. they are needed for the biosynthesis of both CPS and LPS.

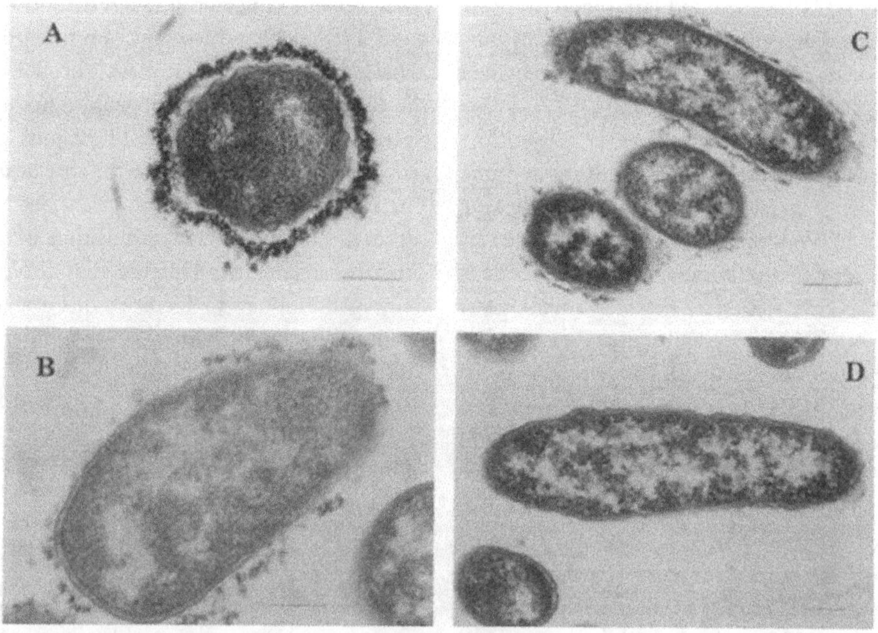

Figure 6.2 Thin sections of non-O1 *V. cholerae* stained with polycationic ferritin. (A) NRT36S/O; (B) NRT36S/T; (C) JVB52; (D) A5. Bar, 200 nm. (Figure is reprinted from the Infect. Immun. (Johnson *et al.*, 1992) with permission of the publisher.)

Vibrio cholerae O139

Several initial reports (Albert, 1994; Albert *et al.*, 1993; Ramamurthy *et al.*, 1993) suggested that pathogenic *V. cholerae* O139 strains originated in the Bay of Bengal. In addition, molecular and epidemiological analyses as well as whole-genome studies revealed that O139 strains are very similar to O1 El Tor strains, which supports the hypothesis that the pathogenic O139 isolates evolved from the recent seventh pandemic El Tor strains (Berche *et al.*, 1994; Dziejman *et al.*, 2002; Hall *et al.*, 1993). Despite the high degree of genetic similarity between El Tor and O139 strains, they have differences in the genes (*wb**) encoding surface polysaccharide. In O139 strains, the genes required for O1 antigen biosynthesis have been deleted and replaced by the O139 antigen locus (*wbf*) encoding the CPS and the LPS's O-side chain (Stroeher *et al.*, 1998). The structure of the CPS and the LPS's O side chains are identical, and one side chain is linked to the serotype's lipid A-core oligosaccharide (OS); however, the LPS-unlinked CPS is a polymer of side chain repeat units (Knirel *et al.*, 1995; 1997; Waldor *et al.*, 1994). Consistent with the structural data, mutations in the putative O139 side chain's biosynthesis genes yielded LPS O side chain- and CPS-negative mutants (Comstock *et al.*, 1996; Waldor *et al.*, 1994). On the other hand, strains with mutations in the putative capsular translocation genes expressed the O side chain but they were CPS-negative (Bik *et al.*, 1996). Taken together, these results suggest that the genes involved with LPS and CPS biosynthesis by *V. cholerae* O139 are interspersed in the *wbf* region. Subsequent studies (Nesper *et al.*, 2002) created defined mutations in the genes required for biosynthesis of (i) LPS's O side chain, (ii) CPS, and (iii) LPS's core OS by *V. cholerae* O139. Deletion of *waaL* (encodes WaaL, a lipid A core-surface polymer ligase) yielded an O side chain-negative strain but did not affect capsule biogenesis. However, inactivation of *wbfF* (encodes WbfF/OtnA, which is involved in CPS translocation) yielded a CPS-negative, O side chain-positive phenotype (Nesper *et al.*, 2002). The biosynthesis of *Escherichia coli* and *V. cholerae* O139 capsules appears to have similarities. The biosynthetic model proposed for the synthesis of CPS and the lipid A-core OS is as follows: individual repeat units are assembled on a lipid carrier in the cytoplasm and then transferred across the inner membrane by a polysaccharide exporter protein. The repeat units can serve as a substrate for the polymerization and capsule translocation machinery or for the Waal enzyme, which ligates one repeating unit onto the lipid A-core OS (Whitfield and Roberts, 1999).

The rugose variant of *Vibrio cholerae*

Overview

The ability of *V. cholerae* to switch reversibly from a smooth to a wrinkled, corrugated, colonial phenotype (designated as the 'rugose' variant phenotype) (Fig. 6.3) was reported in the early 1920s and 1930s (Baltenau, 1926; White, 1938). White noted that the rugose variant spontaneously appeared in old alkaline peptone water (APW) cultures and that it was resistant to adverse environmental conditions (White, 1938, 1940). Based on his observations, he suggested that the rugose variant of *V. cholerae* plays an important role in the bacterium's persistence in its aquatic reservoirs.

Recently, it was observed that rugose variants of *V. cholerae* strains are more resistant to harsh conditions (e.g. exposure to chlorine and UV light) and to serum/complement-mediated killing than are the strains' smooth variants (Morris *et al.*, 1996; Rice *et al.*, 1993). In addition, the rugose variant was shown to be more resistant to osmotic and oxidative stresses (Wai *et al.*, 1998). Rice and Morris suggested that the increased resistance of rugose variants of *V. cholerae* to harsh conditions is due to aggregation of cells in the polysaccharide matrix (Morris *et al.*, 1996; Rice *et al.*, 1992; 1993). Intriguingly, rugose variants produce as much cholera toxin and are as virulent for human volunteers as smooth variants (Ali *et al.*, 2000a, Morris *et al.*, 1996). Other bacterial species, including *Salmonella enterica* Typhimurium (Anriany *et al.*, 2001), *Vibrio vulnificus* (Grau *et al.*, 2005), *Grimontia* (*Vibrio*) *hollisae* (Curtis *et al.*, 2007), *Pseudomonas aeruginosa* (D'Argenio *et al.*, 2002) and *Salmonella enterica* Enteritidis (Peter, 1993) also have been reported to switch to a rugose or rugose-like colonial phenotype. The hallmark of

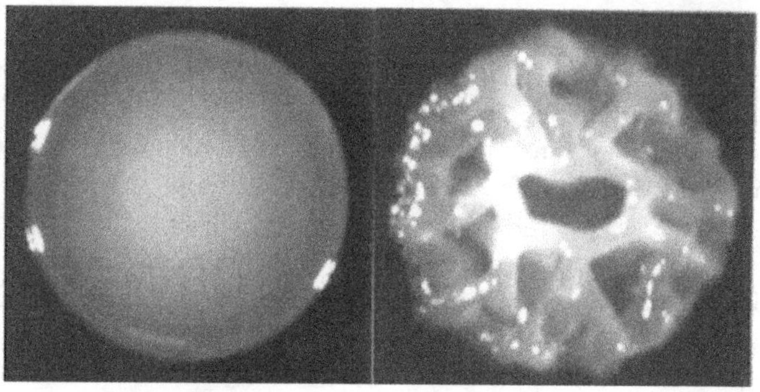

(A) (B)

Figure 6.3 Phenotypes of *V. cholerae* serogroup O1 (N16961) strain on L-agar. (A) smooth and (B) rugose.

this phenotype is its occurrence in old culture conditions and in nutrient-poor growth conditions, which suggests that nutrient starvation of the smooth variant elicits the phenomenon.

Physiology of the rugose variant of *Vibrio cholerae*

Usually, *V. cholerae* O1 strains are non-encapsulated but a majority of *V. cholerae* non-O1 strains possess a capsule (Johnson *et al.*, 1992). However, *V. cholerae* O1 and non-O1 strains can switch to a rugose colonial variant that produces copious amounts of exopolysaccharide (EPS) containing nearly equal amounts of glucose and galactose and small quantities of N-acetylglucosamine and mannose (Yildiz and Schoolnik, 1999). The precise mechanism for expression of the rugose variant/rugose EPS synthesis is not fully understood. However, as mentioned above, growing the smooth variant in a nutrient-poor medium for a prolonged period of time appears to elicit the phenomenon. In contrast, the culture medium's pH and osmolarity does not affect expression (Ali *et al.*, 2005). Many sugars, including sucrose, dextrose, arabinose, fructose and maltose, inhibit expression of the rugose phenotype. However, polyhydric alcohols; e.g. mannitol and glycerol, do not affect it (Ali *et al.*, 2005). Thus, more studies are needed to elucidate the precise cue(s) that stimulate(s) the conversion of the smooth to rugose variant of *V. cholerae*.

Public health significance of the rugose variant of *Vibrio cholerae*

The rugose variant of *V. cholerae* has public health importance for two reasons. First, the variant's EPS enables it to resist various public health measures used to kill *V. cholerae*; e.g. chlorination of drinking water. Second, since thiosulfate citrate bile salts sucrose (TCBS) medium (the primary selective medium for *V. cholerae*) inhibits expression of the rugose phenotype but not the smooth phenotype, the true prevalence of the former in *V. cholerae*'s aquatic reservoirs may be severely underestimated (Ali *et al.*, 2002; 2005). Alternatively, laboratory personnel not familiar with the rugose variant of *V. cholerae* may misidentify rugose *V. cholerae* colonies as microbial contaminants on non-selective diagnostic media.

Genetics of the rugose variant of *Vibrio cholerae*

Structural genes required for biosynthesis of the rugose EPS

The genetic elements of *V. cholerae*, which regulate the smooth to rugose conversion that occurs under nutrient-poor growth conditions, have not yet been fully characterized. However, Yildiz and Schoolnik (1999) have identified a gene cluster, designated $\text{VPS}^{\text{ElTor}}$ (for *Vibrio* polysaccharide of the El Tor biotype), encoding enzymes required for rugose EPS biosynthesis. They (i) subjected a rugose variant of *V. cholerae* to mutagenesis with the transposon mini-Tn5 km2, (ii) mapped each transposon insertion junction in the smooth mutants, and (iii) reported that the transposon inactivated several genes (in a single region on chromosome I) encoding proteins with high homologies to proteins involved in EPS and capsule biosynthesis and transport by other bacterial species. Introduction of a cosmid clone (30.7-kb), carrying the *V. cholerae* gene cluster responsible for rugose EPS biosynthesis, into each rugose mutant confirmed the role of each gene that was inactivated by the transposon (Yildz and Schoolnik, 1999). However, their report did not describe the *vps* region's precise organization. Subsequently, Ali *et al.* (2000b) used the recently released whole genome sequence of *V. cholerae* (Heidelberg *et al.*, 2000) to analyse the *vps* region. They found that the entire region has 26,458-bp (consisting of 22 ORFs) located adjacent to *acrD* (encodes for acriflavine resistance) at the left junction. At the right junction, the gene immediately downstream of the *vps* region shows very weak similarity (24%) to hypothetical proteins in *Bacillus subtilis* (*yitA* and *yuxA*), and it is followed by *glyA* (encodes a serine hydroxymethyltransferase). The ORFs within the *vps* region and the predicted functions of the proteins are shown in Fig. 6.4. The region has two clusters (ORFs VC0916-VC0927 and VC0934-VC0937) that are homologous to known polysaccharide biosynthesis genes, which are interrupted by a cluster (VC0928-VC0933) containing genes with seemingly-unrelated functions. Despite its having no known polysaccharide synthesis-related function, ORF VC0930 sustained three Tn*phoA* insertions resulting in the loss of the rugose phenotype, which suggests that it is involved in biosynthesis of the rugose EPS. A subsequent gene expression study (Yildiz *et al.*, 2004) confirmed the presence of a higher level of the VC0930 transcript in a rugose strain than in an isogenic smooth strain, thus corroborating the earlier finding that VC0930 has a role in rugose EPS synthesis (Ali *et al.*, 2000). However, the genes downstream of VC0937 do not appear to be involved in EPS biosynthesis. Upstream of the *vps* cluster, between *acrD* (VC0914) and VC0916, is an intergenic region (426-bp) containing putative σ^{70} promoter elements. In summary, the genetic element responsible for biosynthesis of *V. cholerae*'s rugose EPS consists of three gene clusters located on chromosome I: *vpsI* (ORFs VC0916-VC0927, 12 genes), *vpsII* (ORFs VC0934-VC0937, 4 genes), and an intervening region between *vpsI* and *vpsII* (ORFs VC0928-VC0933, 6 genes) (Fig. 6.4). While the roles of *vpsI* and *vpsII* gene clusters in rugose polysaccharide synthesis were substantiated early (Yildz and Schoolnik, 1999), the function of intervening genes between *vpsI* and *vpsII* remained elusive until recently. A recent study has demonstrated that the proteins encoded by genes located between intervening region of *vpsI* and *vpsII* are involved in the biofilm matrix production and in the maintenance of biofilm structure and its stability (Fong and Yildiz, 2007).

Regulatory genes involved in biosynthesis of the rugose EPS

hapR

Before the gene cluster (VPS^{ElTor}) required for biosynthesis of the rugose EPS was characterized, Jobling and Holmes (1997) identified a transcriptional activator in *V. cholerae*, which they designated HapR (encoded by the 1.7-kb *hapR* gene). HapR negatively regulates the rugose colony phenotype in El Tor *V. cholerae* strains, but not in classical O1 strains. HapR contains 203 amino acids with 71% identity to LuxR, the positive regulator for the *lux* operon in *V. harveyi*. Two lines of genetic evidence suggested that the *hapR* gene product (HapR) negatively regulates expression of the rugose EPS by El Tor *V. cholerae*: (i) a null mutant of *hapR* yielded rugose colonies by El Tor *V. cholerae*, and introducing $hapR^+$ into the mutant reverted the rugose colonies to the smooth phenotype, and (ii) introducing $hapR^+$ into a spontaneously-generated rugose strain reverted the rugose strain to a smooth variant.

hydG/vpsR

Ali *et al.* (2000) used transposon-induced and subsequent in-frame deletion mutagenesis to identify a gene (ORF VC0665) involved in positively-regulating expression of *V. cholerae*'s rugose phenotype. VC0665 (1,335-bp-long) encodes a protein containing 444 amino acids, and it has a high degree of homology to the HydG-encoding gene found in other gram-negative bacteria, including *E. coli* (Stoker, *et al.*, 1989). Thus, the authors considered the gene to be a *V. cholerae* homolog of *hydG*. The gene product (HydG) belongs to the two-component signal transduction system of the NtrBC family, and, in *E. coli*, HydG is involved in the expression of a labile hydrogenase (Stoker, *et al.*, 1989). The authors found that (i) inactivating ORF VC0665 resulted in colonies possessing a stable, smooth

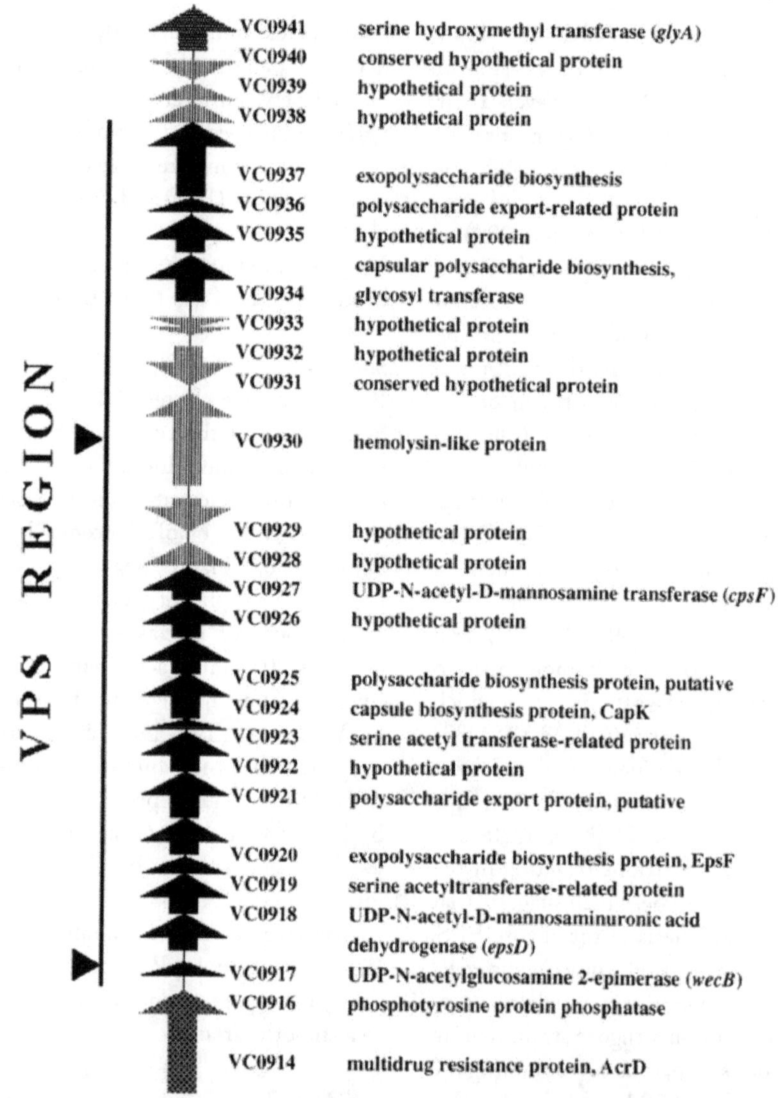

Figure 6.4 Genetic organization of the VPS region. The DNA sequence of the *V. cholerae* VPS region was obtained from the TIGR microbial genome database. The entire region in the map is 33,024 bp, and the vps region is 26,458 bp. Sequence analysis was done by the NCI BLAST server program, and the recently published annotation of the vps region has been used to define the products of the ORFs. *acrD* (VC0914) is at the left junction of the *vps* cluster, and *yitA* (VC0940) and *glyA* (VC0941) are at the right junction. There is a 426-bp intergenic region containing the *vps* promoter. There are 22 ORFs arranged in three clusters: ORFs VC0916 to VC0927 and ORFs VC0934 to VC0937, separated by the cluster of ORFs from VC0927 to VC0933. The middle cluster genes apparently encode functions unrelated to polysaccharide biosynthesis. The arrowheads indicate the positions of the Tn*phoA* insertions in the vps region. The three hly::Tn*phoA* insertions were in close proximity to each other. (Figure is reprinted from the Infection and Immunity (Ali *et al.*, 2000b) with permission of the publisher.)

phenotype, and (ii) introducing *hydG*$^+$ into the *hydG* mutants reverted the mutants' smooth phenotype to the rugose phenotype.

Subsequent studies by Yildiz *et al.* (2001) confirmed that HydG is a positive transcriptional regulator of *vps* genes involved in rugose EPS production and biofilm formation. During their study, HydG was renamed VpsR, a positive transcriptional regulator for EPS^{ElTor} biosynthesis. RT-PCR revealed that much larger amounts of cDNAs specific for *vpsA* (the first gene in the *vpsI* cluster) and *vpsL* (the first gene in the *vpsII* cluster) were present in the rugose variant than in the isogenic smooth variant of *V. cholerae* El

Tor. Interestingly, the rugose and smooth variants contained approximately equal amounts of *vpsR*-specific cDNA, which suggested that VpsR is not required for rugose EPS synthesis. However, a *vpsR* mutant exhibited very low levels of *vpsA* and *vpsL*, and complementing the mutant with *vpsR*$^+$ restored *vpsA* and *vpsL* expression.

eps operon

An extracellular protein secretion pathway (*eps*) is known to be required for the secretion of cholera toxin, chitinase and protease by *V. cholerae* (Sandkvist *et al.*, 1993, 1997). In addition, Ali *et al.*, (2000) discovered that the *eps* operon, which contains 12 genes designated *eps*C through *epsN*, is involved in the production of rugose EPS by *V. cholerae*. Transposon mutagenesis of *epsD*, and defined deletion mutations in *epsD* and *eps*C in a rugose strain resulted in the loss of the rugose phenotype. Complementing the mutants with the respective wild-type genes restored the rugose phenotype, confirming the importance of *epsD* and *epsE* in the expression of the rugose phenotype. Yildiz *et al.* (2004) provided further evidence that the *eps* operon of *V. cholerae* is involved in expression of the rugose phenotype. They found that genes of *eps* operon of a rugose variant were significantly more up-regulated than were the same genes in a smooth variant. Although the *eps* operon is involved in rugose EPS production by *V. cholerae*, further studies are needed to elucidate the mechanism(s) by which it inhibits/regulates the process.

vpsT

During their whole-genome expression profiling studies, Yildiz *et al.* (2004) identified a gene (ORF VCA 0952, designated *vpsT*) whose expression by a rugose variant was four-fold higher than by a smooth variant. *vpsT* (671-bp) encodes VpsT, a 25.8-kDa protein that belongs to the UhpA (FixJ) family of transcriptional response regulators and is homologous (44% homology and 65% similarity) to the transcriptional regulator CsgD of *E. coli*. Targeted disruption of *vpsT* (Casper-Lindley and Yildiz, 2004) in a rugose variant yielded a smooth colonial variant, and complementing the *vpsT*-defective mutant with *vpsT*$^+$ reverted the smooth mutant to the rugose phenotype. Moreover, the introduction of plasmids carrying constructs of *vpsA::lacZ* and *vpsL::lacZ* into the smooth, *vpsT* mutant significantly decreased its β-galactosidase activity. Thus, the authors demonstrated a transcriptional role for *vpsT* in the expression of *vps* genes by a rugose variant of *V. cholerae*.

rbmA

During their studies differentiating the secreted protein profiles of smooth and rugose variants of *V. cholerae*, Fong *et al.*, (2006), used 2D-gel electrophoresis analysis to identify a protein that was exclusively secreted by a rugose variant during its exponential and stationary growth phases. The authors then determined the sequence of 12 amino acids corresponding to the internal part of the protein. By Blast search, they determined that the sequence aligned to an amino acid sequence in a hypothetical protein encoded by ORF VC0928 (located between the *vpsI* and *vpsII* regions) in the genome of *V. cholerae* biotype El Tor (Fig. 6.4). The strategic position of the VC0928 (between two *vps* regions) and the exclusive secretion of its encoded protein by a rugose variant prompted the authors to examine further the protein's possible role in rugose EPS production. Targeted disruption of VC0928 (Fong *et al.*, 2006) revealed that it is required for expression of the rugose phenotype and biofilm formation, and it was designated *rbmA* (for rugosity and biofilm structure modulation). Like the *vps* genes, two positive transcriptional regulators, (VpsR and VpsT) positively regulated the expression of RbmA. Also, the authors demonstrated that RbmA is predominantly present in the culture supernatant fluids of *V. cholerae*. Based on their analysis, the authors concluded that the *rbmA*'s product may assist in maintaining the structural integrity and stability of the rugose EPS of *V. cholerae*.

Transcriptome analysis and the rugose variant

To gain better understanding of the molecular basis of the rugose variant, Yildiz *et al.* (2004) undertook microarray expression profiling studies of the rugose and smooth variants of the same *V. cholerae* strain. This study resulted in the identification of 124 genes differentially expressed in the rugose variant. Of the differentially expressed genes, 77 genes were up-regulated whereas 47 genes were down-regulated. The major groups of genes that were differentially expressed can be organized as follows: (i) as expected, *vps* region genes that encode enzymes required for the biosynthesis of rugose exopolysaccharide were markedly induced. Also, *vps* region genes including *vpsI*, *vpsII* and genes located between *vpsI* and *vpsII* are all upregulated. Interestingly, expression of *vpsII* region genes was much higher (7.9- to 20-fold) than that of *vpsI* region genes (3.7- to 6.3 fold), suggesting that *vpsI* and *vpsII* genes may be under the subject of differential regulation. Alternatively, the difference in the expression of *vpsII* and *vpsI* genes could be linked to stochiometric requirement of the gene products of those *vps* genes in the production of rugose exopolysaccharide, (ii) four of the 12 genes in extracellular protein secretion system (*eps*) were up-regulated, corroborating an earlier finding by Ali *et al.* (2000a) that *epsD* and *epsE* are involved in the production of rugose exopolysaccharide, (iii) expression of several flagellar genes were markedly lower in rugose variant relative to smooth variant, suggesting that genes encoding flagellar proteins and *vps* proteins are regulated divergently. Indeed, motility of rugose variant is much lower than the motility of smooth variant on soft motility agar, despite the presence of flagella in both rugose and smooth variants as determined by electron microscopy (Ali *et al.*, 2002), and (iv) microarray analysis also yielded 12 genes that were expressed in rugose variant and predicted to encode regulatory functions (Yildiz *et al.*, 2004). All but one gene was induced, including *vpsR*, a transcriptional activator required for the transcription of *vps* operon, and, thus, required for the biosynthesis of rugose exopolysaccharide. The role of other regulatory genes (induced in this study) in rugose variant is yet to be determined.

Flagellar genes and the rugose variant

As discussed above, the rugose variant of *V. cholerae* El Tor strain displayed lower motility than the smooth variant of the same strain. Watnick *et al* (2001) have demonstrated that mutations in flagellar genes in *V. cholerae* O139 strain (MO10), but not in *V. cholerae* O1 El Tor strain, resulted in the production of a rugose colony variant. This observation was only attributed to O139 mutants defective in the complete flagellar filament. However, the mutants defective in *motB* and *motY* (encoding flagellar motor proteins that confer motility to the flagella) that possess completed flagellar filament but lacking motility did not display a rugose colony variant (Watnik *et al.*, 2001). Taken together, these observations suggest that the complete absence of flagellum provides *V. cholerae* O139 with an appropriate signalling pathway leading to the production of higher level of exopolysaccharide needed to express the rugose variant. This report also suggests that a multitude of signalling pathways may be required to switch from smooth to a rugose variant and that signalling pathways leading to the expression of rugose variant may vary in *V. cholerae* strains (Watnik *et al.*, 2001).

C-diGMP and the rugose variant of *Vibrio cholerae*

Cyclic di-guanylic acid (c-diGMP) is a second messenger that modulates the cell surface properties in microorganisms, including intracellular aggregation, exopolysaccharide production, biofilm formation and motility (Jenal, 2004; Lim *et al.*, 2006; Romling *et al.*, 2005; Simm *et al.*, 2004). The intracellular concentration of c-diGMP is controlled and regulated by the opposing activities of diguanylate cyclase and phosphodiesterase that contain GGDEF and EAL domains, respectively (Romling *et al.*, 2005). Rashid *et al.* (2003) first reported the role of c-diGMP protein in the production of rugose exopolysaccharide by a *V. cholerae* O1 El Tor strain (N16961). From their transposon library, these authors identified a transposon mutant that was mapped to the ORF VC0653 encoding a protein with regulatory functions. This regulatory protein has a GGDEF and EAL domain, a hallmark for c-diGMP proteins. The authors designated this gene as *rocS* (for regulation of cell signalling). A mutant with a transposon-induced inactivated *rocS* in the smooth background of N16961 was unable to switch to a rugose variant, even under growth conditions promoting high-frequency rugose exopolysaccharide production in *V. cholerae* (Ali *et al.*, 2002). Based on their observation, the authors suggested that *rocS* plays an important role in the smooth to rugose conversion. Additionally, a motility assay showed that the *rocS* mutant's motility was 50% less than that of the parent strain (N16961) (Rashid *et al.*, 2003).

Recent comprehensive studies by Lim *et al.* (2006) have elucidated how c-diGMP plays a role in the rugose variant and motility. This report confirmed that the genome of *V. cholerae* harbours 53 genes encoding proteins that have GGDEF and/or EAL domains, a hallmark for c-diGMP proteins. Comparative genomic expression profiles aided the authors in identifying 5 c-diGMP genes induced in the rugose variant of *V. cholerae* compared to that in its smooth variant. These genes were designated *cdgA* (VCA0074), *cdgB* (VC1029), *cdgC* (VCA0785), *cdgD* (VCA0697) and *cdgE* (VC1367). Disruptions of these five genes along with previously described C-diGMP genes (*rocS* and *mbaA*), and complementation of these mutants *in trans* in the background of both rugose and isogenic smooth variants led the authors to conclude that three genes, including *cdgC*, *rocS*, and *mbaA* negatively regulated the rugose colony phenotype. In contrast, CdgA positively regulated the expression of rugose colony phenotype. As expected, the biofilm formation of these mutants (CdgC, RocS and MbaA) was increased relative to their parents as measured by crystal violet staining and confocal scanning laser microscopy (CSLM) (Lim *et al.*, 2006).

Concluding remarks and future directions

Based on its ability to resist harsh conditions, including its resistance to chlorine and oxidative and osmotic stresses, the rugose variant has been predicted (Ali *et al.*, 2002; Morris *et al.*, 1996; Rice *et al.*, 1993; Wai *et al.*, 1998; Yildiz *et al.*, 1999) to be the survival prototype of *V. cholerae.* The variant produces a copious amount of an amorphous exopolysaccharide that allows the bacterium to resist harsh environmental conditions. It was possible to switch smooth variants of *V. cholerae* to rugose variants by starving the smooth cells into media deficient in nutrients, suggesting that smooth *V. cholerae* strain switches to the rugose variant in response to nutrient deficiency in aquatic environments. Although aquatic reservoirs of *V. cholerae* are constantly changing and are deficient in nutrients, studies have not yet been undertaken to determine the prevalence of the rugose phenotype in aquatic environments. TCBS, a primary selective medium used to isolate *V. cholerae*, inhibits the expression of the rugose phenotype, and may have resulted in underestimation of the prevalence of rugose variants. In order to examine the prevalence of rugose *V. cholerae* in the aquatic environments in a cholera-endemic country (Bangladesh), we plated water samples (collected from various sites at every 14 days for 2 years) directly onto L-agar plates and looked for rugose colonies. Out of 1,585 *V. cholerae* isolates obtained from L-agar plates, 56 (3.5%) were found to be rugose variants (Ali *et al.*, *unpublished observations*). While our currently unpublished observation suggests that smooth *V. cholerae* can switch to the rugose variant in aquatic environments, only a small fraction of *V. cholerae* strains were able to switch from smooth to rugose colonial variants. Our earlier *in vitro* assay also mimicked the current observation; i.e. only a small fraction of smooth *V. cholerae* were able to shift to the rugose variant under high-frequency, rugose-inducing growth conditions (Ali *et al.*, 2002). These observations led us to raise a question, i.e. what is the significance of the majority of *V. cholerae* strains that were unable to switch to rugose variant? Is there any difference in persistence between *V. cholerae* strains with ability to undergo rugose variant or unable to undergo rugose variant? Further studies are needed to address these questions.

V. cholerae produces the rugose exopolysaccharide only in response to nutrient starvation. Even under high-frequency rugose-producing conditions, *V. cholerae* was able to switch from the smooth to rugose variant at or after 16 h of incubation (Ali *et al.*, 2002), suggesting that smooth *V. cholerae* induces rugose variant in response to stationary growth phase conditions and/or in response to severe nutrient deficiency. This observation suggests that, at stationary growth phase, smooth cells may sense an environmental cue(s) leading to its conversion to rugose variant. Such an environmental cue(s) has not yet been determined, and future studies should be directed to understanding the associated environmental cue(s).

The genes encoding the enzymes required for the biosynthesis of the rugose exopolysaccharide (VPS^{Eltor}) have been identified. In addition, several positive and negative transcriptional regulators affecting the transcription of *vps* expression have been identified. In a recent study, Beyhan and Yildiz (2007) reported that a single nucleotide change of a T to C at the 718th bp of *vpvC* (VC2454) gene (encodes a diguanylate cyclase) led the conversion of a smooth to a rugose variant of *V. cholerae*. This transition leads to the conversion of a tryptophan residue in a smooth variant to an arginine residue in the rugose variant, leading to the conversion of a smooth variant to a rugose variant. However, the key master genetic element(s) regulating the reversible switch between the smooth to rugose variant remains mysterious. Understanding the nature and modulation of the regulatory element(s) will expand our knowledge on rugosity and its implication on the persistence of the rugose variant of *V. cholerae* in the aquatic environment, a research area that remains in its infancy.

Acknowledgements

I thank Glenn Morris, Jr., and Arnold Kreger for their critical reading and valuable comments on this chapter. Part of the work, presented in this chapter, was supported by a National Institute of Health grant (RO1 123456) to J. G. Morris, Jr.

References

Albert, M.J. (1994). *Vibrio cholerae* O139 Bengal. J. Clin. Microbiol. *32*, 2345–2349.

Albert, M.J., Siddique, A.K., Islam, M.S., Faruque, A.S.G., Ansaruzzaman, M., Faruque, S.M. and Sack, R.B. (1993). A large outbreak of clinical cholera due to *Vibrio cholerae* non-O1 in Bangladesh. Lancet *341*, 704.

Aldova, E., Laznickova, K., Stepankova, E. and Lietava, J. (1968). Isolation of nonagglutinable vibrio from an enteritis outbreak in Czechoslovakia. J. Infect. Dis. *118*, 25–31.

Ali, A., Johnson, J.A., Franco, A.A., Metzger, D.J., Connell, T.D., Morris, J.G.J. and Sozhamannan, S. (2000a). Mutations in the extracellular protein secretion pathway genes (*eps*) interfere with rugose polysaccharide production in and motility of *Vibrio cholerae*. Infect. Immun. *68*, 1967–1974.

Ali, A., Mahmud, Z.H., Morris, JR., J.G., Sozhamannan, S. and Johnson, J.A. (2000b). Sequence analysis of Tn*phoA* insertion sites in *Vibrio cholerae* mutants defective in rugose polysaccharide production. Infect. Immun. *68*, 6857–6864.

Ali, A., Morris, Jr., J.G. and Johnson, J.A. (2005). Sugars inhibit expression of the rugose phenotype of *Vibrio cholerae*. J. Clin. Microbiol. *43*, 1426–1429.

Ali, A., Rashid, M.H. and Karaolis, D.K.R. (2002). High-frequency rugose exopolysaccharide production by *Vibrio cholerae*. Appl. Environ. Microbiol. *68*, 5773–5778.

Anriany, Y.A., Weiner, R.M., Johnson, J.A., De Rezende, C.E. and Joseph, S.W. (2001). *Salmonella enterica* Serovar Typhimurium DT104 displays a rugose phenotype. Appl. Environ. Microbiol. *67*, 4048–4056.

Bailey, M.J.A., Hughes, C. and Koronakis, V. (1997). Rfah and the ops element, components of a novel system controlling bacterial transcription elongation. Mol. Microbiol. *26*, 845–852.

Baltenau, I. (1926). The receptor structure of *Vibrio comma* with observations on cholera and cholera-like organisms. J. Pathol. Bacteriol. *29*, 251–277.

Baum, J.A. (1994). Tn5401, a new class II transposable element from *Bacillus thuringiensis*. J. Bacteriol. *176*, 2835–2845.

Berche, P., Poyart, C., Abachin, E., Latievre, H., Vandepitte, J., Dodin, A. and Fournier, J.M. (1994). The novel epidemic strain O139 is closely related to pandemic strain O1 of *Vibrio cholerae*. J. Infect. Dis. *170*, 701–704.

Beyhan, S., and Yildiz, F.H. (2007). Smooth to rugose phase variation in *Vibrio cholerae* can be mediated by a single nucleotide change that targets c-di-GMP signalling pathway. Mol. Microbiol. *63*, 995–1007.

Bik, E.M., Bunschoten, A.E., Gouw, R.D. and Mooi, F.R. (1995). Genesis of the novel epidemic *Vibrio cholerae* O139 strain: evidence for horizontal transfer of genes involved in polysaccharide synthesis. Eur. Mol. Biol. Org. J. *14*, 209–216.

Bik, E.M., Bunscoten, A.E., Willems, R.J.L., Chang, A.C.Y. and Mooi, F.R. (1996). Genetic organization and functional analysis of the otn DNA essential for cell-wall polysaccharide synthesis in *Vibrio cholerae* O139. Mol. Microbiol. *20*, 799–811.

Blake, P.A., Weaver, R.E. and Hollis, D.G. (1980). Diseases of humans (other than cholera) caused by vibrios. Annu. Rev. Microbiol. *34*, 341–367.

Casper-Lindley, C. and Yildiz, F.H. (2004). VpsT Is a transcriptional regulator required for expression of vps biosynthesis genes and the development of rugose colonial morphology in *Vibrio cholerae* O1 El Tor. J. Bacteriol. *186*, 1574–1578.

Chatterjee, S.N. and Chaudhuri, K. (2003). Lipopolysaccharide of *Vibrio cholerae*: I. Physical and chemical characterization. Bioche. Biophy. Acta. *1639*, 65–79.

Chatterjee, S.N. and Chaudhuri, K. (2004). Lipopolysaccharides of *Vibrio cholerae* II. Genetics of biosynthesis. Bioche. Biophy. Acta. *1690*, 93–109.

Chen, Y., Bystricky, P., Adeyeye, J., Panigrah, P., Ali, A., Johnson, J.A., Bush, C.A., Morris, Jr., J.G. and Stine, O.C. (2007). The capsule polysaccharide structure and biogenesis for non-O1 *Vibrio cholerae* NRT36S: genes are embedded in the LPS region. BMC Microbiol. *15*, 7–20.

Comstock, L.E., Johnson, J.A., Michalski, J.M., Morris, Jr., J.G. and Kaper, J.B. (1996). Cloning and sequence of a region encoding a surface polysaccharide of *Vibrio cholerae* O139 and characterization of the insertion site in the chromosome of *V. cholerae* O1. Mol. Microbiol. *19*, 815–826.

Comstock, L.E., Maneval, Jr., D., Panigrahi, P., Joseph, A., Levine, M.M., Kaper, J.B., Morris, Jr., J.G. and

Johnson, J.A. (1995). The capsule and O antigen in *Vibrio cholerae* O139 Bengal are associated with a genetic region not present in *Vibrio cholerae* O1. Infect. Immun. *63*, 317–323.

Curtis, S.K., Kothary, M.H., Blodgett, R.J., Raybourne, R.B., Ziobro, G.C. and Tall, B.D. (2007). Rugosity in *Grimontia hollisae*. Appl. Environ. Microbiol. *73*, 1215–1224.

D'Argenio, D.A., Calfee, M.W., Rainey, P.B. and Pesci, E.C. (2002). Autolysis and autoaggregation in *Pseudomonas aeruginosa* colony morphology mutants. J. Bacteriol. *184*, 6481–6489.

Das, B., Ghosh, R.K., Sharma, L., Vasin, N. and Ghosh, A. (1993). Tandem repeats of cholera toxin gene in *Vibrio cholerae* O139. Lancet *342*, 1173–1174.

Dumontier, S. and Berche, P. (1998). *Vibrio cholerae* O22 might be a putative source of exogenous DNA resulting in the emergence of the new strain of *Vibrio cholerae* O139. FEMS Microbiol. Lett. *164*, 91–98.

Dziejman, M., Balon, E., Boyd, D., Fraser, C.M., Heidelberg, J.F. and Mekalanos, J.J. (2002). Comparative genomic analysis of *Vibrio cholerae*: genes that correlate with cholera endemic and pandemic disease. Proc. Natl. Acad. Sci. USA *99*, 1556–1561.

Fallarino, A., Mavrangelos, C., Stroeher, U.H. and Manning, P.A. (1997). Identification of additional genes required for O-antigen biosynthesis in *Vibrio cholerae* O1. J. Bacteriol. *179*, 2147–2153.

Faruque, S.M., Ahmed, K.M., Abdul Alim, A.R., Qadri, F., Siddique, A.K. and Albert, M.J. (1997). Emergence of a new clone of toxigenic *Vibrio cholerae* O1 biotype El Tor displacing *Vibrio cholerae* O139 Bengal in Bangladesh. J. Clin. Microbiol. *35*, 624–630.

Faruque, S.M., Asadulghani, Rahman, M.M., Waldor, M.K. and Sack, D.A. 2000. Sunlight-Induced propagation of the lysogenic phage encoding cholera toxin. Infect. Immun. *68*, 4795–4801.

Faruque, S.M., Sack, D.A., Sack, R.B., Colwell, R.R., Takeda, Y. and Nair, G.B. (2003). Inaugural Article: Emergence and evolution of *Vibrio cholerae* O139. Proc. Natl. Acad. Sci. *100*, 1304–1309.

Fong, J.C.N., Karplus, K., Schoolnik, G.K. and Yildiz, F.H. (2006). Identification and characterization of RbmA, a novel protein required for the development of rugose colony morphology and biofilm structure in *Vibrio cholerae*. J. Bacteriol. *188*, 1049–1059.

Fong, J.C.N. and Yildiz, F.H. (2007). The *rbmBCDEF* gene cluster modulates development of rugose colony morphology and biofilm formation in *Vibrio cholerae*. J. Bacteriol. *189*, 2319–2330.

Grau, B.L., Henk, M.C. and Pettis, G.S. (2005). High-frequency phase variation of *Vibrio vulnificus* 1003: Isolation and characterization of a rugose phenotypic variant. J. Bacteriol. *187*, 2519–2525.

Hall, R.H., Khambaty, F.M., Kothary, M. and Keasle, S.P. (1993). Non-O1 *Vibrio cholerae*. Lancet *342*, 430.

Heidelberg, J.F., Eisen, J.A., Nelson, W.C., Clayton, R.A., Gwinn, M.L., Dodson, R.J., Haft, D.H., Hickey, E.K., Peterson, J.D., Umayam, L., Gill, S.R., Nelson, K.E., Read, T.D., Tettelin, H., Richardson, D., Ermolaeva, M.D., Vamathevan, J., Bass, S., Qin, H., Dragoi, I., Sellers, P., McDonald, L., Utterback, T., Fleishmann, R.D., Nierman, W.C., White, O., Salzberg, S.L., Smith, H.O., Colwell, R.R., Mekalanos, J.J., Venter, J.C. and F.C.M. (2000). DNA sequence of both chromosomes of the cholera pathogen *Vibrio cholerae*. Nature. *406*, 477–483.

Heinriches, D.E., Yethon, J.A. and Whitefield, C. (1998). Molecular basis for structural diversity in the core regions of the lipopolysaccharides of *Escherichia coli* and *Salmonella enterica*. Mol. Microbiol. *30*, 221–232.

Hill, C.W., Sandt, C.H. and Vlazny, D.A. (1994). Rhs elements of *Escherichia coli*: a family of genetic composites each encoding a large mosaic protein. Mol. Microbiol. *12*, 865–871.

Hisatsune, K., Kondo, S., Isshiki, Y., Iguchi, T., Kawamata, Y. and Shimada, T. (1993). O-antigenic LPS of *Vibrio cholerae* O139 Bengal, a new epidemic strain for recent cholera in the Indiand subcontinent. Biochem. Biophys. Res. Commun. *196*, 1309–1315.

Hobbs, M., and Reeves, P.R. (1994). The JUMPstart sequence: a 39 bp element common to several polysaccharide gene clusters. Mol. Microbiol. *12*, 855–856.

Isshiki, Y., Kondo, S., Iguchi, T., Sano, Y., Shimada, T. and Hisatsune, K. 1996. An immunological study of serological cross-reaction between lipopolysaccharides from *Vibrio cholerae* O22 and O139. Microbiology *142*, 1499–1504.

Jedani, K.E., Stroeher, U.H. and Manning, P.A. (1998). Distribution of IS1358 and linkage to *rfb*-related genes in *Vibrio cholerae* and *Vibrio anguillarum*. Gene. 223, 269–282.

Jenal, U. (2004). Cyclic di-guanosine-monophosphate comes of age: a novel secondary messenger involved in modulating cell surface structures in bacteria? Curr. Opin. Microbiol. *7*, 185–191.

Jobling, M.G. and Holmes, R.K. (1997). Characterization of *hapR*, a positive regulator of the *Vibrio cholerae* HA/protease gene *hap*, and its identification as a functional homologue of the *Vibrio harveyi luxR* gene Mol. Microbiol. *26*, 1023–1034.

Johnson, J.A., Panigrahi, P. and Morris, Jr., J.G. (1992). Non-O1 *Vibrio cholerae* NRT36S produces a polysaccharide capsule that determines colony morphology, serum resistance, and virulence in mice. Infect. Immun. *60*, 864–869.

Johnson, J.A., Salles, C.A., Panigrahi, P., Albert, M.J., Wright, A.C., Johnson, R.J. and Morris, Jr., J.G. (1994). *Vibrio cholerae* O139 synonym Bengal is closely related to *Vibrio cholerae* El Tor but has important differences. Infect. Immun. *62*, 2108–2110.

Kamal, A.M. (1971). Outbreak of gastro-enteritis by non-agglutinable (NAG) vibrios in the republic of Sudan. J. Egypt. Public Health Assoc. XL*VI*, 125–159.

Kaper, J.B., Morris, Jr., J.G. and Levine, M.M. (1995). Cholera. Clin. Microbiol. Rev. *8*, 48–86.

Knirel, Y.A., Paredes, L., Jansson, P.E., Weintraub, A., Widmalm, G. and Albert, M.J. (1995). Structure of the capsular polysaccharide of *Vibrio cholerae* O139 synonym Bengal containing D-galactose 4,6-cyclophosphate. Eur. J. Biochem. *232*, 391–396.

Knirel, Y.A., Widmalm, G., Senchenkova, S.N., Jansson, P.E. and Weintraub, A. (1997). Structural studies on the short-chain lipopolysaccharide of *Vibrio cholerae* O139 Bengal. Eur. J. Biochem. *247*, 402–410.

Li, M., Shimada, T., Morris, Jr., J.G. Sulakvelidze, A. and Sozhamannan, S. (2002). Evidence for the emergence of non-O1 and non-O139 *Vibrio cholerae* strains with pathogenic potential by exchange of O-antigen biosynthesis regions. Infect. Immun. *70*, 2441–2453.

Lim, B., Beyhan, S., Meir, J. and Yildiz, F.H. (2006). Cyclic-diGMP signal transduction systems in *Vibrio cholerae*: modulation of rugosity and biofilm formation. Mol. Microbiol. *60*, 331–348.

Manning, P.A., Heuzenroeder, M.W., Yeadon, J., Leavesley, D.I., Reeves, P.R. and Rowley, D. (1986). Molecular cloning and expression in *Escherichia coli* K-12 of the O-antigens of the Inaba and Ogawa serotypes of the *Vibrio cholerae* O1 lipopolysaccharides and their potential for vaccine development. Infect. Immun. *53*, 272–277.

Manning, P.A., Stroeher, U.H., Karageorgos, L.E. and Morona, R. (1995). Putative O-antigen transport genes within the *rfb* region of *Vibrio cholerae* O1 are homologous to those of capsule transport. Gene *158*, 1–7.

Manning, P.A., Stroeher, U.H. and Morona, R. (1993). Molecular basis for O-antigen biosynthesis in *Vibrio cholerae* O1: Ogawa-Inaba switching, p. in press. In I. K. Wachsmuth, P. Blake, and O. Olsvik (ed.), *Vibrio cholerae* and cholera. American Society for Microbiology, Washington DC.

Manning, P.A., Stroeher, U.W. and Morona, R. (1994). Molecular basis for O-antigen biosynthesis in *Vibrio cholerae* O1: Ogawa-Inaba switching, p. 77–94. In I. K. Wachsmuth, P. A. Blake, and Ø. Olsvik (ed.), *Vibrio cholerae* and cholera. American Society for Microbiology, Washington DC.

Maris, A.E., Sawaya, M.R., Kaczor-Grzeskowiak, M., Jarvis, M.R., Bearson, S.M., Kopka, M.L., Schroder, I., Gunsalus, R.P. and Dickerson, R.E. (2002). Dimerization allows DNA target site recognition by the NarL response regulator. Nat. Struct. Biol. *9*, 771–778.

Mazel, D., Dychinco, B., Webb, V.A. and Davies, J. (1998). A distinctive class of integron in the *Vibrio cholerae* genome. Science *280*, 605–608.

Mitra, R., Basu, A., Dutta, D., Nair, G.B. and Takeda, Y. (1996). Resurgence of *Vibrio cholerae* O139 Bengal with altered antibiogram in Calcutta, India. Lancet *348*, 1181.

Mooi, F.R. and Bik, E.M. (1997). The evolution of epidemic *Vibrio cholerae* strains. Trends Microbiol. *5*, 161–165.

Morris Jr., J.G. (1994). Cholera and other vibrios, p. 753–762. In P. D. Hoeprich, M. C. Jordan, and A. R. Ronald (ed.), Infectious Diseases. J. B. Lippincott, Philadelphia, USA.

Morris Jr., J.G., Sztein, M.B., Rice, E.W., Nataro, J.P., Losonsky, G.A., Panigrahi, P., Tacket, C.O. and Johnson, J.A. (1996). *Vibrio cholerae* O1 can assume a chlorine-resistant rugose survival form that is virulent for humans. J. Infect. Dis. *174*, 1364–1368.

Morris Jr., J.G., Takeda, T., Tall, B.D., Losonsky, G.A., Bhattacharya, S.K., Forrest, B.D., Kay, B.A. and Nishibuchi, M. (1990). Experimental non-O group 1 *Vibrio cholerae* gastroenteritis in humans. J. Clin. Invest. *85*, 697–705.

Morris Jr., J.G., Wilson, R., Davis, B.R., Wachsmuth, I.K., Riddle, C.F., Wathen, H.G., Polard, R.A. and Blake, P.A. (1981). Non-O1 group 1 *Vibrio cholerae* gastroenteritis in the United States: clinical, epidemiologic, and laboratory aspects of sporadic cases Ann. Intern. Med. *94*, 656–658.

Mukhopadhyay, A.K., Basu, A., Garg, P., Bag, P.K., Ghosh, A., Bhattacharya, S.K., Takeda, Y. and Nair, G.B. (1998). Molecular epidemiology of reemergent *Vibrio cholerae* O139 Bengal in India. J. Clin. Microbiol. *36*, 2149–2152.

Nesper, J., Krai, A., Schild, S., Bla, J., Klose, K.E., Bockemuhl, J. and Reidl, J. (2002). Comparative and genetic analyses of the putative *Vibrio cholerae* lipopolysaccharide core oligosaccharide biosynthesis (*wav*) gene cluster. Infect. Immun. *70*, 2419–2433.

Peter, J.G. (1993). Detection of two smooth colony phenotypes in a *Salmonella enterica* Enteritidis isolate which vary in their ability to contaminate eggs. Appl. Environ. Microbiol. *59*, 2884–2890.

Popovic, T., Fields, P.I., Olsvik, O., Wells, J.G., Evins, G.M., Cameron, D.N., Farmer, III J.J., Bopp, C.A., Wachsmuth, I.K., Sack, R.B., Albert, M.J., Nair, G.B., Shimada, T. and Feeley, J.C. (1995). Molecular subtyping of toxigenic *Vibrio cholerae* O139 causing epidemic cholera in India and Bangladesh, 19992–1993. J. Infect. Dis. *171*, 122–127.

Ramamurthy, T., Garg, S., Sharma, R., Bhattacharya, S.K., Nair, G.B., Shimada, T., Takeda, T., Karasawa, T., Kurazano, H., Pal, A. and Takeda, Y. (1993). Emergence of a novel strain of *Vibrio cholerae* with epidemic potential in Southern and Eastern India. Lancet *341*, 703–704.

Rashid, M.H., Rajanna, C., Ali, A. and Karaolis, D.K.R. (2003). Identification of genes involved in the switch between the smooth and rugose phenotypes of *Vibrio cholerae*. FEMS Microbiol. Lett. *227*, 113–119.

Recchia, G.D. and Hall, M.R. (1995). Gene cassettes: a new class of mobile element. Microbiology *141*, 3015–3027.

Reeves, P.R., Hobbs, M., Valvano, M.A., Skurnik, M., Whitefield, C., Coplin, D., Kido, N., Klena, J., Maskell, D., Raetz, C.R. and Rick, P.D. (1996). Bacterial polysaccharide synthesis and gene nomenclature. Trends Microbiol. *4*, 495–503.

Rice, E.W., Johnson, C.H., Clark, R.M., Fox, K.R., Reasoner, D.J., Dunnigan, M.E., Panigrah, P., Johnson, J.A. and Morris, J.G.J. (1993). Chlorine and survival of 'rugose' *Vibrio cholerae*. Int. J. Environ. Health Res. *3*, 89–98.

Rice, E.W., Johnson, C.J., Clark, R.M., Fox, K.R., Reasoner, D.J., Dunnigan, M.E., Panigrah, P., Johnson, J.A. and Morris, Jr., J.G. (1992). Chlorine and survival of 'rugose' *Vibrio cholerae*. Lancet *340*, 740.

Romling, U., Gomelsky, M. and Galperin, M.Y. (2005). C-di-GMP: the drawing of a novel bacterial signalling system. Mol. Microbiol. *57*, 629–639.

Sandkvist, M., Michel, L.O., Hough, L.P., Morales, V.M., Bagdasarian, M., Koomey, M., DiRita, V.J. and Bagdasarian, M. (1997). General secretion pathway (eps) genes required for toxin secretion and outer membrane biogenesis in *Vibrio cholerae*. J. Bacteriol. *179*, 6994–7003.

Sandkvist, M., Morales, V. and Bagdasarian, M. (1993). A protein required for secretion of cholera toxin through the outer membrane. Gene. *123*, 81–86.

Simm, R., Morr, M., Kader, A., Nimtz, M. and Romling, U. (2004). GGDEF and EAL domains inversely regulate cyclic di-GMP levels and transition from sessility to motility. Mol. Microbiol. *53*, 1123–1134.

Sozhamannan, S., Deng, Y.K., Li, M., Sulakvelidze, A., Kaper, J.B., Johnson, J.A., Nair, G.B. and Morris, J.G.J. (1999). Cloning and sequencing of the genes downstream of the *wbf* gene cluster of *Vibrio cholerae* serogroup O139 and analysis of the junction genes in other serogroups. Infect. Immun. *67*, 5033–5040.

Stoker, K., Reijnders, W.N., Oltmann, L.F. and Stouthamer, A.H. (1989). Initial cloning and sequencing of *hydHG*, an operon homologous to *ntrBC* and regulating the labile hydrogenase activity in *Escherichia coli* K-12. J. Bacteriol. *171*, 4448–4456.

Stroeher, U.H., Jedani, K.E., Dredge, B.K., Morona, R., Brown, M.H., Karageorgos, L.E., Albert, M.J. and Manning, P.A. (1995a). Genetic rearrangements in the *rfb* regions of *Vibrio cholerae* O1 and O139. Proc. Natl. Acad. Sci. USA 92, 10374–10378.

Stroeher, U.H., Jedani, K.E. and Manning, P.A. (1998). Genetic organization of the genes associated with surface polysaccharide synthesis in *Vibrio cholerae* O1, O139 and *Vibrio anguillarum* O1 and O2: a review. Gene. *223*, 269–282.

Stroeher, U.H., Karageorgos, L.E., Brown, M.H., Morona, R. and Manning, P.A. (1995b). A putative pathway for perosamine biosynthesis is the first function encoded within the *rfb* region of *Vibrio cholerae* O1 Gene. *166*, 33–42.

Stroeher, U.H., Karageorgos, L.E., Morona, R. and Manning, P.A. (1995c). In *Vibrio cholerae* serogroup O1, *rfaD* is closely linked to the *rfb* operon. Gene. *155*, 67–72.

Stroeher, U.H., Karageorgos, L.E., Morona, R. and Manning, P.A. (1992). Serotype conversion in *Vibrio cholerae* O1. Proc. Natl. Acad. Sci. USA *89*, 2566–2570.

Stroeher, U.H. and Manning, P.A. (1997a). *Vibrio cholerae* serotype O139: swapping genes for surface polysaccharide biosynthesis. Trends Microbiol. *5*, 178–180.

Stroeher, U.H., Parasivam, G., Dredge, B.K. and Manning, P.A. (1997b). Novel *Vibrio cholerae* O139 genes involved in lipopolysaccharide biosynthesis. J. Bacteriol. *179*, 2740–2747.

Wai, S.N., Mizunoe, Y., Takade, A., Kawabata, S.I. and Yoshida, S.I. (1998). *Vibrio cholerae* O1 strain TSI-4 produces the exopolysaccharide materials that determine colony morphology, stress resistance, and biofilm formation. Appl. Environ. Microbiol. *64*, 3648–3655.

Waldor, M.K., Colwell, R. and Mekalanos, J.J. (1994). The *Vibrio cholerae* O139 serogroup antigen includes an O-antigen capsule and lipopolysaccharide virulence determinants. Proc. Natl. Acad. Sci. USA *91*, 11388–11392.

Waldor, M.K. and Mekalanos, J.J. (1996). Lysogenic conversion by a filamentous phage encoding cholera toxin. Science. *272*, 1910–1914.

Ward, H.M. and Mannning, P.A. (1989). Mapping of chromosomal loci associated with lipopolysaccharide synthesis and serotype specificity in *Vibrio cholerae* O1 by transposon mutagenesis using Tn5 and Tn2680. Mol. Gen. Genet. *218*, 367–370.

Ward, H.M., Morelli, G., Kamke, M., Morona, R., Hackett, J.A. and Manning, P.A. (1987). A physical map of the chromosomal region determining O-antigen biosynthesis in *Vibrio cholerae* O1. Gene. *55*, 197–204.

Watnick, P.I., Lauriano, C.M., Klose, K.E., Croal, L. and Kolter, R. (2001). The absence of a flagellum leads to altered colony morphology, biofilm development and virulence in *Vibrio cholerae* O139. Mol. Microbiol. *39*, 223–235.

Weintraub, A., Widmalm, G., Jansson, P.E., Jansson, M., Hultenby, K. and Albert, M.J. (1994). *Vibrio cholerae* O139 Bengal possesses a capsular polysaccharide which may confer increased virulence. Microbial Pathogen. *16*, 235–241.

White, P.B. (1940). The characteristic hapten and antigen of rugose races of cholera and El Tor vibrios. J. Pathol. Bacteriol. *50*, 160–164.

White, P.B. (1938). The rugose variant of vibrios. J. Pathol. Bacteriol. *46*, 1–6.

Whitfield, C. and Roberts, I.S. (1999). Structure, assembly and regulation of expression of capsules in *Escherichia coli*. Mol. Microbiol. *31*, 1307–1319.

Xiang, S.H., Hobbs, M. and Reeves, P.R. (1994). Molecular analysis of the *rfb* gene cluster of a group D2 *Salmonella enterica* strain: evidence for its origin from an insertion sequence-mediated recombination event between group E and D1 strains. J. Bacteriol. *176*, 4357–4365.

Yamasaki, S., Shimizu, T., Hoshino, K., Ho, S.T., Shimada, T., Nair, G.B. and Takeda, Y. (1999). The genes responsible for O-antigen synthesis of *Vibrio cholerae* O139 are closely related to those of *Vibrio cholerae* O22. Gene *237*, 321–332.

Yildiz, F.H., Dolganov, N.A. and Schoolnik G.K. (2001). VpsR, a member of the response regulators of the two-component regulatory systems, is required for expression of vps biosynthesis genes and EPS^{ETr}-associated phenotypes in *Vibrio cholerae* O1 El Tor. J. Bacteriol. *183*, 1716–1726.

Yildiz, F.H., Liu, X.S., Heydorn, A. and Schoolnik, G.K. (2004). Molecular analysis of rugosity in a *Vibrio cholerae* O1 Eltor phase variant. Mol. Microbiol. *53*, 497–515.

Yildiz, F.H. and Schoolnik, G.K. (1999). *Vibrio cholerae* O1 El Tor: identification of a gene cluster required for the rugose colony type, exopolysaccharide production, chlorine resistance, and biofilm formation. Proc. Natl. Acad. Sci. USA *96*, 4028–4033.

Zhao, S., Sandt, C.H., Feuler, G., Valzany, D.A., Gray, J.A. and Hill, C.W. (1993). Rhs elements of *Escherichia coli* K-12: complex composites of shared and unique components that have different evolutionary history. J. Bacteriol. *175*, 2799–2808.

Zhou, D.X., Massenet, O., Quingly, F., Marion, M.J., Moneger, F., Huber, P. and Mache, R. (1988). Characterization of a large inversion in the spinach chloroplast genome relative to *Marchantia*, a possible transposon-mediated origin. Curr. Genet. *13*, 433–439.

Genetics and Microbiology of Biofilm Formation by *Vibrio cholerae*

7

Fitnat H. Yildiz and Roberto Kolter

Abstract

In nature, most bacteria grow as matrix-enclosed, surface-associated communities known as biofilms. *Vibrio cholerae,* the causative agent of the disease cholera, forms biofilms on diverse surfaces. This ability to form biofilms appears to be critical for the environmental survival and the transmission of *V. cholerae*. The molecular mechanisms utilized by *V. cholerae* to form and maintain biofilms are being investigated through a combination of molecular genetic and microscopic approaches. A better understanding of the life cycle of this important human pathogen should prove useful in the development of future strategies for predicting and controlling cholera epidemics.

Introduction

Biofilms are surface attached microbial communities composed of microorganisms and the extra-polymeric substances they produce (Costerton *et al.*, 1995). Biofilms play a significant role in environmental, industrial, and clinical settings (Davey and O'Toole G, 2000; Donlan and Costerton, 2002; Parsek and Singh, 2003). The biofilm mode of growth is the preferred lifestyle in the microbial world as it enhances survival in natural settings. Because surfaces adsorb and concentrate nutrients that are usually scarce, growth on surfaces is favoured. In addition, biofilms provide the constituent microbes protection from predators, such as protozoa and viruses, and from toxic compounds, such as antimicrobial agents (Davey and O'Toole G, 2000; Donlan and Costerton, 2002).

V. cholerae, the causative agent of the disease cholera, is a facultative human pathogen (Kaper *et al.*, 1995). The life cycle of this bacterium involves rapid growth within the human intestine and prolonged survival in aquatic habitats (Faruque *et al.*, 1998; Kaper *et al.*, 1995). The pathogen is found in coastal and estuarine environments either as individual cells in the water column, or as biofilm-associated cells attached to surfaces (Colwell and Huq, 1994; Colwell *et al.*, 2003; Faruque *et al.*, 1998; Huq *et al.*, 1983; Huq *et al.*, 1986; Huq *et al.*, 1995; Tamplin *et al.*, 1990). It has been proposed that the ability of *V. cholerae* to form biofilms is crucial for its survival in aquatic habitats in between epidemics and advantageous in its host-to-host transmission during epidemics. Recent studies suggest that *V. cholerae* may form biofilms within the intestine (Faruque *et al.*, 2006; Nielsen *et al.*, 2006). However, the role that biofilms play in the progression and pathology of the disease, if any, remains unknown. Regardless, it does appear that the ability to form biofilms is an important part of the natural history of this important human pathogen. In this chapter, we summarize our current understanding of biofilm formation by *V. cholerae* and point to areas that are still poorly understood.

Vibrio cholerae can form biofilms on many different surfaces

The biofilm forming capacity of *V. cholerae,* both in natural habitats and under laboratory conditions, is well documented. It has been reported that removal of particles >20 μm in diameter from water reduced cholera incidence by 48%, suggesting that biofilms are critical for transmission of *V. cholerae* (Colwell *et al.*, 2003). Field studies have shown that *V. cholerae* can be recovered from the surfaces of phytoplankton, zooplankton, aquatic plants, crustaceans, insects and sediments (Broza *et al.*, 2005; Halpern *et al.*, 2004; Huq *et al.*, 1986; Huq *et al.*, 1995; Islam *et al.*, 2007). Microcosm studies that monitored attachment of *V. cholerae* to zooplankton and phytoplankton from rivers and ponds showed that the organism can bind to a surprisingly wide range of plankton species (Hood and Winter, 1997; Huq *et al.*, 1983; Huq *et al.*, 1984). The bacterium is able to attach to living surfaces such as estuarine algae and sea grasses as well as to purified plant and animal surface compounds such as cellulose and chitin (Chiavelli *et al.*, 2001; Hood and Winter, 1997; Reguera and Kolter, 2005; Watnick *et al.*, 1999). In addition, it attaches to artificial compounds with widely different surface properties, e.g. silicone, glass and Teflon (Hood and Winter, 1997; Watnick and Kolter, 1999; Yildiz *et al.*, 2004). In summary, *V. cholerae* can attach to hydrophilic and hydrophobic surfaces and optimal rates of attachment are observed at temperatures and salinities typical of semitropical and tropical estuarine environments.

Biofilm formation appears to increase survival of *V. cholerae* in natural settings. The ability of heterotrophic bacteria such as *V. cholerae* to grow in aquatic habitats is often limited by the low availability of carbon sources. Because photosynthetic microorganisms (i.e. phytoplankton) excrete a variety of organic compounds, attachment of *V. cholerae* to phytoplankton can provide access to those compounds and thus facilitate its growth and survival. Similarly, attachment to the chitinaceous surfaces of zooplankton, rich in N-acetyl glucosamine, provides close access to a compound that can serve as both a carbon and a nitrogen source. In support of this notion, Huq *et al.* showed that survival of *V. cholerae* was enhanced when co-cultured with copepods (Huq *et al.*, 1984). Hood *et al.* showed that bacteria growing as biofilms on abiotic surfaces survive longer than freely suspended cells (Hood and Winter, 1997). In aquatic environments, abiotic surfaces contain absorbed substrates which the attached bacteria can readily utilized as nutrients. Thus, attachment of *V. cholerae* to surfaces and subsequent biofilm formation facilitate its growth survival and persistence in aquatic ecosystems.

Aside from biofilms playing a role in environmental survival, it now appears that *V. cholerae* biofilms may also be important in the disease process. Stool samples from cholera patients contain *V. cholerae* not only as planktonic cells but also in biofilm-like aggregates (Faruque *et al.*, 2006). This result was somewhat surprising, as biofilm formation by *V. cholerae* was thought to occur mainly in the environment. This prompted the study of the role that the biofilm-like aggregates found in stools might have in the transmission of the disease. Indeed, using competition assays it was possible to show that the average infectivity of the aggregate form was significantly higher than that of the planktonic cells (Faruque *et al.*, 2006). Interestingly, *V. cholerae* found in the surface waters of areas where cholera is endemic exists in biofilm-like aggregates (Faruque *et al.*, 2006). Within these aggregates the bacterial cells are in what has been called a 'metabolically impeded' state. That is, they are viable but their metabolic state is kept low by virtue of their being entrapped in the aggregate. These 'conditionally viable environmental cells', or CVEC, can regain metabolic activity under some special laboratory conditions or by inoculation into ligated rabbit ileal loops (Faruque *et al.*, 2006). This suggests that CVEC of *V. cholerae* might play a role in transmission. The source of CVEC is thought to be the biofilm-like aggregates of *V. cholerae* present in human stools. But it is not only in stools where *V. cholerae* populations are found as multicellular aggregates within an infected host. The cells that colonize the mucus membranes of the small intestine are also predominantly in biofilms, i.e. clusters or microcolonies encased in a matrix (Nielsen *et al.*, 2006). We know very little about the genes and processes that allow *V. cholerae* to form these biofilms *in vivo*, or the role of this process in *V. cholerae* pathogenesis.

Molecular analyses of biofilm formation

Biofilm formation is thought to be a developmental process that begins when a bacterium reaches and attaches to a surface. The process of reaching the surface is influenced by both fluid flow and bacterial motility. Once at the surface, attachment to it depends on the bacterial cell surface properties, defined by the outer membrane proteins, lipopolysaccharides, exopolysaccharides, flagella, and pili (Brading *et al.*, 1995; Danese *et al.*, 2000; Jackson *et al.*, 2004; Ma *et al.*, 2006; O'Toole and Kolter, 1998; Pratt and Kolter, 1998; Sauer and Camper, 2001; Watnick *et al.*, 1999; Watnick and Kolter, 1999). Attachment also depends on the substratum surface properties such as the net surface charge, the critical surface charge, and the surface free energy (Brading *et al.*, 1995; Dalton *et al.*, 1994). After initial attachment, colonization of a surface is mediated by movement and growth of attached bacteria. Surface colonization behaviour leads to formation of microcolonies where cells are held together by extracellular polymeric substances (EPS). Growth of bacteria and continued production of EPS leads to development of mature biofilms. (Danese *et al.*, 2000; Friedman and Kolter, 2004a, 2004b; Jackson *et al.*, 2004; Watnick and Kolter, 1999; Yildiz and Schoolnik, 1999). When these biofilms form on a submerged solid surface they attain elaborate architectures characterized by pillars and channels. Fig. 7.1 illustrates the developmental cycle of biofilm formation on a submerged solid surface.

Before entering into a detailed presentation of the molecular basis of biofilm formation it is important to remind the reader that not all *V. cholerae* are alike and therefore not all *V. cholerae* biofilms are the same. Different strains and even phenotypic variants derived from a single strain can generate profoundly different biofilms (Watnick and Kolter, 1999; Watnick *et al.*, 2001; Yildiz and Schoolnik, 1999). While there are over 200 different serotypes of *V. cholerae* known, only the O1 and O139 serotypes have endemic and epidemic potential (Kaper *et al.*, 1995). Strains of the O1 serotype are classified further as belonging to classical or El Tor biotypes on the basis of their biochemical properties and phage susceptibility. Given their importance in human health, it is not surprising that most of our understanding of *V. cholerae* biofilm biology is derived from studies performed on O1-classical, O1-El Tor and O139 strains.

As is the case for many bacteria, growth of a *V. cholerae* strain can result in populations that contain a genotypically and phenotypically diverse individuals (Morris *et al.*, 1996; Wai *et al.*, 1998; Yildiz *et al.*, 2004). This ability to generate diversity during growth is thought to increase the adaptive capabilities of the organism in rapidly changing environments. One example of such diversity in *V. cholerae* is its ability to switch colonial morphology types between smooth and rugose forms (Fig. 7.2) (Morris *et al.*, 1996; Wai *et al.*, 1998; Yildiz *et al.*, 2004). Certain environmental stresses, such as carbon limitation and

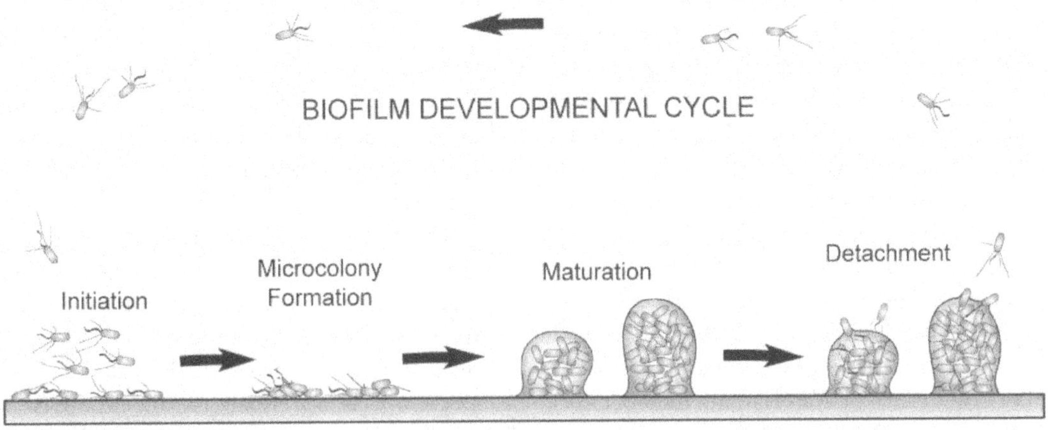

Figure 7.1 A model of biofilm developmental cycle in *V. cholerae*. Biofilm formation begins with transport of the organism to a surface using flagellar motility. Attachment to surfaces requires type IV pili. Attachment is followed by the formation of microcolonies and mature biofilm structures, which require the presence of VPS as well as matrix proteins. The biofilm developmental cycle is completed with detachment of bacteria from mature biofilms that is mediated in part by the VPS lyase.

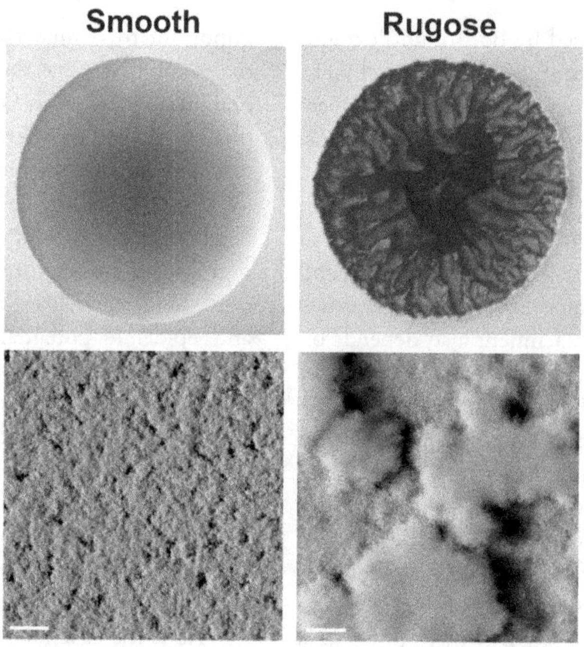

Figure 7.2 Colony morphology and biofilm formation phenotypes of smooth and rugose variants. Colony morphology of the smooth and rugose variants grown on LB agar plates at 30°C for 48 hours (top panels). Biofilm structures of the variants formed after 24 inoculation in once-through flow cell. Images were acquired with CSLM and top-down view of biofilms are shown. Bar = 30 μm (bottom panels).

treatment with bactericidal agents, along with biofilm growth appear to enhance the generation of phenotypic variants. Although both smooth and rugose variants are capable of forming biofilms, the biofilm forming capacity of the rugose variants is superior to that of smooth variants (Fig. 7.2) (Wai *et al.*, 1998; Yildiz and Schoolnik, 1999). Furthermore, the dynamics of biofilm development and the biofilm's architecture differ markedly between the two variants (Yildiz and Schoolnik, 1999). The genetic bases of biofilm formation have been studied using both smooth and rugose variants and we will discuss their differences and similarities in this chapter.

Not only have different strains been used to study *V. cholerae* biofilms, different experimental set-ups have also been used. In many instances, particularly those that require ease of high-throughput in the analysis (e.g. mutant screens), static culture conditions in microtitre dishes have been utilized (Watnick and Kolter, 1999). In this setting, the extent of biofilm formation is most easily followed by staining of the surface-associated biofilm with crystal violet. When more detail of the biofilm architecture is desired, the biofilms that form in static cultures can be analysed using laser scanning confocal microscopy (LSCM) (Watnick and Kolter, 1999; Yildiz and Schoolnik, 1999). A second setting commonly used to analyse biofilms is the 'once-through' flow cell reactor which is thought to simulate riverine and estuarine environments. In these flow cell reactors individual bacteria are allowed to attach to the submerged surface, flow is started and then the growth and maintenance of the biofilm can be followed in real time with LSCM usually utilizing strains that express fluorescent proteins (Yildiz *et al.*, 2004). Results obtained from static and flow culture conditions – their similarities and differences – will therefore also be discussed in this chapter.

The process of biofilm development can be divided into distinct stages: attachment to the surface, colonization of that surface to form a monolayer of cells, and formation of a mature biofilm with its characteristic three dimensional architecture as the result of the synthesis of the extracellular matrix (Brading *et al.*, 1995; Costerton *et al.*, 1995; O'Toole *et al.*, 2000). The following sections review what is known about each of these stages in the process of biofilm formation by *V. cholerae*.

Initiation of biofilm formation: surface attachment and colonization factors

The cellular factors involved in the early steps of biofilm development were discovered by combining genetic screens with microscopic observation (Watnick and Kolter, 1999). In this way it was possible to show that under static culture conditions the mannose-sensitive haemagglutinin type IV pilus (MSHA) and the flagellum facilitated early attachment to abiotic surfaces (Watnick and Kolter, 1999). Flagellar motility was also shown to be required to generate movement towards and along the surface. Under conditions of nutrient flow the MSHA pilus was also shown to be important for attachment because very few cells of an *mshA* mutant attached to the surface (Fig. 7.3). However, the few cells that attach go on to form well-developed biofilms that are even thicker than those formed by the wild-type strain. In contrast, the role of the flagellum in biofilm formation appears to be different under static and flow conditions. Under static conditions, non-flagellated mutants showed extremely delayed kinetics of biofilm formation (Watnick and Kolter, 1999) while under flow conditions they formed biofilms that were more developed than those formed by a wild-type strain (Fig. 7.3) (Bilecen and Yildiz, unpublished).

The factors involved in the attachment and colonization of biotic surfaces by *V. cholerae* have also been extensively investigated. Under laboratory conditions, the MSHA pili contribute to attachment to chitin beads and the chitinaceous exoskeleton of the crustacean zooplankton *Daphnia pulex* (Chiavelli *et al.*, 2001; Meibom *et al.*, 2004). The analyses of the roles of pili in attachment, however, are complicated by the fact that *V. cholerae* is capable of producing three different type IV pili (Fullner and Mekalanos, 1999; Jonson *et al.*, 1991; Taylor *et al.*, 1987). In addition to MSHA, there are two other type IV pili produced under different conditions: the toxin co-regulated pili (TCP) (Taylor *et al.*, 1987) and the chitin regulated pilus (ChiRP) (Meibom *et al.*, 2004). In addition to MSHA, TCP and ChiRP have also been shown to contribute to chitin colonization (Chiavelli *et al.*, 2001; Meibom *et al.*, 2004; Reguera and Kolter, 2005). The colonization of chitin is considered ecologically significant as the capacity of *V. chol-*

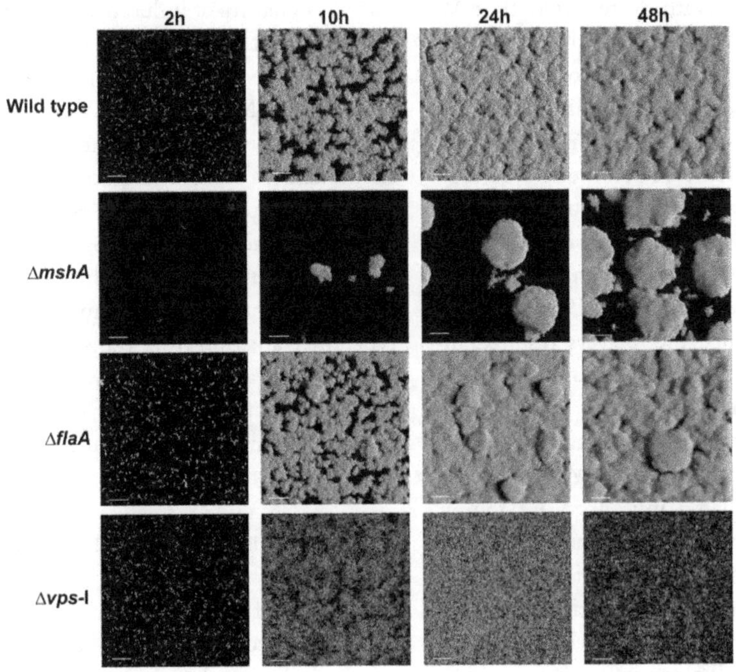

Figure 7.3 Comparison of biofilm formation by wild type and mutants unable to produce MSHA pili, flagellum, and VPS. Three dimensional biofilm structures of wild type, Δ*mshA*, Δ*flaA*, Δ*vps*-I cluster mutants formed after 2, 10, 24 and 48 h post-inoculation in once-through flow cell. Images were acquired with CSLM and top-down views of biofilms are shown. Bar= 30μm.

erae to colonize chitin and utilize it as a carbon and nitrogen source contribute to the environmental survival of the pathogen (Meibom *et al.*, 2004; Nalin *et al.*, 1979). It should be noted that the contribution of both the MSHA and the flagellum to biofilm formation in both abiotic and biotic surfaces appears to vary from strain to strain. For example, lack of MSHA leads to a reduction of colonization of *V. cholerae* O1 El Tor on crustacean zooplankton, while it completely eliminates such colonization if the parent strain is *V. cholerae* O139 (Chiavelli *et al.*, 2001). In contrast, adherence of *V. cholerae* O1 classical is unaffected by a mutation in *mshA*, which is to be expected because classical biotype strains do not express the MSHA pilus, at least under the conditions investigated.

The genetic requirements for surface colonization during biofilm development have also been identified and characterized. Recent studies have shown that, after individual cells attach, formation of a monolayer of cells is a distinct stage in biofilm development (Moorthy and Watnick, 2004, 2005). Growth of *V. cholerae* in a minimal medium that lacks monosaccharides arrests biofilm development at the monolayer stage. Moorthy and Watnick reported that *V. cholerae* O139 required MSHA for monolayer formation while the contribution of the flagellum and the *V. cholerae* polysaccharide (VPS) during this stage was negligible. Consistent with the requirement of MSHA for monolayer formation, supplementation of the biofilm growth medium with alpha-methylmannoside (AMM), a non-metabolizable analogue of mannose, interfered with monolayer formation. But, addition of AMM to preformed monolayers resulted in detachment of only a subset of monolayer bacteria, suggesting that monolayers contain both AMM-sensitive and AMM-resistant populations (Moorthy and Watnick, 2004). As monolayers of non-flagellated mutants were found to be AMM-resistant, the investigators propose that AMM-resistant population of monolayers of wild type may not have an active flagellum and thus may be unable to break their association with the surface. This suggests that in order for a cell to become permanently attached to a surface it may have to lose flagellar motility. Interestingly, transcription of the gene encoding the flagellin subunit, *flaA*, was found to be significantly decreased in monolayers compared to the cells in the planktonic state (Moorthy and Watnick, 2004). In addition, *flaA* transcription was also reduced in planktonic cells of an *mshA* mutant (Moorthy and Watnick, 2004). Taken together these observations led to the hypothesis that MSHA, both by acting as an adhesin and by somehow down-regulating flagellar motility, is required for surface attachment and monolayer formation. However, the mechanism by which MSHA regulates *flaA* transcription and whether the interaction of MSHA with a surface leads to repression of *flaA* transcription is yet to be determined.

There is growing evidence that cells undergo large physiological changes as they transit from the planktonic state to being members of a monolayer. Most of these insights come from whole genome transcriptional profiling studies that have been performed in minimal medium with and without mannose (Moorthy and Watnick, 2005). The results obtained suggest that the transcriptome of each population is distinct, furthering the notion that biofilm formation is a developmental process and that each step has its own signature of induced and repressed genes. Expression of a set of genes coding for methyl-accepting chemotaxis proteins was induced in monolayers (Moorthy and Watnick, 2005). To more directly investigate the possible role of chemotaxis in biofilm formation, the capacity of a mutant lacking *cheY*-3, the regulator of flagellum-based chemotaxis, to form monolayer and biofilm was investigated. While biofilms of the *cheY*-3 mutant were similar to those of the wild type, the transition from transient to permanent attachment was found to be retarded, suggesting that the role for chemotaxis may be limited to the stage of monolayer formation (Moorthy and Watnick, 2005).

Biofilm maturation: the extracellular matrix

Production of mature biofilms requires extracellular matrix components (the so-called extracellular polymeric substances or EPS) which hold the cells together and the biofilm attached to the surface. A major component of the *V. cholerae* biofilm matrix is the VPS exopolysaccharide

(Wai *et al.*, 1998; Watnick and Kolter, 1999; Yildiz and Schoolnik, 1999). VPS production is essential for development of three dimensional biofilm structures under both static and flow conditions (Wai *et al.*, 1998; Watnick and Kolter, 1999; Yildiz and Schoolnik, 1999). VPS was first isolated from rugose strains of *V. cholerae* that produce in much greater quantities. In fact, the main reason for the corrugated appearance of rugose colonies is this increase in the levels of VPS produced (Wai *et al.*, 1998; Yildiz and Schoolnik, 1999). The genes required for VPS synthesis, designated as *vps* for *Vibrio* polysaccharide, were identified through a transposon mutagenesis using rugose to smooth conversion as a screen (Yildiz and Schoolnik, 1999). The *vps* genes are clustered in two regions, *vpsU-vpsA-K*, VC0916–27 (*vps*-I cluster) and *vpsL-Q*, VC0934–9 (*vps*-II cluster) on the large chromosome of *V. cholerae*.

Although the chemical structure of VPS is yet to be determined, initial characterization of VPS isolated from different strains revealed that the glycosyl composition of the VPS varies among different isolates (Wai *et al.*, 1998; Yildiz and Schoolnik, 1999). For example, the VPS produced by the *V. cholerae* O1 El Tor strain A1552 was found to contain nearly equal amounts of glucose and galactose, with smaller amounts of *N*-acetylglucosamine and mannose (Yildiz and Schoolnik, 1999). The chemical linkage analysis of the VPS produced by this strain revealed that it contains nearly equal amounts of 4-linked galactose and 4-linked glucose and these therefore may comprise the backbone of the polysaccharide. Branching is indicated by the presence of 3,4- and 4,6-linked galactose and glucose and by 2,4-linked galactose. VPS composition analysis of another *V. cholerae* O1 El Tor strain TSI-4 revealed that it lacked glucose and contained *N*-acetylglucosamine, D-mannose, 6-deoxy-D-galactos, and D-galactose (Wai *et al.*, 1998). Whether these differences in glycosyl composition of the VPS have any impact on biofilm structure and function is not known. In addition to the *vps* genes *galU*, encoding for UDP-glucose pyrophosphorylase – responsible for the production of UDP-glucose, and *galE*, which encodes for UDP-galactose epimerase that catalyses the production of UDP-galactose, are also required for synthesis of VPS and in turn biofilm formation (Nesper *et al.*, 2001).

Protein components of the extracellular matrix of *V. cholerae* biofilms have also been recently identified. The first one of these proteins, termed RbmA (rugosity and biofilm structure modulator), is a secreted protein which does not have any homologues in the protein sequence databases (Fong *et al.*, 2006). The *rbmA* mutant forms less differentiated biofilms than those of the wild-type rugose variant. Microscopic analyses have shown that *rbmA* mutant biofilms are less organized and more disperse, and exhibit an increase in average thickness, compared to the wild type. Structure prediction of RbmA using different programs suggests that RbmA may fold into either the FnIIID or beta-trefoil tertiary structures (Fong *et al.*, 2006). Other proteins that contain these folds include proteins capable of binding carbohydrates, suggesting a similar role for RbmA in biofilms. RbmA is required for the maintenance of rugose colonial morphology and for the development of wild-type biofilm architecture. Since both of these processes involve the VPS, one possibility is that RbmA may be a VPS-binding protein. It is also possible that RbmA acts like an agglutinin or adhesin, binding cells together or anchoring the biofilm to a surface that contains carbohydrates.

RbmC and Bap1 (biofilm-associated protein), which exhibit 46.7% sequence similarity to each other, are two additional proteins found the biofilm matrix (Fong and Yildiz, 2007). Mutations in *bap1* in either *V. cholerae* classical strain O395 (Hung *et al.*, 2006) or *V. cholerae* O139 (Moorthy and Watnick, 2005) lead to defects in biofilm formation. It was also reported that a transposon insertion into *rbmC* (VC0930) in the chromosome of a *V. cholerae* O1 El Tor C6706 rugose variant, converts it to a smooth morphology (Ali *et al.*, 2000). In *V. cholerae* O1 El Tor strain A1552, colony morphology of *rbmC* mutant is not different from that of the rugose variant parent, while *bap1* and a double deletion Δ*rbmC* Δ*bap1* form colonies that are flat and less wrinkled compared to the rugose parent (Fong and Yildiz, 2007). Similarly, biofilm structures of single *rbmC* and *bap1* deletions in the rugose genetic background were found to be similar to those of rugose variant. In contrast, the *rbmC bap1* double mutant formed a thin biofilm, which detached after 24 hours of development (Fig. 7.4). The total biomass, biofilm thickness and substrate coverage of the double mutant were found to be dramatically lower 24 hours post inoculation than those of the rugose parent. Analysis of floating biofilms (pellicles) formed by these mutants further confirmed the role of RbmC and Bap1 in biofilm formation. The pellicle formed by Δ*rbmC* mutant was not significantly different from that formed by the rugose parent while Δ*bap1* mutant formed a pellicle with a slightly altered structure. In contrast, the double mutant Δ*rbmC* Δ*bap1* was incapable of forming pellicles. The defects in pellicle and solid surface biofilm formation of the *rbmC bap1* double mutant could be complemented by introducing the wild-type copy of either *rbmC* or *bap1* on a multicopy plasmid, suggesting that RbmC and Bap1 may have a similar functions.

Computational analyses revealed that both RbmC and Bap1 contain FG-GAP and carbohydrate-binding domains, including three putative *N*-acetylglucosamine binding sites (Fong and Yildiz, 2007). The InterPro database indicates that both RbmC and Bap1 contain one putative calcium binding site, two and one FG-GAP domains respectively, and one galactose mutarotase-like domain. The FG-GAP domains are found in the extracellular N-terminal region of the integrin α chain which has been reported to be involved in ligand recognition and binding to extracellular matrix proteins of eukaryotic cells. The galactose mutarotase-like domains, on the other hand, are described in the Structural Classification of Proteins (SCOP) database as probable carbohydrate-binding domains in enzymes acting on sugars. Three putative *N*-acetylglucosamine binding sites were also predicted in both RbmC and Bap1 by ModBase software. All of these computational predictions are consistent with the possible roles of RbmC and Bap1 in maintaining rugosity of colony morphology and stabilizing pellicle and biofilm structures.

Cellular detachment from biofilms: VPS lyase

One of the least studied aspects of biofilm biology is cell detachment. Since the matrix is required to keep cells associated to the biofilm, any process or factor that is able to modify or degrade matrix components can lead to cells detaching from the biofilm. Enzymes capable of degrading exopolysaccharides, or altering its chemical structure, should in principle be able to modulate biofilm detachment. Similarly, proteases capable of degrading matrix proteins should be able to alter the structural integrity of biofilms and thus facilitate detachment. In the particular case of *V. cholerae*, recent studies suggest that RbmB, a protein with sequence homology to polysaccharide hydrolases, could be a VPS lyase (Fong and Yildiz, 2007). In a rugose background, an *rbmB* mutant accumulates more VPS and exhibits an increase in colony corrugation and subtle structural changes in biofilm architecture. Overexpression of *rbmB* from an arabinose inducible promoter results in a decrease in colony corrugation and less compact biofilms under flow conditions. Two putative *N*-acetylglucosamine and seven putative mannose binding sites are predicted to be present in

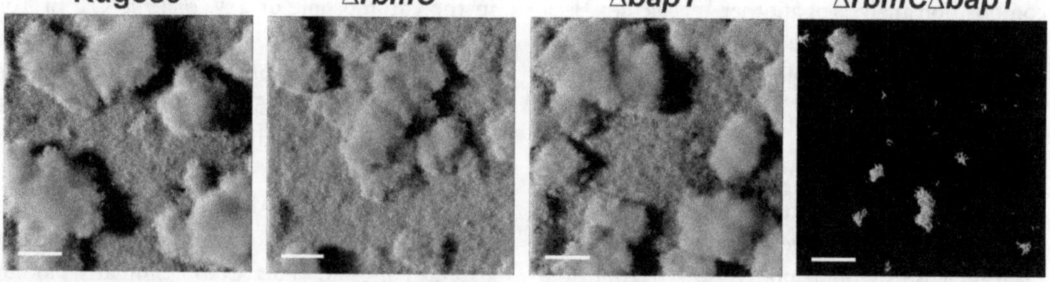

Figure 7.4 Comparison of biofilm formation by rugose variant and mutants unable to produce matrix proteins RbmC and Bap1. Three dimensional biofilm structures of rugose, Δ*rbmC*, Δ*bap1*, Δ*rbmC* Δ*bap1* formed after 24 h post inoculation in one-through flow cell. Images were acquired with CSLM and top-down views of biofilms are shown. Bar = 40 μm.

RbmB. These predictions are consistent with the possible role of RbmB as a polysaccharide lyase that breaks down VPS.

Regulation of biofilm formation

Regulation of biofilm formation in *V. cholerae* appears to be effected at multiple levels and involves numerous positive and negative transcriptional regulators as well as alternative sigma factors. In the following sections we summarize what is currently known about this complex regulatory network.

Positive regulation

A positive transcriptional regulator of the *vps* genes, denoted as VpsR (*Vibrio* polysaccharide regulator) is required for VPS synthesis and appears to be the central regulator of biofilm formation in *V. cholerae* (Yildiz *et al.*, 2001). VpsR is a member of the NtrC subclass of response regulators of the two-component signal transduction systems. While disruption of *vpsR* prevents expression of *vpsA* and *vpsL*, it is yet to be determined if VpsR directly binds to the regulatory regions of these genes. In these two-component systems response regulators possesses an N-terminal domain that can be phosphorylated by a cognate sensor histidine kinase. While the conserved aspartate residue of VpsR (Asp-59) needs to be phosphorylated for its function, the kinase responsible has not been identified (Lauriano *et al.*, 2004).

Members of the NtrC subclass of response regulators act in concert with the alternative sigma factor RpoN. This suggests that transcription of the *vps* genes might require both VpsR and RpoN. Whole genome transcriptional profiling of *rpoN* mutant and wild-type strains indicate that RpoN is indeed involved in the expression of the *vps* genes (Yildiz *et al.*, 2004). As expected, *rpoN* mutant biofilms are less developed, compared to those formed by the wild type.

Another response regulator involved in *V. cholerae* biofilm formation is VpsT (Casper-Lindley and Yildiz, 2004). Whole-genome expression profiling of exponentially grown smooth and rugose variants and mutational analyses of regulatory genes that are differentially expressed between the two variants led to the identification of VpsT (Yildiz *et al.*, 2004). This protein is similar to proteins that belong to the UhpA (FixJ) family of transcriptional response regulators. VpsT is homologous to the transcriptional regulators CsgD and AgfD from *Escherichia coli* and *Salmonella enterica*, respectively (Casper-Lindley and Yildiz, 2004). CsgD and AgfD are required for the production of extracellular matrix components, cellulose and curli fimbriae (Romling *et al.*, 1998, 2000; Uhlich *et al.*, 2001). As a consequence, *E. coli* and *S. enterica* mutants lacking these regulators produce flat colonies in contrast to the characteristically wrinkled colonies of wild-type strains from both species. Consistent with the notion that VpsT is functionally homologous to CsgD and AgfD, mutants lacking VpsT yield smooth colonies (Casper-Lindley and Yildiz, 2004).

VpsT and VpsR are required for *vps* gene expression and, in turn, for formation of rugose colonies and biofilms (Beyhan *et al.*, 2007; Casper-Lindley and Yildiz, 2004; Yildiz *et al.*, 2001). The processes that are controlled by each of the positive regulators were determined by comparing whole genome expression profiles of *vpsR* and *vpsT* mutants to those of the rugose parent during exponential growth (Beyhan *et al.*, 2007). This study showed that the VpsR and VpsT regulons are largely identical. However, the magnitude of induction or repression of the genes regulated by them was, for the most part, greater for the *vpsR* mutant than for the *vpsT* mutant. This study also showed that the overall expression profile of the *vpsT vpsR* double mutant is most similar to that of the *vpsR* single mutant suggesting that *vpsR* is epistatic to *vpsT*. Whether the actions of VpsT and VpsR are indeed mediated by the direct binding to the *vpsL* and *vpsA* promoter regions or indirectly through other regulatory proteins remains unknown. Adding to the complexity of the regulation of *vps* gene expression, VpsR and VpsT positively regulate their own and each other's expression (Casper-Lindley and Yildiz, 2004).

VpsT and VpsR have different roles in determining biofilm architecture (Beyhan *et al.*, 2007). The biofilm structures formed by a *vpsT* mutant in a flow cell are markedly different from those formed by its rugose parent. But the *vpsT* mutant is at least still able to from biofilms. In contrast, the *vpsR* and *vpsT vpsR* mutants produce only attached single cells or small microcolonies attached to the substratum. Under static conditions, the rugose variant forms well-developed biofilms. In contrast, the *vpsT* mutant yields mostly attached single cells and microcolonies, while the *vpsR* and *vpsTvpsR* mutants attach to surfaces mainly as single cells, not even forming microcolonies. Taken together the results indicate that in the once-through flow cell system, where fresh nutrients are continuously provided and waste materials and extracellular signalling molecules are removed from the system, the structure of *vpsT* biofilms differ greatly from the ones formed under static conditions. In contrast, both *vpsR* and *vpsT vpsR* are unable to form biofilms under either condition.

Negative regulation

Population density-dependent regulatory systems, known as quorum sensing systems, negatively regulate biofilm formation in *V. cholerae* (Hammer and Bassler, 2003; Yildiz *et al.*, 2004; Zhu and Mekalanos, 2003). In quorum sensing, bacteria produce secreted signalling molecules termed autoinducers (AIs) that accumulate in the medium in proportion to population density. When the AIs reach a critical concentration they are detected by the producing cells, resulting in extensive changes in gene expression and, in turn, the physiology and behaviour of the organism can be dramatically altered (Miller and Bassler, 2001). *V. cholerae* produces two AIs, known as CAI-1 and AI-2 (Miller *et al.*, 2002; Zhu *et al.*, 2002). These are produced by the CqsA and LuxS autoinducer synthases, respectively. CAI-1 and AI-2 are detected via CqsS and LuxQ/P sensors, respectively. Information from the sensors is transduced through a phosphorelay, first to the LuxU phosphotransfer protein and then to LuxO which is an *RpoN*-dependent response regulator. LuxO activity is defined by its phosphorylation state. At low cell density, when the concentrations of AIs are low, LuxO is phosphorylated and represses expression of the major quorum sensing regulator, HapR. It does this indirectly by activating the expression of a set of small regulatory RNAs (Lenz *et al.*, 2004). By contrast, at high cell density, LuxO is de-phosphorylated and expression of HapR is increased.

Jobling and Holmes showed that *hapR* mutants of *V. cholerae* form rugose colonies (Jobling and Holmes, 1997). Subsequently, it was shown that expression of the *vps* genes is increased in *hapR* mutants and that *hapR* mutants form biofilms of increased thickness (Hammer and Bassler, 2003; Yildiz *et al.*, 2004; Zhu and Mekalanos, 2003). HapR could affect *vps* gene expression directly, by binding to the upstream regulatory regions of the *vps* structural genes, or indirectly by regulating either expression or activation of the *vps* regulatory proteins. In contrast, *luxO* mutants are impaired in biofilm formation as they are 'locked' in a regulatory state mimicking high cell density (Hammer and Bassler, 2003). Zhu and Mekalanos have shown that deletion of *luxS* does not affect biofilm formation while a *cqsA* mutant forms thicker biofilms (Zhu and Mekalanos, 2003). This suggests that CAI-1 signals are critical for repression of biofilm formation via the quorum sensing regulatory circuitry. Interestingly, the concentration of CAI-1 was found to be higher in biofilms than in planktonic cultures (Zhu and Mekalanos, 2003). In addition, the quorum sensing system was activated faster in biofilm bacteria when analysed in a static experimental set-up. It was further shown that biofilm thickness is modulated by the timing of quorum sensing induction. The investigators suggest that quorum sensing ensures development of 'normal' biofilm structures that permit rapid dispersion of bacteria from the biofilm, thereby facilitating transmission. It has also been shown that the VarA/VarS two component regulatory system acts parallel with of the quorum sensing regulatory circuit to relay information to LuxO which controls *hapR* transcription and therefore negatively regulates *vps* gene expression (Lenz *et al.*, 2005). It was recently shown that VpsT and VpsR, besides being positive transcriptional activators of *vps* genes, can also contribute to transcription of *vps* genes by causing a reduction in *hapR* message levels. It is yet to be determined whether VpsT and VpsR modulate *hapR* transcription through LuxO or by another pathway (Beyhan *et al.*, 2007).

Another global regulator that affects biofilm formation negatively is the alternative sigma factor RpoS (Yildiz *et al.*, 2004). RpoS facilitates adaptation of bacteria to numerous stresses by controlling the expression of many genes during both the exponential and stationary phases of growth (Hengge-Aronis, 2002). Expression of the genes of the *vps* cluster is increased in an *rpoS* mutant relative to the wild type (Nielsen *et al.*, 2006; Yildiz *et al.*, 2004). As predicted, the biofilm forming capacity of the *rpoS* mutant was found to be 3.6-fold greater than that of its smooth parent under static conditions (Yildiz *et al.*, 2004). However, when biofilms of the *rpoS* mutant formed under flow conditions were compared to those of the smooth parent it was observed that the biofilm structures of the two strains were similar, though not identical (Yildiz *et al.*, 2004). Interestingly, *rpoS* is required for transcription of *hapR* (Yildiz *et al.*, 2004). Thus, RpoS can negatively regulate biofilm formation by being a positive regulator of *hapR* expression (which in turn negatively regulates *vps* expression) or by other mechanisms that affect the positive transcriptional regulators VpsR and VpsT. Indeed, it has been shown that RpoS negatively regulates expression of VpsT (Yildiz *et al.*, 2004).

In a screen to identify increased biofilm forming mutants, Haugo and Watnick discovered *cytR* and showed that smooth variants lacking *cytR* form rugose colonies (Haugo and Watnick, 2002). They further showed that in *V. cholerae* O139, CytR is not only a regulator of nucleoside uptake and catabolism (where it functions as a repressor of these functions under nucleoside limited conditions) but it also negatively regulates the expression of the *vpsL* gene. In contrast *cytR* mutant of *V. cholerae* O1 El Tor A1552 do not develop rugose colonies (Yildiz, unpublished).

At present it is not known if the negative regulators of biofilm formation described in the foregoing paragraphs function independently or whether they form a hierarchical network whose end effect is the regulation of *vps* gene expression. A model of regulation of biofilm formation is shown in Fig. 7.5.

c-diGMP signalling

Cyclic dimeric guanosine monophosphate (c-diGMP) has emerged as a ubiquitous second messenger in bacteria that controls the transi-

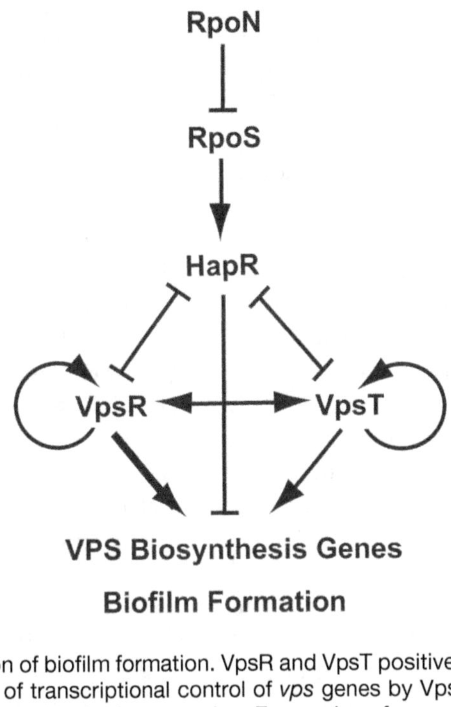

Figure 7.5 A Model of regulation of biofilm formation. VpsR and VpsT positively control transcription of *vps* genes; however, the magnitude of transcriptional control of *vps* genes by VpsR is greater. VpsR and VpsT positively regulate their own and each other's expression. Expression of *vps* genes and the regulators VpsR and VpsT are negatively controlled by HapR. VpsT and VpsR negatively regulate either the transcription or the message abundance of *hapR*. The expression of *vps* genes is positively controlled by the alternative sigma factor RpoN and negatively regulated by the alternative sigma factor RpoS. It is yet to be determined if the interactions discussed above are direct or indirect.

tion from the free-living, motile lifestyle to the biofilm lifestyle (Jenal, 2004; Romling *et al.*, 2005; Romling and Amikam, 2006). c-diGMP production and degradation is controlled by proteins containing the GGDEF and EAL domains, respectively. The GGDEF and EAL domains, named for a short segment of their conserved amino acid sequence, were recently shown to harbour intrinsic diguanyl cyclase (DGC) and c-diGMP phosphodiesterase (PDE) activity, respectively (Ryjenkov *et al.*, 2005; Schmidt *et al.*, 2005). Recent studies have also shown that proteins containing an HD-GYP domain also have c-diGMP phosphodiesterase (PDE) activity (Ryan *et al.*, 2006).

Microbial genome sequencing projects have revealed that the GGDEF, EAL and HD-GYP proteins are nearly ubiquitous in bacteria (Galperin *et al.*, 2001). These enzymatic domains are linked to various sensory domains, suggesting that cells make a significant effort to monitor various environmental and intracellular signals in order to adjust their c-diGMP levels. *V. cholerae* has 62 genes predicted to encode GGDEF/EAL/HD-GYP domain proteins in its genome (Heidelberg *et al.*, 2000).

c-diGMP was initially identified in *Gluconacetobacter xylinus* as an allosteric activator of cellulose synthase (Ross *et al.*, 1990; Ross *et al.*, 1991). The physiological role of c-diGMP, beyond its known effect on cellulose synthetic activity in G. *xylinus*, has started to emerge from the observation of phenotypes of mutants unable to produce proteins with GGDEF and EAL domains. These mutants generally show defects in biofilm formation, motility, cell differentiation and virulence. Recently, genome-wide mutational analysis of all genes encoding proteins with GGDEF and EAL domains were carried out in *Salmonella enterica* serovar Typhimurium and *Pseudomonas aeruginosa* (Garcia *et al.*, 2004; Kulasakara *et al.*, 2006) From these studies, the following simplified scheme emerges: increased c-diGMP levels (achieved by the inactivation of selected PDEs or overexpression of selected DGCs) inhibit flagellar motility and type IV pili-dependent twitching motility, and/or stimulate synthesis of exopolysaccharide that constitute biofilm matrices. Therefore, in general terms, c-diGMP appears to regulate bacterial transition from the planktonic, single-cellular lifestyle to the multicellular (biofilm) lifestyle.

GGDEF and EAL domains play important roles in biofilm formation in *V. cholerae*. This was first indicated by Bomchil *et al.* in a study showing that mutants of *mbaA*, a GGDEF-EAL module encoding gene, had increased capacity to produce extracellular matrix and consequently formed increased biofilms that lacked the typical biofilm architecture (Bomchil *et al.*, 2003). Subsequently, VieA, a response regulator harbouring an EAL module, was shown to repress transcription of *vps* genes in *V. cholerae* O1 classical biotype in defined minimal medium supplemented with amino acids (Tischler and Camilli, 2004). The EAL module of VieA is capable of hydrolyzing c-diGMP and in the *vieA* mutant the levels of c-diGMP are greatly increased compared to the wild type. Increased c-diGMP levels activate expression of the *vps* genes through a mechanism yet to be identified. Interestingly, VieA does not play a role in *vps* gene regulation in the El Tor biotype (Beyhan *et al.*, 2006). *RocS* (regulation of cell signalling), which was the first identified regulator of the smooth to rugose switch and which encodes a protein containing both GGDEF and EAL domains, also regulates VPS production, motility, and biofilm formation in *V. cholerae* (Lim *et al.*, 2006; Rashid *et al.*, 2003).

A whole genome expression profiling study designed to gain insights into responses of *V. cholerae* to the increased intracellular c-diGMP level revealed that the genes regulated by c-diGMP are likely to participate in regulating diverse cellular functions (Beyhan *et al.*, 2006). For example, transcription of the *vps* and *msh* genes were found to be increased while expression of the genes required for flagellar biogenesis was decreased, further confirming the notion that in *V. cholerae* c-diGMPs signalling regulates bacterial transition from a planktonic state to a biofilm state.

Whole genome expression profile comparison of smooth and rugose variants of *V. cholerae* O1 El Tor A1552 revealed that, during exponential growth expression of five genes encoding proteins with GGDEF and GGDEF plus EAL modules are differentially expressed between the smooth and rugose variants (Yildiz *et al.*, 2004). These genes were designated *cdgA-E* for cyclic-di-guanylate (Lim *et al.*, 2006). Of these five genes *cdgA*, encoding a single GGDEF module, positively regulates *vps* transcription. In contrast, *cdgC*, which encodes both a GGDEF and a EAL module, negatively regulates *vps* transcription. Deletion of *cdgD* causes an increase in motility and alters biofilm development. The *cdgD* mutant exhibited a decrease in the surface coverage during the early stages of biofilm formation suggesting that CdgG may play its key a role during the initial stages of biofilm formation.

More recent studies have shown that a novel protein domain, designated PilZ, is involved in c-diGMP binding (Amikam and Galperin, 2006; Ryjenkov *et al.*, 2006). *V. cholerae* contains four predicted PilZ domain proteins: VCA0042, VCA0735, VC0697 and VC2344. It was recently reported that several of these proteins can bind c-diGMP (VC2344/plzC and (VCA0042/plzD) while others were unable to bind under the conditions tested (Pratt *et al.*, 2007). VC2344/PlzC and another extremely distant PilZ homolog (does not contain the conserved residues), VC1885/PlzB, are positive regulators of motility and biofilm formation, while PlzD negatively regulates motility (Pratt *et al.*, 2007). This study showed that PilZ domain proteins can function as c-diGMP receptors and regulate c-diGMP dependent process in *V. cholerae*.

At present, we have a limited understanding of the functions of GGDEF and EAL proteins in *V. cholerae* and in other bacteria, as only a few of them have been studied in any detail. Many questions remain unanswered: How is the transcription of GGDEF/EAL encoding genes regulated? How are the activities of GGDEF/EAL proteins regulated? How are the cellular levels c-diGMP altered? Is there a single concentration of intracellular c-diGMP that is critical or is there compartmentalization of the second messenger such that local concentrations can be sensed? How PilZ domains regulate c-diGMP dependent process and whether there are other types of c-diGMP binding proteins?

Environmental signals

Biofilm formation, as part of the overall life cycle of *V. cholerae*, must be responsive to changes in the environment if it is to provide the organism with a selective advantage. Natural aquatic habitats are constantly subject to changes in many physicochemical parameters, including nutrient availability, salinity, and the sheer forces generated by flowing water. For *V. cholerae* to be adequately responsive, flexible, and robust, some environmental signals should trigger biofilm formation, whereas others should inhibit the process and initiate biofilm dissolution.

We know very little about how the physical and chemical factors that are likely to be encountered by *V. cholerae* in natural settings might affect biofilm development. *V. cholerae* grown under different conditions is likely to have different cell surface properties which may lead the organism to use different biofilm development pathways. Analyses of biofilm formation by *V. cholerae* grown under conditions that simulate its natural habitat (fresh, estuarine, and marine waters) have shown that it has the capacity to form biofilms through both VPS-dependent and VPS-independent pathways (Kierek and Watnick, 2003a,b). The VPS-independent pathway is preferred in a seawater environment, and requires Ca^{2+} and the O-chain of the outer membrane lipopolysaccharide (LPS) (Kierek and Watnick, 2003). In VPS-independent biofilms, intercellular interactions are thought to be mediated by Ca^{2+} via bridging of the negative charges present on the O-chains of the LPS on the surfaces of the cells. VPS-dependent biofilms form in growth medium supplemented with monosaccharides and, as expected, those monosaccharides activate *vps* gene transcription. Taken together, these studies have shown that the signals and regulatory networks that are necessary for biofilm formation by *V. cholerae* grown under conditions that simulate its natural ecosystems are different to those observed in standard laboratory media commonly used for the growth of this bacterium. It has been suggested that colonization of surfaces that are composed of polymers of monosaccharides, such as those produced by phytoplankton and chitin surfaces of zooplankton, and subsequent degradation of these surfaces by hydrolytic enzymes, may trigger biofilm formation. Interestingly the polysaccharide composition of the biofilm matrix formed by cells grown in rich nutrient broth and in artificial seawater was found to be similar (Kierek and Watnick, 2003), suggesting that a novel exopolysaccharide is not produced in biofilms of cells grown in artificial seawater.

During its life cycle *V. cholerae* experiences fluctuations in salinity and osmolarity (Colwell and Huq, 1994; Huq *et al.*, 2005; Lipp *et al.*, 2002). Studies on the influence of osmoadaptation on biofilm formation in *V. cholerae* have shown that addition of glycine betaine, a compatible solute that facilitates osmoadaptation, to a growth medium of high osmolarity increases biofilm formation by activating *vpsL* gene expression (Kapfhammer *et al.*, 2005). This activation depends on the glycine betaine transporter OpuD. Because *V. cholerae* is unable to synthesize glycine betaine it is possible that this compound serves as an environmental signal that induces biofilm formation in *V. cholerae*.

Perspectives

There is still much to be learned regarding the genes and processes that allow *V. cholerae* to form biofilms *in vivo* and the role of this process in *V. cholerae* pathogenesis. In nature, *V. cholerae* is found both in a free-living state and attached to surfaces. However, we know practically nothing about the differences in disease progression and outcome that result from ingestion of the biofilm versus planktonic forms of *V. cholerae*. For example, are the biofilm bacteria dispersed during infection? How would the disease progress if the biofilm-associated bacteria were unable to disperse due to the lack of VPS lyase? Does VPS lyase have any role in *in vivo* biofilm dynamics? What are the roles, if any, of VPS and matrix proteins in *in vivo* biofilm formation and pathogenesis? Answers to these questions may allow us to define targets to combat *V. cholerae* in specific aquatic ecosystems and to develop tools to predict and/or control cholera epidemics and may facilitate development of novel treatment and prevention strategies against cholera.

Acknowledgements

Work in our laboratories are supported by grants from NIH (AI055987) to F.H.Y. and NIH (GM582132) to RK. The authors thank Ates Gurcan, Sinem Beyhan and Kivanc Bilecen for providing the figures.

References

Ali, A., Mahmud, Z.H., Morris, J.G., Sozhamannan, Jr., S. and Johnson, J.A. (2000). Sequence analysis of Tn*phoA* insertion sites in *Vibrio cholerae* mutants defective in rugose polysaccharide production. Infect. Immun. *68*, 6857–6864.

Amikam, D. and Galperin, M.Y. (2006). PilZ domain is part of the bacterial c-di-GMP binding protein. Bioinformatics *22*, 3–6.

Beyhan, S., Tischler, A.D., Camilli, A. and Yildiz, F.H. (2006). Transcriptome and phenotypic responses of *Vibrio cholerae* to increased cyclic di-GMP level. J. Bacteriol. *188*, 3600–3613.

Beyhan, S., Tischler, A.D., Camilli, A. and Yildiz, F.H. (2006). Differences in gene expression between the classical and El Tor biotypes of *Vibrio cholerae* O1. Infect. Immun. *74*, 3633–3642.

Beyhan, S., Bilecen, K., Salama, S.R., Casper-Lindley, C. and Yildiz, F.H. (2007). Regulation of rugosity and biofilm formation in *Vibrio cholerae*: comparison of VpsT and VpsR regulons and epistasis analysis of *vpsT*, *vpsR*, and *hapR*. J. Bacteriol. *189*, 388–402.

Bomchil, N., Watnick, P. and Kolter, R. (2003). Identification and characterization of a *Vibrio cholerae* gene, mbaA, involved in maintenance of biofilm architecture. J. Bacteriol. *185*, 1384–1390.

Brading, M.G., Jass, J. and Lappin-Scott, H.M. (1995). Dynamics of bacterial biofilm formation. In H. M. Lappin-Scott and J. W. Costerton (ed.), Microbial Biofilms, Cambridge University Press, Cambridge, pp. 47–63.

Broza, M., Gancz, H., Halpern, M. and Kashi, Y. (2005). Adult non-biting midges: possible windborne carriers of *Vibrio cholerae* non-O1 non-O139. Environ. Microbiol. *7*, 576–585.

Casper-Lindley, C. and Yildiz, F.H. (2004). VpsT is a transcriptional regulator required for expression of *vps* biosynthesis genes and the development of rugose colonial morphology in *Vibrio cholerae* O1 El Tor. J. Bacteriol. *186*, 1574–1578.

Chiavelli, D.A., Marsh, J.W. and Taylor, R.K. (2001). The mannose-sensitive hemagglutinin of *Vibrio cholerae* promotes adherence to zooplankton. Appl Environ Microbiol. *67*, 3220–3225.

Colwell, R.R. and Huq, A. (1994). Environmental reservoir of *Vibrio cholerae*. The causative agent of cholera. Ann. NY Acad. Sci. *740*, 44–54.

Colwell, R.R., Huq, A., Islam, M.S., Aziz, K.M., Yunus, M., Khan, N.H., Mahmud, A., Sack, R.B., Nair, G.B., Chakraborty, J., Sack, D.A. and Russek-Cohen, E. (2003). Reduction of cholera in Bangladeshi villages by simple filtration. Proc. Natl. Acad. Sci. USA *100*, 1051–1055.

Costerton, J.W., Lewandowski, Z., Caldwell, D.E., Korber, D.R. and Lappin-Scott, H.M. (1995). Microbial biofilms. Annu. Rev. Microbiol. *49*, 711–745.

Dalton, H.M., Poulsen, L.K., Halasz, P., Angles, M.L., Goodman, A.E. and Marshall, K.C. (1994). Substratum-induced morphological changes in a marine bacterium and their relevance to biofilm structure. J. Bacteriol. *176*, 6900–6906.

Danese, P.N., Pratt, L.A., Dove, S.L. and Kolter, R. (2000). The outer membrane protein, antigen 43, mediates cell-to-cell interactions within *Escherichia coli* biofilms. Mol. Microbiol. *37*, 424–432.

Danese, P.N., Pratt, L.A. and Kolter, R. (2000). Exopolysaccharide production is required for development of *Escherichia coli* K-12 biofilm architecture. J. Bacteriol. *182*, 3593–3596.

Davey, M.E. and O'Toole, G.A. (2000). Microbial biofilms: from ecology to molecular genetics. Microbiol. Mol. Biol. Rev. *64*, 847–867.

Donlan, R.M. and Costerton, J.W. (2002). Biofilms: survival mechanisms of clinically relevant microorganisms. Clin. Microbiol. Rev. *15*, 167–193.

Faruque, S.M., Albert, M.J. and Mekalanos, J.J. (1998). Epidemiology, genetics, and ecology of toxigenic *Vibrio cholerae*. Microbiol. Mol. Biol. Rev. *62*, 1301–1314.

Faruque, S.M., Biswas, K., Udden, S.M., Ahmad, Q.S., Sack, D.A., Nair, G.B. and Mekalanos, J.J. (2006). Transmissibility of cholera: *in vivo*-formed biofilms and their relationship to infectivity and persistence in the environment. Proc. Natl. Acad. Sci. USA *103*, 6350–6355.

Fong, J.C., Karplus, K., Schoolnik, G.K. and Yildiz, F.H. (2006). Identification and characterization of RbmA, a novel protein required for the development of rugose colony morphology and biofilm structure in *Vibrio cholerae*. J. Bacteriol. *188*, 1049–1059.

Fong, J.C. and Yildiz, F.H. (2007). The *rbmBCDEF* Gene cluster modulates development of rugose colony morphology and biofilm formation in *Vibrio cholerae*. J. Bacteriol. *189*, 2319–2330.

Friedman, L. and Kolter, R. (2004). Two genetic loci produce distinct carbohydrate-rich structural components of the *Pseudomonas aeruginosa* biofilm matrix. J. Bacteriol. *186*, 4457–4465.

Friedman, L. and Kolter, R. (2004). Genes involved in matrix formation in *Pseudomonas aeruginosa* PA14 biofilms. Mol. Microbiol. *51*, 675–690.

Fullner, K.J. and Mekalanos, J.J. (1999). Genetic characterization of a new type IV-A pilus gene cluster found in both classical and El Tor biotypes of *Vibrio cholerae*. Infect. Immun. *67*, 1393–1404.

Galperin, M.Y., Nikolskaya, A.N. and Koonin, E.V. (2001). Novel domains of the prokaryotic two-component signal transduction systems. FEMS Microbiol. Lett. *203*, 11–21.

Garcia, B., Latasa, C., Solano, C., Garcia-del Portillo, F., Gamazo, C. and Lasa, I. (2004). Role of the GGDEF protein family in Salmonella cellulose biosynthesis and biofilm formation. Mol. Microbiol. *54*, 264–277.

Halpern, M., Broza, Y.B., Mittler, S., Arakawa, E. and Broza, M. (2004). Chironomid egg masses as a natural reservoir of *Vibrio cholerae* non-O1 and non-O139 in freshwater habitats. Microb. Ecol. *47*, 341–349.

Hammer, B.K. and Bassler, B.L. (2003). Quorum sensing controls biofilm formation in *Vibrio cholerae*. Mol. Microbiol. *50*, 101–104.

Haugo, A.J. and Watnick, P.I. (2002). *Vibrio cholerae* CytR is a repressor of biofilm development. Mol. Microbiol. *45*, 471–483.

Heidelberg, J.F., Eisen, J.A., Nelson, W.C., Clayton, R.A., Gwinn, M.L., Dodson, R.J., Haft, D.H., Hickey, E.K., Peterson, J.D., Umayam, L., Gill, S.R., Nelson, K.E., Read, T.D., Tettelin, H., Richardson, D., Ermolaeva, M.D., Vamathevan, J., Bass, S., Qin, H., Dragoi, I., Sellers, P., McDonald, L., Utterback, T., Fleishmann, R.D., Nierman, W.C., White, O., Salzberg, S.L., Smith, H.O., Colwell, R.R., Mekalanos, J.J., Venter, J.C. and Fraser, C.M. (2000). DNA sequence of both chromosomes of the cholera pathogen *Vibrio cholerae*. Nature *406*, 477–483.

Hengge-Aronis, R. (2002). Signal transduction and regulatory mechanisms involved in control of the sigma(S) (RpoS) subunit of RNA polymerase. Microbiol Mol Biol Rev. *66*, 373–395.

Hood, M.A. and Winter, P.A. (1997). Attachment of *Vibrio cholerae* under various environmental conditions and to selected substrates. FEMS Microbiol. Ecol. *22*, 215–223.

Hung, D.T., Zhu, J., Sturtevant, D. and Mekalanos, J.J. (2006). Bile acids stimulate biofilm formation in *Vibrio cholerae*. Mol Microbiol, *59*, 193–201.

Huq, A., Small, E.B., West, P.A., Huq, M.I., Rahman, R. and Colwell, R.R. 1983. Ecological relationships between *Vibrio cholerae* and planktonic crustacean copepods. Appl. Environ. Microbiol. *45*, 275–283.

Huq, A., West, P.A., Small, E.B., Huq, M.I. and Colwell, R.R. (1984). Influence of water temperature, salinity, and pH on survival and growth of toxigenic *Vibrio cholerae* serovar 01 associated with live copepods in laboratory microcosms. Appl. Environ. Microbiol. *48*, 420–424.

Huq, A., Huq, S.A., Grimes, D.J., O'Brien, M., Chu, K.H., Capuzzo, J.M. and Colwell, R.R. (1986). Colonization of the gut of the blue crab (Callinectes sapidus) by *Vibrio cholerae*. Appl. Environ. Microbiol. *52*, 586–588.

Huq, A., Colwell, R.R., Chowdhury, M.A., Xu, B., Moniruzzaman, S.M., Islam, M.S., Yunus, M. and Albert, M.J. (1995). Coexistence of *Vibrio cholerae* O1 and O139 Bengal in plankton in Bangladesh. Lancet *345*, 1249.

Huq, A., Sack, R.B., Nizam, A., Longini, I.M., Nair, G.B., Ali, A., Morris, J.G., Khan, Jr. M.N., Siddique, A.K., Yunus, M., Albert, M.J., Sack, D.A. and Colwell, R.R. (2005). Critical factors influencing the occurrence of *Vibrio cholerae* in the environment of Bangladesh. Appl Environ Microbiol. *71*, 4645–4654.

Islam, M.S., Jahid, M.I., Rahman, M.M., Rahman, M.Z., Islam, M.S., Kabir, M.S., Sack, D.A. and Schoolnik, G.K. (2007). Biofilm Acts as a Microenvironment for Plankton-Associated *Vibrio cholerae* in the Aquatic Environment of Bangladesh. Microbiol. Immunol. *51*, 369–379.

Jackson, K.D., Starkey, M., Kremer, S., Parsek, M.R. and Wozniak, D.J. (2004). Identification of *psl*, a locus encoding a potential exopolysaccharide that is essential for *Pseudomonas aeruginosa* PAO1 biofilm formation. J. Bacteriol. *186*, 4466–4475.

Jenal, U. (2004). Cyclic di-guanosine-monophosphate comes of age: a novel secondary messenger involved in modulating cell surface structures in bacteria? Curr. Opin. Microbiol. 7, 185–191.

Jobling, M.G. and Holmes, R.K. (1997). Characterization of *hapR*, a positive regulator of the *Vibrio cholerae* HA/protease gene *hap*, and its identification as a functional homologue of the Vibrio harveyi luxR gene. Mol. Microbiol. *26*, 1023–1034.

Jonson, G., Holmgren, J. and Svennerholm, A.M. (1991). Identification of a mannose-binding pilus on *Vibrio cholerae* El Tor. Microb Pathog. *11*, 433–441.

Kaper, J.B., Morris, Jr., J.G. and Levine, M.M. (1995). Cholera. Clin. Microbiol. Rev. *8*, 48–86.

Kapfhammer, D., Karatan, E., Pflughoeft, K.J. and Watnick, P.I. (2005). Role for glycine betaine transport in *Vibrio cholerae* osmoadaptation and biofilm formation within microbial communities. Appl. Environ. Microbiol. *71*, 3840–3847.

Kierek, K. and Watnick, P.I. (2003). The *Vibrio cholerae* O139 O-antigen polysaccharide is essential for Ca^{2+}-dependent biofilm development in sea water. Proc. Natl. Acad. Sci. USA *100*, 14357–14362.

Kierek, K. and Watnick, P.I. (2003). Environmental determinants of *Vibrio cholerae* biofilm development. Appl. Environ. Microbiol. *69*, 5079–5088.

Kulasakara, H., Lee, V., Brencic, A., Liberati, N., Urbach, J., Miyata, S., Lee, D.G., Neely, A.N., Hyodo, M., Hayakawa, Y., Ausubel, F.M. and Lory, S. (2006). Analysis of *Pseudomonas aeruginosa* diguanylate cyclases and phosphodiesterases reveals a role for bis-(3′-5′)-cyclic-GMP in virulence. Proc. Natl. Acad. Sci. USA *103*, 2839–2844.

Lauriano, C.M., Ghosh, C., Correa, N.E. and Klose, K.E. (2004). The sodium-driven flagellar motor controls exopolysaccharide expression in *Vibrio cholerae*. J. Bacteriol. *186*, 4864–4874.

Lenz, D.H., Mok, K.C., Lilley, B.N., Kulkarni, R.V., Wingreen, N.S. and Bassler, B.L. (2004). The small RNA chaperone Hfq and multiple small RNAs control quorum sensing in *Vibrio harveyi* and *Vibrio cholerae*. Cell *118*, 69–82.

Lenz, D.H., Miller, M.B., Zhu, J., Kulkarni, R.V. and Bassler, B.L. (2005). CsrA and three redundant small RNAs regulate quorum sensing in *Vibrio cholerae*. Mol. Microbiol. *58*, 1186–1202.

Lim, B., Beyhan, S., Meir, J. and Yildiz, F.H. (2006). Cyclic-diGMP signal transduction systems in *Vibrio cholerae*: modulation of rugosity and biofilm formation. Mol. Microbiol. *60*, 331–348.

Lipp, E.K., Huq, A. and Colwell, R.R. (2002). Effects of global climate on infectious disease: the cholera model. Clin. Microbiol. Rev. *15*, 757–770.

Ma, L., Jackson, K.D., Landry, R.M., Parsek, M.R. and Wozniak, D.J. (2006). Analysis of *Pseudomonas aeruginosa* conditional *psl* variants reveals roles for the *psl* polysaccharide in adhesion and maintaining biofilm structure postattachment. J. Bacteriol. *188*, 8213–8221.

Meibom, K.L., Li, X.B., Nielsen, A.T., Wu, C.Y., Roseman, S. and Schoolnik, G.K. (2004). The *Vibrio cholerae* chitin utilization program. Proc. Natl. Acad. Sci. USA *101*, 2524–2529.

Miller, M.B. and Bassler, B.L. (2001). Quorum sensing in bacteria. Annu. Rev. Microbiol. *55*, 165–199.

Miller, M.B., Skorupski, K., Lenz, D.H., Taylor, R.K. and Bassler, B.L. (2002). Parallel quorum sensing systems converge to regulate virulence in *Vibrio cholerae*. Cell *110*, 303–314.

Moorthy, S. and Watnick, P.I. (2004). Genetic evidence that the *Vibrio cholerae* monolayer is a distinct stage in biofilm development. Mol. Microbiol. *52*, 573–587.

Moorthy, S. and Watnick, P.I. (2005). Identification of novel stage-specific genetic requirements through whole genome transcription profiling of *Vibrio cholerae* biofilm development. Mol. Microbiol. *57*, 1623–1635.

Morris, J.G., Sztein, Jr. M.B., Rice, E.W., Nataro, J.P., Losonsky, G.A., Panigrahi, P., Tacket, C.O. and Johnson, J.A. (1996). *Vibrio cholerae* O1 can assume a chlorine-resistant rugose survival form that is virulent for humans. J. Infect. Dis. *174*, 1364–1368.

Nalin, D.R., Daya, V., Reid, A., Levine, M.M. and Cisneros, L. (1979). Adsorption and growth of *Vibrio cholerae* on chitin. Infect. Immun. *25*, 768–770.

Nesper, J., Lauriano, C.M., Klose, K.E., Kapfhammer, D., Kraiss, A. and Reidl, J. (2001). Characterization of *Vibrio cholerae* O1 El Tor galU and galE mutants: influence on lipopolysaccharide structure, colonization, and biofilm formation. Infect. Immun. *69*, 435–445.

Nielsen, A.T., Dolganov, N.A., Otto, G., Miller, M.C., Wu, C.Y. and Schoolnik, G.K. (2006). RpoS controls the *Vibrio cholerae* mucosal escape response. PLoS Pathog. *2*, e109.

O'Toole, G., Kaplan, H.B. and Kolter, R. (2000). Biofilm formation as microbial development. Annu. Rev. Microbiol. *54*, 49–79.

O'Toole, G.A. and Kolter, R. (1998). Flagellar and twitching motility are necessary for *Pseudomonas aeruginosa* biofilm development. Mol. Microbiol. *30*, 295–304.

Parsek, M.R. and Singh, P.K. (2003). Bacterial biofilms: an emerging link to disease pathogenesis. Annu. Rev. Microbiol. *57*, 677–701.

Pratt, J.T., Tamayo, R., Tischler, A.D. and Camilli, A. (2007). PilZ domain proteins bind cyclic diguanylate and regulate diverse processes in *Vibrio cholerae*. J Biol Chem. *282*, 12860–70.

Pratt, L.A. and Kolter, R. (1998). Genetic analysis of *Escherichia coli* biofilm formation: roles of flagella, motility, chemotaxis and type I pili. Mol. Microbiol. *30*, 285–293.

Rashid, M.H., Rajanna, C., Ali, A. and Karaolis, D.K. (2003). Identification of genes involved in the switch between the smooth and rugose phenotypes of *Vibrio cholerae*. FEMS Microbiol Lett. *227*, 113–119.

Reguera, G. and Kolter, R. (2005). Virulence and the environment: a novel role for *Vibrio cholerae* toxin-coregulated pili in biofilm formation on chitin. J. Bacteriol. *187*, 3551–3555.

Romling, U., Sierralta, W.D., Eriksson, K. and Normark, S. (1998). Multicellular and aggregative behaviour of *Salmonella typhimurium* strains is controlled by mutations in the *agfD* promoter. Mol. Microbiol. *28*, 249–264.

Romling, U., Rohde, M., Olsen, A., Normark, S. and Reinkoster, J. (2000). AgfD, the checkpoint of multicellular and aggregative behaviour in *Salmonella typhimurium* regulates at least two independent pathways. Mol. Microbiol. *36*, 10–23.

Romling, U., Gomelsky, M. and Galperin, M.Y. (2005). C-di-GMP: the dawning of a novel bacterial signalling system. Mol. Microbiol. *57*, 629–639.

Romling, U. and Amika, D. (2006). Cyclic di-GMP as a second messenger. Curr. Opin. Microbiol. *9*, 218–228.

Ross, P., Mayer, R., Weinhouse, H., Amikam, D., Huggirat, Y., Benziman, M., de Vroom, E., Fidder, A., de Paus, P., Sliedregt, L.A. and *et al.* (1990). The cyclic diguanylic acid regulatory system of cellulose synthesis in *Acetobacter xylinum*. Chemical synthesis and biological activity of cyclic nucleotide dimer, trimer, and phosphothioate derivatives. J. Biol. Chem. *265*, 18933–18943.

Ross, P., Mayer, R. and Benziman, M. (1991). Cellulose biosynthesis and function in bacteria. Microbiol. Rev. *55*, 35–58.

Ryan, R.P., Fouhy, Y., Lucey, J.F., Crossman, L.C., Spiro, S., He, Y.W., Zhang, L.H., Heeb, S., Camara, M., Williams, P. and Dow, J.M. (2006). Cell-cell signaling in *Xanthomonas campestris* involves an HD-GYP domain protein that functions in cyclic di-GMP turnover. Proc. Natl. Acad. Sci. USA *103*, 6712–6717.

Ryjenkov, D.A., Tarutina, M., Moskvin, O.V. and Gomelsky, M. (2005). Cyclic diguanylate is a ubiquitous signaling molecule in bacteria: insights into biochemistry of the GGDEF protein domain. J. Bacteriol. *187*, 1792–1798.

Ryjenkov, D.A., Simm, R., Romling, U. and Gomelsky, M. (2006). The PilZ domain is a receptor for the second messenger c-di-GMP: the PilZ domain protein YcgR controls motility in enterobacteria. J. Biol. Chem. *281*, 30310–30314.

Sauer, K. and Camper, A.K. (2001). Characterization of phenotypic changes in *Pseudomonas putida* in response to surface-associated growth. J. Bacteriol. *183*, 6579–6589.

Schmidt, A.J., Ryjenkov, D.A. and Gomelsky, M. (2005). The ubiquitous protein domain EAL is a cyclic diguanylate-specific phosphodiesterase: enzymatically active and inactive EAL domains. J. Bacteriol. *187*, 4774–4781.

Tamplin, M.L., Gauzens, A.L., Huq, A., Sack, D.A. and Colwell, R.R. (1990). Attachment of *Vibrio cholerae* serogroup O1 to zooplankton and phytoplankton of Bangladesh waters. Appl. Environ. Microbiol. *56*, 1977–1980.

Taylor, R.K., Miller, V.L., Furlong, D.B. and Mekalanos, J.J. (1987). Use of *phoA* gene fusions to identify a pilus colonization factor coordinately regulated with cholera toxin. Proc. Natl. Acad. Sci. USA *84*, 2833–2837.

Tischler, A.D. and Camilli, A. (2004). Cyclic diguanylate (c-di-GMP) regulates *Vibrio cholerae* biofilm formation. Mol. Microbiol. *53*, 857–869.

Uhlich, G.A., Keen, J.E. and Elder, R.O. (2001). Mutations in the *csgD* promoter associated with variations in curli expression in certain strains of *Escherichia coli* O157:H7. Appl. Environ. Microbiol. *67*, 2367–2370.

Wai, S.N., Mizunoe, Y., Takade, A., Kawabata, S.I. and Yoshida, S.I. (1998). *Vibrio cholerae* O1 strain TSI-4 produces the exopolysaccharide materials that determine colony morphology, stress resistance, and biofilm formation. Appl. Environ. Microbiol, *64*, 3648–3655.

Watnick, P.I., Fullner, K.J. and Kolter, R. (1999). A role for the mannose-sensitive hemagglutinin in biofilm formation by *Vibrio cholerae* El Tor. J. Bacteriol. *181*, 3606–3609.

Watnick, P.I. and Kolter. R. (1999). Steps in the development of a *Vibrio cholerae* El Tor biofilm. Mol. Microbiol. *34*, 586–595.

Watnick, P.I., Lauriano, C.M., Klose, K.E., Croal, L. and Kolter, R. (2001). The absence of a flagellum leads to altered colony morphology, biofilm development and virulence in *Vibrio cholerae* O139. Mol. Microbiol. *39*, 223–235.

Yildiz, F.H. and Schoolnik, G.K. (1999). *Vibrio cholerae* O1 El Tor: identification of a gene cluster required for the rugose colony type, exopolysaccharide production, chlorine resistance, and biofilm formation. Proc. Natl. Acad. Sci. USA *96*, 4028–4033.

Yildiz, F.H., Dolganov, N.A. and Schoolnik. G.K. (2001). VpsR, a member of the response regulators of the two-component regulatory systems, is required for expression of *vps* biosynthesis genes and EPS(ETr)-associated phenotypes in *Vibrio cholerae* O1 El Tor. J. Bacteriol. *183*, 1716–1726.

Yildiz, F.H., Liu, X.S., Heydorn, A., and Schoolnik, G.K. (2004). Molecular analysis of rugosity in a *Vibrio cholerae* O1 El Tor phase variant. Mol Microbiol, *53*, 497–515.

Zhu, J., Miller, M.B., Vance, R.E., Dziejman, M., Bassler, B.L. and Mekalanos, J.J. (2002). Quorum-sensing regulators control virulence gene expression in *Vibrio cholerae*. Proc. Natl. Acad. Sci. USA *99*, 3129–3134.

Zhu, J. and Mekalanos, J.J. (2003). Quorum sensing-dependent biofilms enhance colonization in *Vibrio cholerae*. Dev. Cell. *5*, 647–656.

Molecular Ecology of *Vibrio cholerae* 8

Shah M. Faruque and John J. Mekalanos

Abstract

Although *Vibrio cholerae* cause human disease, aquatic ecosystems are major habitats of *V. cholerae*, and all *V. cholerae* are not pathogenic for humans. *V. cholerae* represents a paradigm for origination of pathogenic bacteria from environmental non-pathogenic progenitor strains by horizontal transfer of genes. Besides environmental factors which are not precisely defined, bacteriophages, and horizontally transmissible genetic elements have a significant role in the epidemiology and evolution of the pathogen. Recent studies are beginning to reveal the mechanisms associated with the occurrence of seasonal epidemics in endemic areas, water-borne spread of cholera, and the factors that enable the organisms to survive unfavourable conditions in the aquatic environment. The emergence of new epidemic strains, and their enrichment during epidemics of cholera appear to constitute a natural system for the evolution of *V. cholerae* and genetic elements that mediate horizontal transfer of genes among bacterial strains.

Introduction

Cholera epidemics caused by toxigenic *Vibrio cholerae* are a major public health problem in developing countries where water quality and sanitation are poor. In Southern Asia and parts of Africa and Latin America, cholera is endemic with seasonal epidemics occurring widely (Kaper *et al.*, 1995; Faruque *et al.*, 1998). Although some strains of *V. cholerae* cause disease in humans, the species is part of the normal bacterial flora of the aquatic environments. *V. cholerae* belonging to the O1 and O139 serogroups which are associated with human disease are commonly known to carry a set of virulence genes necessary for pathogenesis in humans (Faruque *et al.*, 1998). However, recent studies have indicated that virulence genes or their homologues are also dispersed among environmental strains of *V. cholerae* belonging to diverse serogroups, and these strains often possess a lower virulence potential than the epidemic strains. The eco-system for *V. cholerae* (comprising the aquatic environment, elements mediating horizontal gene transfer, and the mammalian host) presumably support the clustering of critical virulence genes in rare combinations leading to the origination of new strains with epidemic potential. A sudden outbreak of cholera can occur either due to introduction of pathogenic strains for the first time into a region or by the *de novo* emergence of a new pathogenic strain through rare genetic recombination events. Thus, pathogenic strains emerge most efficiently in regions of the world where transmission between humans occurs and most often this emergence is associated with the introduction of a fully virulent strain into a susceptible population, for example, the Latin American epidemic of 1991 (Levine, 1991). Occasionally, strains of low pathogenicity acquire virulence genes and thus emerge as a unique new pathogenic clone. The emergence of the El Tor biotype of *V. cholerae* to cause disease in Indonesia in 1961, is an example of such events (Faruque *et al.*, 1998).

Once endemicity is established in a region, cholera outbreaks appear to occur in a seasonal pattern. For example, in the Ganges Delta region of Bangladesh and India, two epidemics occur every year. Factors associated with the maintenance of the periodicity of cholera epidemics are not clear, but a multitude of environmental, genetic, ecological and socioeconomic factors may be involved. Interestingly, the evolution of *V. cholerae* and the selection of a pathogenic clone are linked to the occurrence of disease in humans, and thus the epidemics of cholera also enhances the origination and emergence of new epidemic strains. While the issues of virulence genes, their regulation and pathogenesis have been addressed in the other chapters of this book, here we discuss the factors affecting occurrence of seasonal cholera epidemics, transmission and mode of persistence of *V. cholerae* in the environment, and the relationships between evolution of *V. cholerae* and its selection through human disease.

Ecology of *Vibrio cholerae*

Over the last decade many aspects of the ecology of *V. cholerae* have been revealed and some of these clearly play a role in the seasonal pattern of outbreaks of the disease in endemic areas. *V. cholerae* has been regarded as a member of a group of organisms whose major habitats are aquatic ecosystems (Colwell and Spira, 1992; Colwell *et al.*, 2004). Although, *V. cholerae* is part of the normal, free-living bacterial flora in riverine and estuarine areas, non-pathogenic strains are more commonly isolated from the environment than the strains capable of causing cholera epidemics. This is apparently because there are more than 200 serogroups of *V. cholerae*, of which only two serogroups are able to cause epidemic disease.

The structure of the major virulence gene clusters in *V. cholerae* indicates that these are capable of being propagated horizontally (Faruque and Mekalanos, 2003), and indeed the genes encoding cholera toxin (CT) have been demonstrated to be transferred through the filamentous phage CTXϕ (Waldor and Mekalanos, 1996). This is also suggestive of the possibility that environmental strains of *V. cholerae* may develop the ability to adapt to the intestinal environment through acquisition of virulence genes. It is therefore apparent that the ecosystem for *V. cholerae* should consists of a number of components. These include (a) the bacterium and its aquatic environment, (b) CTXϕ and other genetic elements and/or unidentified phages involved in the transfer of virulence genes, (c) the intestinal environment of the host population, and (d) phages that modulate the concentration of epidemic strains both in the environment and possibly also under *in vivo* conditions. It is important to understand the role of each of these components, to identify the factors which enhance water-borne spread of cholera epidemics and sustain the epidemic strain in nature.

Environmental survival and persistence of *Vibrio cholerae*

The physicochemical conditions for the survival of *V. cholerae* O1 have been studied and the possibility of survival of the organism in an estuarine environment and other brackish waters is widely accepted (Colwell and Spira, 1992). However, the nature of the survival and persistence of toxigenic *V. cholerae* O1 or O139 in aquatic ecosystems as well as the factors controlling their periodic appearance causing epidemics of cholera in endemic areas are being investigated. The survival may be dependent on a number of factors, such as occurrence of particular physicochemical conditions, specific association of the bacteria with aquatic plants or animals, and /or the existence of specific ecological association involving several components of the aquatic environment. It was postulated that under stress conditions the vibrios are converted to a viable but non-culturable (VBNC) form, that cannot be recovered by culture techniques, and that such forms are able to produce infection and can revert to the culturable form (Colwell and Huq, 1994). However, contrary to this proposition, studies on a marine *Vibrio* strain ANT-300 had shown that the organism responded to starvation by reducing metabolic activities and inducing morphological changes, e.g. rod shaped to a coccoid shape, and producing progeny cells significantly decreased in volume, but still remained culturable (Novitsky and Morita, 1976, 1977, 1978). The ecological and public health importance of the possible survival forms depends on whether these forms are re-convertible to normal infectious bacteria.

Conditionally viable environmental cells (CVEC)

Until recently, there was very little evidence to conclusively establish that *V. cholerae* exist in a culturable form in the environment during inter-epidemic periods when there is almost no incidence of cholera. It was recently confirmed that environmental water samples contain significantly more viable pathogenic *V. cholerae* cells than are recovered by usual enrichment culture techniques (Faruque *et al.*, 2005, 2006a,b). A significant proportion of epidemic *V. cholerae* O1 cells in the environment was found to exist in a 'conditionally viable' state requiring special culture techniques for their recovery. Since these cells were in fact 'culturable' using a modified technique (Faruque *et al.*, 2005), for clarity these cells were designated as conditionally viable environmental cells (CVEC), to distinguish them from the previously described VBNC. It was demonstrated that *V. cholerae* in the environment can be efficiently cultured after pre-enriching the water in a nutrient broth, and plating on a medium containing an appropriate antibiotic to which the bacteria are known to be resistant (Faruque *et al.*, 2005, 2006a). The improved method of *V. cholerae* enrichment termed AST for 'antibiotic selection technique' constitutes a breakthrough in understanding the ecology and epidemiology of cholera. In further studies involving AST and fluctuation analysis, CVEC appear to be clumps of partially dormant cells that can be revived under appropriate *in vivo* and *in vitro* conditions. It was demonstrated that the CVEC could be revived into infectious bacteria when inoculated in the intestines of rabbits (Faruque *et al.*, 2006b), as well as when enriched in nutrient medium. As described later in this chapter, CVEC may originate from biofilm-like clumps of cells excreted in human cholera stools (Faruque, *et al.*, 2006). Critically, the role of CVEC in the epidemiology of cholera needs to be more precisely demonstrated. There is much to elucidate regarding *V. cholerae* CVEC, including their metabolic status, innate infectivity, phage sensitivity, and the changes in gene regulation that occur during formation of CVEC and their reactivation as viable planktonic cells.

Interactions of *Vibrio cholerae* and other environmental organisms

The concept of an environmental reservoir of *V. cholerae* O1 or O139 implies not only that the vibrios survive, in whatever form, but that they form an essential component of the ecosystem. Laboratory studies of microcosms have illustrated the ability of *V. cholerae* O1 to associate with a variety of zooplankton, phytoplankton and algae (Islam *et al.*, 1994a, 1994b). The associations prolong survival and presumably the vibrios gain nutrients from the host. For example, *V. cholerae* is known to produce enzymes such as chitinase, which allows the bacteria to use chitin present in the exoskeleton of aquatic crustaceans as a source of nutrients.

In addition to the genes encoding the major virulence factors, i.e. cholera toxin (CT), and toxin-coregulated pili (TCP), *V. cholerae* are known to carry the genes for a number of putative accessory virulence factors. These factors include the mannose-sensitive haemagglutinin (MSHA) pilus, the RTX toxin, haemolysins, as well as a few other accessory toxins (Faruque and Mekalanos, 2003; Faruque *et al.*, 1998; Faruque *et al.*, 2004; Kaper *et al.*, 1995). However, the roles of the accessory virulence factors in cholera pathogenesis are not well established, and recent studies are beginning to reveal that at least some of these putative virulence factors also play a role in environmental fitness of the pathogen (Chiavelli *et al.*, 2001; Watnick *et al.*, 1999). Although the role of MSHA pilus in *V. cholerae* virulence is controversial (Thelin and Taylor, 1996), recent studies have indicated that the MSHA pilus is used by environmental *V. cholerae* to adhere to plankton or to form biofilms on abiotic surfaces (Chiavelli *et al.*, 2007; Watnick *et al.*, 1999). Given the role of MSHA in environmental persistence, possession of *mshA* gene would provide a strong selective advantage to any *V. cholerae* recipient in the aquatic environment. This assumption is reflected in the widespread prevalence of *mshA* in *V. cholerae* (Faruque and Mekalanos, unpublished). This finding together with recent studies showing that MSHA can also act as a receptor for diverse filamentous phages (Faruque *et al.*, 2005; Jouravlev *et al.*, 1998) further suggest that possession of *mshA* gene might well have been a significant event in early *V. cholerae* evolution, since filamentous phages can facilitate horizontal gene transfer among *V. cholerae* (Faruque and Mekalanos, 2003).

While *V. cholerae* is autochthonous to the aquatic environment, it has been suggested that acquisition of virulence genes including the TCP pathogenicity island and the CTX prophage encoding CT allowed specific strains of *V. cholerae* to become adapted to the human intestinal environment (Faruque *et al.*, 1998). Although, virulence genes are believed to be carried by pathogenic clinical isolates, recent studies have revealed that virulence genes or their homologues are also dispersed among environmental strains of *V. cholerae* belonging to diverse serogroups (Chakraborty, *et al.*, 2000; Rivera, *et al.*, 2001; Mukhopadhay *et al.*, 2001). Environmental isolates of *V. cholerae* analysed for the presence of *ctxA*, *hlyA*, *ompU*, *stn/sto*, *tcpA*, *tcpI*, *toxR*, and *zot* genes, using multiplex PCR showed that *toxR*, *hlyA*, and *ompU* genes were present in 100%, 98.6%and 87.01% of the isolates, respectively (Rivera *et al.*, 2001). The CTX prophage and toxin-coregulated pilus were present in all toxigenic *V. cholerae* O1 and *V. cholerae* O139 strains examined in this study.

In another study, (Chakraborty *et al.*, 2000) the presence of five structural and regulatory genes involved in virulence, including *ctxAB*, *tcpA*, *toxR* and *toxT* in environmental strains of *V. cholerae* cultured from three different freshwater lakes and ponds in the eastern part of Calcutta, India were investigated. PCR analysis revealed the presence of these virulence genes or their homologues among diverse serotypes and ribotypes of environmental *V. cholerae*. Sequencing of a part of the *tcpA* gene carried by an environmental strain showed 97.7% homology to the *tcpA* gene of the classical biotype of *V. cholerae* O1. Strains carrying the *tcpA* gene expressed the toxin coregulated pilus (TCP), demonstrated by both autoagglutination analysis and electron microscopy of the TCP pili. Strains carrying *ctxAB* genes also produced CT, as determined by GM_1 ELISA and by passage in the ileal loops of rabbits. Thus this study demonstrated the presence and expression of critical virulence genes, or their homologues, in diverse environmental strains of *V. cholerae*, which appear to constitute an environmental reservoir of virulence genes, thereby providing new insights into the ecology of *V. cholerae* (Chakraborty, *et al.*, 2000).

The existence of different *tcpA* alleles in vibrios of various non-epidemic serogroups, including toxigenic and non-toxigenic strains from clinical and environmental sources have also been documented recently. The different *tcpA* alleles may have evolved in response to selective pressures that vary between the environment and the host. In view of a recent report that TCP pili can promote biofilm formation on chitin surfaces (Reguera and Kolter, 2005), the function of TCP in these strains could also be to promote environmental persistence by using TCP to adhere to aquatic flora. The virulence genes carried by pathogenic vibrios might also have been derived from the environmental strains, and a crucial combination of different genes may be necessary for the origination of a potential pathogenic strain.

The role in virulence of a newly discovered type III secretion system (T3SS) in *V. cholerae* non-O1 is not clearly known, but it has been suggested that the T3SS might also have functions in the environment (Dziejman *et al.*, 2005), since *V. cholerae* are found in contact with eukaryotic cells not only within the human host but also in the marine environment where the bacteria can associate with insects, plankton, copepods, and shrimp. *Vibrio cholerae*, that reside in environmental reservoirs, also survive predation by unicellular eukaryotes, and a recently described 'type VI' secretion system (T6SS) in *V. cholerae* appear to be involved in the process (Pukatzki *et al.*, 2006). The genes of the T6SS referred to as virulence-associated secretion (VAS) genes, are required for cytotoxicity of *V. cholerae* cells toward an amoebae *Dictyostelium discoideum*, by a contact-dependent mechanism (Pukatzki *et al.*, 2006).

Since, cholera is a water-borne disease and the species is part of the normal aquatic flora, increased survival in the environment may also favour epidemic or pandemic spread of the pathogen. Many of the accessory virulence genes are also present in environmental strains, which do not carry the major virulence genes, i.e. CTX and TCP island genes. Thus the accessory genes may also have a role in surviving unfavourable environmental conditions. Therefore, possession of these genes may provide a selective advantage to the bacteria in the aquatic environment. This also raises questions whether the factors which have been described as virulence factors in the context of human infection with *V. cholerae* have other roles when the bacterium resides in its environmental habitat.

Enrichment of pathogenic strains in the host intestine

Although we now have some understanding of the pathogenic mechanism of toxigenic *V. cholerae*, we do not yet fully understand whether the role of the virulence factors is simply to cause diarrhoea and thus disseminate the organism to its next victim or could the virulence property providing a more crucial function for the enrichment and continued existence of the bacteria. Studies directed towards the development of attenuated *V. cholerae* mutants altered in toxin production for use as live oral cholera vaccines provided a means to investigate the role of CT in the intestine. Non-toxigenic mutants of strain 569B by mutagenesis with nitrosoguanidine were unable to induce a secretory response in the rabbit intestinal loop model, and did not survive or multiply in the intestinal environment (Howard, 1971). It was also observed that the ability of various mutants to grow in the intestinal environment correlates with the ability of the mutants to induce a residual secretory response in the infant rabbit model (Finkelstein *et al*, 1974; Holmes *et al.*, 1975). Moreover, several hypotoxigenic mutants were noted to be unstable in the rabbit intestinal loop model; during passage they produced toxigenic revertants which eventually displaced the mutant strains *in vivo* (Baselski *et al.*, 1978, 1979). A variety of different toxin deficient mutants of *V. cholerae* tested in rabbit and infant mouse models also suggested that the mutants showed enhanced killing and mechanical clearance in the intestinal environment compared to the toxigenic parental strain. It seems possible that many of these early mutants were altered in TCP as well as CT expression. The selection for their reversion *in vivo* may have, therefore, been driven by the need to up-regulate the expression of TCP more than that of CT.

Mekalanos (1983) demonstrated that significant amplification of the genes encoding CT occurred in hypertoxigenic variants of *V. cholerae* selected during intestinal passage in rabbits. The selective pressure causing this enrichment of hypertoxigenic variants *in vivo* is unknown but is probably related to the *in vivo* selection process that was involved in the reversion of hypotoxigenic mutants observed in earlier animal studies (Holmes *et al.*, 1975; Mekalanos, *et al.*, 1978). Induction of CTX phage *in vivo* (Lazar and Waldor, 1998) might have contributed to the process. In agreement with these observations, many clinical isolates of *V. cholerae* O1 or O139 carry multiple copies of CTX genetic praphage and hence the *ctxAB* operon (Mekalanos, 1983; Faruque *et al.*, 1998). It was also demonstrated that *ctxAB* deletion mutants colonized rabbit intestines about 10- to 100-fold less efficiently than the parental strain (Mekalanos, 1985). At least two possible mechanisms have been proposed to explain how CT might enhance intestinal colonization by *V. cholerae*. These are by relief of nutritional deprivation (Freter and Brein, 1981), and by inhibition of bactericidal activity produced by epithelial cells (Knop and Rowley, 1975; Freter and Brein, 1981).

Studies so far suggest that causation of cholera in humans is also linked with a natural process of enrichment of toxigenic *V. cholerae*, and partly explains the benefit imparted to the pathogen during the disease in humans. However, in order to understand the general epidemiological behaviour of *V. cholerae*, which includes mechanisms leading to seasonal pattern of epidemics, transient appearance and disappearance of different clones, and emergence of new epidemic clones, it is important to study the interactions among the bacteria, genetic elements mediating the transfer of virulence genes, the human host and possible environmental factors.

Evolution of toxigenic *Vibrio cholerae*

V. cholerae represents a group of bacteria which is autochthonous to riverine, coastal, and estuarine ecosystems, but at the same time, pathogenic for humans (Colwell *et al.*, 2004). Over the last few decades, dramatic progress has been made in understanding the molecular aspects that account for virulence of this organism and how virulent strains emerge in nature. With the completion of the *V. cholerae* genomic sequence, new methods have been applied to reveal changes that the organism has undergone, and have been linked to its emergence as a pathogen (Dziejman *et al.*, 2002; Raskin *et al.*, 2006). Some of these genetic changes probably also define important virulence altering traits. For example, microarray-based analysis has provide a reliable framework for understanding the emergence of pathogenic *V.*

cholerae strains. The most important conclusions reached from these studies is that pathogenic strains of *V. cholerae* are quite different from environmental strains and acquisition of virulence genes accounts for only a small part of the variation observed between strains. For example, pathogenic strains typically carry genes for virulence factors and the lipopolysaccharide O antigens that are typically absent in environmental strains (Faruque *et al.*, 1998; Faruque and Mekalanos 2003). Thus, while there are over 200 serogroups of *V. cholerae*, only two serogroups (O1 and O139) are associated with epidemic and endemic disease. Virtually all O139 strains carry the genes for CTX and TCP, but many O1 strains (particularly in the environment) do not. The explanation for this resides in the fact that the O139 serogroup is a derivative of the highly successful CTX+ TCP+ O1 El Tor 'seventh pandemic clone' (Waldor and Mekalanos 1994). *V. cholerae* O139 emerged in the coast of the Bay of Bengal in 1992 (Cholera Working Group, 1993), an area that has been the home of cholera for centuries. Both classical and El Tor biotypes of *V. cholerae* O1 carry CTX and TCP, so the explanation for the evolutionary success of the seventh pandemic clone over the preexisting sixth pandemic strain remains largely a mystery. The emergence of the O139 serogroup was assumed to mark the beginning of the 8th pandemic but so far this organism has co-existed with its parental O1 El Tor strains only in South Asia instead of wider geographic spread. Thus, it seems likely that O139 and O1 El Tor strains are here to stay, and comparable in their virulence and evolutionary fitness. Both clones are likely more adpated to cholera 'ecology' than earlier classical strains and the explanation for this may reside in their virulence gene makeup and other innate properties (e.g. phage resistance) as discussed below. TCP and CTX negative O1 strains that exist in the environmental waters of Bangladesh are mostly not closely related to the pathogenic clones causing endemic disease (Faruque *et al.*, 2004). Thus, the emergence of pathogenic classical and El Tor strains occurred long ago coincident with their capture of the genes for TCP and CTX. This is not to say that new clones cannot emerge from environmental *V. cholerae*. Indeed, a CTX+ TCP+ clone of *V. cholerae* that is endemic to the Gulf Coast of the United States appears to be an emergent clone that closely related to a CTX-TCP+ strain that co-exists within this same aquatic environment (Popovic *et al.*, 1993). Despite being pathogenic and in the aquatic environment for decades, this Gulf Coast clone did not cause the epidemic in Peru in 1991; the seventh pandemic clone that did cause this epidemic was possibly introduced from Asia and is clearly more ecologically adapted to causing epidemic and endemic disease.

As described above, environmental strains of *V. cholerae* may carry one or more virulence genes or their homologues, suggesting an environmental origin for pathogenic *V. cholerae* (Chakraborty, *et al.*, 2000; Rivera, *et al.*, 2001; Mukhopadhay, *et al.*, 2001), but emergence of new toxigenic strains 'de novo' just does not seem to happen very often. However, these environmental strains may constitute reservoirs of virulence genes, and may in theory, participate in rare gene transfer events leading to origination of strains carrying a crucial combination of virulence genes. In summary, emergence of pathogenic strains of *V. cholerae* with endemic and pandemic potential has been exclusively linked to the acquisition of the TCP and CTX genes by only a limited spectrum of O1 and O139 strains. This fact alone suggest that the precursors of these pandemic clones probably display traits that are lacking in typical environmentally adapted *V. cholerae* regardless of their serogroup. Analysis of the genome content of environmental and pathogenic *V. cholerae* O1, and non-O1, non-O139 support this conclusion (Dziejman *et al.*, 2002; Dziejman *et al.*, 2005). Unfortunately, the strains that represent these true ancestors of the pandemic *V. cholerae* classical and seventh pandemic O1 clones are missing from the microbiological record at this time.

Factors modulating seasonal cholera epidemics

A distinctive epidemiological feature of cholera is its appearance with seasonal regularity in endemic areas, such as the Ganges Delta region of Bangladesh and India. In Bangladesh, outbreaks usually occur twice during a year, with the highest number of cases just after the monsoon and a somewhat smaller number of cholera cases during the spring, and the timing of the epidemics correlates with increased concentration of the causative *V. cholerae* in water (Khan *et al.*, 1984) The physical, chemical, and biological factors which control the concentration of potential epidemic strains in the aquatic environment leading to initiation and conclusion of seasonal cholera epidemics have been investigated, and various hypotheses have been proposed. These include the hypothesis that ecological conditions such as plankton in water are needed to sustain an epidemic. It has also been postulated that the putative VBNC *V. cholerae* cells exist in an ecological association with 'bystander' aquatic organisms (e.g. copepods, filamentous algae, midges), and that undefined environmental factors affecting the bystander organism triggers multiplication of the dormant bacteria and subsequent cholera outbreaks (Colwell and Spira, 1992; Colwell and Huq, 1994). However, these studies did not explain whether such association is specific for the epidemic serogroups of *V. cholerae* or a general phenomenon for all other *V. cholerae* serogroups. Thus, this environmental resuscitation hypothesis does not explain how a single or limited number of clones with epidemic potential are selectively enriched prior to an epidemic from the array of pathogenic clones that caused previous epidemics in the same geographic locale, as well as from the vast majority of non-pathogenic *V. cholerae* strains found in the environment.

The role of geophysical and climatic factors in cholera epidemiology have also been investigated (Colwell, 1996; Lobitz *et al.*, 2000; Pascual *et al.*, 2000; Huq *et al.*, 2005). It has been shown by remote sensing, employing satellites, that sea surface temperature and sea surface height are correlated with the timings of cholera epidemics (Lobitz, *et al*, 2000). Studies in Lima, Peru, have also suggested that increasing water temperature can serve as a trigger for the occurrence of epidemic cholera (Franco, *et al.*, 1997; Gill, *et al.*, 2004). Once again, these observations do not seem to explain the mechanism for selective enrichment of toxigenic *V. cholerae* O1 or O139 in water, prior to an epidemic.

On the contrary, recent studies have provided more insights regarding how *V. cholerae* pathogenic serogroups might be enriched before the initiation of an outbreak. These studies concluded that while seasonal factors may cause a bloom of diverse *V. cholerae* in the environment, epidemics may be preceded by a gradual enrichment of pathogenic strains through passage in human beings who consume surface water (Faruque *et al.*, 2004). The role of the human host in selective amplification of *V. cholerae* from the majority of environmental non-pathogenic *V. cholerae* prior to an epidemic has been proposed to be an important factor in cholera epidemiology. Thus, asymptomatic infections of humans account for the pre-epidemic build-up of epidemic strains which then initiate the index case of cholera. Cholera then spreads rapidly through water contaminated with the faeces of cholera patients leading to en epidemic outbreak (Fig. 8.1). This proposed model involves both environmental and host factors in the initiation of seasonal epidemics and also explains the limited clonality apparent in the strains causing these epidemics.

Cholera epidemics are known to be self limiting, since the epidemics subside after reaching a peak, even without any active intervention. Among other factors, bacteriophages that can kill *V. cholerae* (cholera phages) have been demonstrated to play a significant role in modulating the course of epidemics (Faruque *et al.*, 2005a, 2006). Over a nearly 3-year period between January 2001 and November 2003 in Dhaka, Bangladesh, the number of cholera patients increased whenever the number of lytic vibriophages in water decreased. Similarly, cholera epidemics tended to end concurrent with large increases in the concentration of these viruses in the water.

The dynamics of the *V. cholerae*-phage interaction was further studied during an epidemic in Dhaka City in 2004 (Faruque *et al.*, 2005b). The changing prevalence in the environment of the epidemic *V. cholerae* O1 strain and a particular lytic cholera phage named JSF4 to which it was sensitive, was measured every 48–72 hours for a period of 17 weeks. The incidence of phage excretion in stools of 387 cholera patients during the course of the epidemic was also monitored. The peak of the epidemic was preceded by high *V. cholerae* prevalence in the environment and was followed by high JSF4 phage levels as the epidemic ended. The build up to the phage peak in the environment coincided with increasing

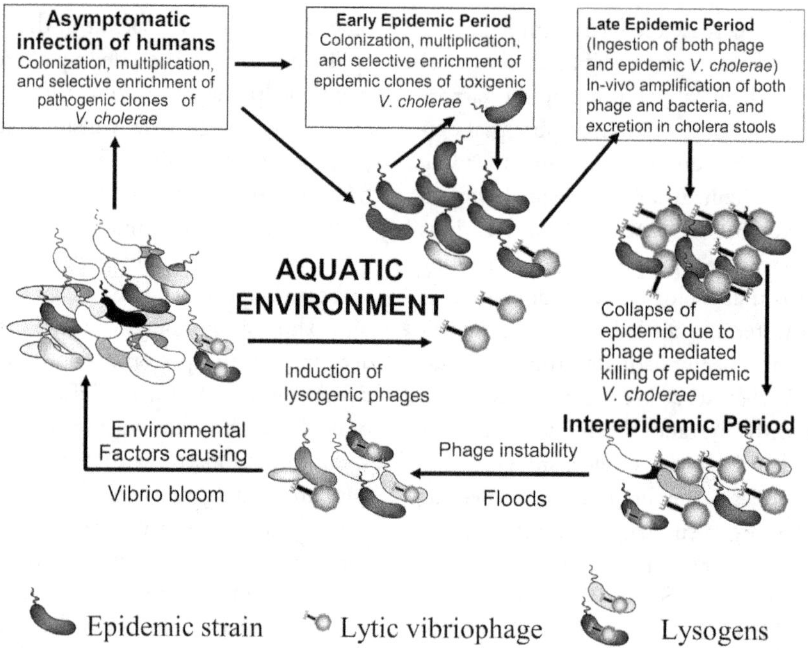

Figure 8.1 Schematic diagram showing involvement of environmental and host factors in the occurrence of seasonal epidemics in an area of endemic cholera. Following blooms of diverse vibrios in the water due to undefined environmental factors, enrichment of pathogenic *Vibrio cholerae* strains occurs in humans who consume the surface water. Epidemics occur only after pre-amplification of pathogenic strains in human victims with mostly asymptomatic infection. At the late stage of the epidemic, amplification of lytic vibriophages occur, and this leads to collapse of the epidemic as a result of rapid reduction of *V. cholerae* concentration in water due to predation by the phages.

excretion of the same phage in the stools of cholera patients (Faruque *et al.*, 2005b). Thus cholera phages in the environment, and their amplification in cholera patients have been shown to remarkably influence the epidemiology of cholera. It has been proposed that when the balance of phage and bacteria tips in favour of the phage, there is a dramatic decline of numbers of bacteria to a point, where they are no longer able to sustain the epidemic (Fig. 8.1). However, a proportion of the epidemic strain are likely to survive by various mechanisms, as described later, and seed the environment before the next epidemic season. Although bacteriophages have been shown to have important roles in cholera dynamics, the understanding of multiple factors and their interactions in cholera is yet to be determined.

Transmissibility of cholera

The factors which enhance water-borne spread of cholera epidemics and sustain the epidemic strain in nature are unclear. Merrell and colleagues (2002) suggested that human colonization creates a hyperinfectious bacterial state, and *V. cholerae* excreted in stools of cholera patients are thus hyperinfectious (Merrel *et al.*, 2002). Furthermore, it was suggested that the hyperinfectious state of *V. cholerae* in stools is maintained after dissemination into the aquatic environment and that this may contribute to the epidemic spread of cholera. More recently, it was shown that stools from cholera patients contain a heterogeneous mixture of biofilm-like aggregates and free swimming planktonic cells of *V. cholerae* (Faruque *et al.*, 2006b). Estimation of relative infectivity of these different forms of *V. cholerae* cells suggested that the enhanced infectivity of *V. cholerae* shed in human stools is largely due to the presence of clumps of cells which disperse *in vivo* providing a high dose of the pathogen.

As mentioned above, conditionally viable environmental cells (CVEC) of pathogenic *V. cholerae* that resist cultivation by conventional techniques was found to exist in surface water as aggregates (biofilms) of partially dormant cells (Faruque *et al.*, 2006b). Such CVEC can be recovered as fully virulent bacteria by inoculation of the water into rabbit intestines. Furthermore, *V. cholerae* shed in stools of cholera patients, when inoculated in environmental water samples in the laboratory, exhibit characteristics similar to CVEC, suggesting that CVEC are the infectious form of *V. cholerae* in water, and that CVEC in nature may have been derived at least in part from human cholera stools. Thus these results support a model of cholera transmission in which *in vivo* formed biofilms contribute to enhanced infectivity and environmental persistence of pathogenic *V. cholerae*.

Further studies proposed that when the apparent concentration of *V. cholerae* in water is low by conventional enrichment and cultivation, certain human victims ingesting clumps of CVEC in the water might still get a fully infectious dose of the pathogen, and may further amplify the strain asymptomatically or even become an index cholera case that begins an epidemic due to that strain. In contrast, towards the end of an epidemic when environmental phage concentration is known to be high even CVEC fails to sustain the epidemic presumably due to predation of the strain when the clumped cells in CVEC disperse *in vivo*. Since the *V. cholerae* strain often re-emerges and causes a subsequent epidemic during the next cholera season, it is obvious that a proportion of the epidemic *V. cholerae* cells survive the phage attack. Therefore, the ability to survive phage attacks in the environment may be an important aspect to the epidemic cycle of cholera. Laboratory-produced biofilms display markedly more resistance to phage predation than planktonic cells (Zhu, Faruque and Mekalanos, unpublished). Furthermore, the metabolic status of cells and the expression of phage receptors can also influence the susceptibility of cells to phage infection. Understanding the complex interactions of many different factors including environmental factors and plankton, the lytic phages, and the role of the human host may lead to the development of a unified model to explain cholera dynamics.

The recent recognition that phage predation may play a role in controlling cholera epidemics (Faruque *et al.*, 2005a; Faruque *et al.*, 2005b) also suggest that changes in this organism may be driven by factors outside of the human host. The emergence of new serogroups (e.g. O139) and the displacement of old biotypes (e.g. the classical by the El Tor) may be driven by environmental forces that include phages, propensity to form biofilms on aquatic matter and even interaction with marine animals (Colwell and Spira 1992; Kirn *et al.*, 2005; Faruque *et al.*, 2006). It seems inevitable that investigation into the history and biology of cholera will continue for years to come, and generate inspiring research questions.

Acknowledgements

Part of the work, presented in this chapter, was supported by the National Institute of Health grants GM068851-5 and AI070963-01A1 under different sub-agreements between the Harvard Medical School and the ICDDR,B. The ICDDR,B is supported by countries and agencies which share its concern for the health problems of developing countries.

References

Baselski, V.S., Medina, R.A., and Parker, C.D. (1978). Survival and multiplication of *Vibrio cholerae* in the upper bowel of infant mice. Infect. Immun. 22, 435–440.

Baselski, V.S., Medina, R.A., and Parker, C.D. (1979). *In vivo* and in vitro characterization of virulence deficient mutants of *Vibrio cholerae*. Infect. Immun. 24, 111–116.

Chakraborty, S., Mukhopadhyay, A.K, Bhadra, R.K, Ghosh, A.N, Mitra, R. Shimada, T., Yamasaki, S., Faruque, S.M., Takeda, Y., Colwell, R.R., and Nair, G.B. (2000). Virulence genes in environmental strains of *Vibrio cholerae*. Appl. Env. Microbiol. 66, 4022–4028.

Colwell, R.R. and Spira, W.M. (1992). The ecology of *Vibrio cholerae*. In Cholera, D. Barua, and W.B. Greenough III eds. (New York, USA; Plenum Medical Book Co.) pp. 107–127.

Colwell, R.R., and Huq, A. (1994). Vibrios in the environment: viable but nonculturable *Vibrio cholerae*. In *Vibrio cholerae* and cholera: molecular to global perspectives. I.K. Wachsmuth, P.A. Blake, and O. Olsvik eds. (Washington DC, USA, ASM Press), pp. 117–133.

Colwell, R.R. (1996). Global climate and infectious disease: the cholera paradigm. Science 274, 2025–2031.

Colwell, R.R. (2004). Infectious disease and environment: cholera as a paradigm for water-borne disease. Int. Microbiol. 7, 285–289.

Dziejman, M., Balon, E., Boyd, D., Fraser, C.M., Heidelberg, J.F. and Mekalanos, J.J. (2002). 'Comparative genomic analysis of *Vibrio cholerae*: genes that correlate with cholera endemic and pandemic disease.' Proc. Natl. Acad. Sci. USA 99; 1556–61.

Cholera Working Group. International Centre for Diarrhoeal Diseases Research, Bangladesh (1993). Large epidemic of cholera-like disease in Bangladesh caused by *Vibrio cholerae* O139 synonym Bengal. Lancet *342*, 387–90.

Faruque, S.M., Albert, M.J. and Mekalanos, J.J. (1998). Epidemiology, genetics, and ecology of toxigenic *Vibrio cholerae*. Microbiol. Mol. Biol. Rev. *62*, 1301–1314.

Faruque, S.M., and Mekalanos, J.J. (2003). Pathogenicity islands and phages in *Vibrio cholerae* evolution. Trends. Microbiology. *11*, 505–510.

Faruque, S.M., Chowdhury, N., Kamruzzaman, M., Dziejman, M., Rahman, M.H., Sack, D.A., Nair, G.B. and Mekalanos, J.J. (2004). Genetic diversity and virulence potential of environmental *Vibrio cholerae* population in a cholera-endemic area. Proc. Natl. Acad. Sci. USA *101*, 2123–2128.

Faruque, S.M., Naser, I.B., Islam, M.J., Faruque, A.S., Ghosh, A.N., Nair, G.B., Sack, D.A. and Mekalanos, J.J. (2005a). Seasonal epidemics of cholera inversely correlate with the prevalence of environmental cholera phages. Proc. Natl. Acad. Sci. USA *102*, 1702–7.

Faruque, S.M., Islam, M.J., Ahmad, Q.S., Faruque, A.S., Sack, D.A., Nair, G.B., and Mekalanos, J.J. (2005b). Self-limiting nature of seasonal cholera epidemics: Role of host-mediated amplification of phage. Proc. Natl. Acad. Sci. USA 2005. *102*, 6119–6124.

Faruque, S.M., Islam, M.J., Ahmad, Q.S., Biswas, K., Faruque, A.S., Nair, G.B., Sack, R.B., Sack, D.A., and Mekalanos, J.J. (2006a). An improved technique for isolation of environmental *Vibrio cholerae* with epidemic potential: monitoring the emergence of a multiple-antibiotic-resistant epidemic strain in Bangladesh. J. Infect. Dis. *193*, 1029–1036.

Faruque, S.M., Biswas, K., Udden, S.M., Ahmad, Q.S., Sack, D.A., Nair, G.B., and Mekalanos, J.J. (2006b). Transmissibility of cholera: *in vivo*-formed biofilms and their relationship to infectivity and persistence in the environment. Proc. Natl. Acad. Sci. USA *103*, 6350–6355.

Finkelstein, R.A., Vasil, M.L., and Holmes, R.K. (1974). Studies on toxinogenesis in *Vibrio cholerae*. I. Isolation of mutants with altered toxinogenicity. J. Infect. Dis. *129*, 117–123.

Franco, A., Fix, A., Prada, A., Paredes, E., Palomino, J., Wright, A., Johnson, J., McCarter, R., Guerra, H., and Morris, J. (1997). Cholera in Lima, Peru correlates with prior isolation of *Vibrio cholerae* from the environment. Am. J. Epidemiol. *146*, 1067–1075.

Freter, R., and Brein, O.P.C.M. (1981). Role of chemotaxis in the association of motile bacteria with intestinal mucosa: fitness and virulence of nonchemotactic *Vibrio cholerae* mutants in infant mice. Infect. Immun. *34*, 222–233.

Gil, A., Louis, V., Rivera, I., Lipp, E., Huq, A., Lanata, C., Taylor, D., Russek-Cohen, E., Choopun, N., Sack, R., and Colwell, R. (2004). Occurrence and distribution of *Vibrio cholerae* in the coastal environment of Peru. Environ. Microbiol. *6*, 699–706.

Howard, B.D. (1971). A prototype live oral cholera vaccine. Nature *230*, 97–99.

Holmes, R.K., M.L. Vasil, and R.A. Finkelstein. 1975. Studies on toxinogenesis in *Vibrio cholerae*. III. Characterization of nontoxigenic mutants in vitro and in experimental animals. J. Clin. Invest. 55:551–556.

Huq, A, Sack R.B., Nizam, A., Longini I.M., Nair, G.B., Ali, A., Morris, J.G., Khan, M.N.H., Siddique, A.K., Yunus, M., Albert, M.J., Sack, D.A., Colwell, R.R. (2005). Critical factors influencing the occurrence of *Vibrio cholerae* in the environment of Bangladesh. Appl. Env. Microbiol. *71*, 4645–4654.

Islam M.S., Drasar, B.S., and Sack, R.B. (1994a). The aquatic environment as reservoir of *Vibrio cholerae*: A review. J. Diarrhoeal Dis. Res. *11*, 197–206.

Islam M.S., Drasar, B.S., and Sack, R.B. (1994b). The aquatic flora and fauna as reservoirs of *Vibrio cholerae*: A review. J. Diarrhoeal Dis. Res. *12*, 87–96.

Kaper, J.B., Morris, J.G., Jr and Levine, M.M. (1995). Cholera. Clin. Microbiol. Rev. *8*, 48–86.

Khan, M.U., Shahidullah, M., Haque, M.S., Ahmed, W.U. (1984). Presence of vibrios in surface water and their relation with cholera in a community. Trop Geogr Med. *36*, 335–40.

Kirn, T.J., Jude, B.A. and Taylor, R.K. (2005). A colonization factor links *Vibrio cholerae* environmental survival and human infection. Nature *438*, 863–6.

Knop, J., and Rowley, D. (1975). Protection against cholera. A bactericidal mechanism on the mucosal surface of the small intestine of mice. Aust. J. Exp. Biol. *53*, 155–165.

Lazar, S. and Waldor, M.K. (1998). ToxR-independent expression of cholera toxin from the replicative form of CTXΦ. Infect. Immun. *66*, 394–397.

Levine, M.M. (1991). South America: the return of cholera. Lancet *338*, 45–46.

Lobitz, B., Beck, L., Huq, A., Wood, B., Fuchs, G., Faruque, A., and Colwell, R. (2000). Climate and infectious disease: use of remote sensing for detection of *Vibrio cholerae* by indirect measurement. Proc. Natl. Acad. Sci. USA *97*, 1438–1443.

Mekalanos, J.J., Collier, R.J., and Roming, W.R. (1978). Affinity filters, a new approach to the isolation of *tox* mutants of *Vibrio cholerae*. Proc. Natl. Acad. Sci. *75*, 941–945.

Mekalanos, J.J. (1983). Duplication and amplification of toxin genes in *Vibrio cholerae*. Cell *35*, 253–263.

Mekalanos, J.J. (1985). Cholera toxin: genetic analysis, regulation and role in pathogenesis. Curr. Top. Microbiol. Immunol. *118*, 97–118.

Merrell, D.S., Butler, S.M. Qadri, F., Dolganov, N.A., Alam, A., Cohen, M.B., Calderwood, S.B. Schoolnik, G.K., and Camilli. A. (2002). Host-induced epidemic spread of the cholera bacterium. Nature. *417*, 642–5.

Mukhopadhay, A.K., Chakraborty, S., Takeda, Y., Nair, G.B. and Berg, D.E. (2001). Characterization of VPI pathogenicity island and CTXΦ prophage in environmental strains of *Vibrio cholerae*. J. Bacteriol. *183*, 4737–4746.

Novitsky, J.S., and Morita, R.Y. (1976). Morphological characterization of small cells resulting from nutrient starvation of a psychrophylic marine vibrio. Appl. Enviorn. Microbiol. *32*, 617–622.

Novitsky, J.S., and Morita, R.Y. (1977). Survival of a psychrophilic marine vibrio under long-term nutrient starvation. Appl. Enviorn. Microbiol. *33*, 635–641.

Novitsky, J.S., and Morita, R.Y. (1978). Possible strategy for the survival of marine bacteria under starvation conditions. Mar. Biol. *48*, 289–295.

Pascual, M., Rodo, X., Ellner, S.P., Colwell, R., and Bouma, M.J. (2000). Cholera dynamics and El Nino-southern oscillation. Science *289*, 1766–1769.

Popovic, T., Bopp, C., Olsvik, O. and Wachsmuth, I.K. (1993). Epidemiologic application of a standardized ribotype scheme for *Vibrio cholerae* O1. J. Clin. Micro. *31*, 2474–2482.

Raskin, D.M., Seshadri, R., Pukatzki, S.U. and Mekalanos, J.J. (2006). Bacterial genomics and pathogen evolution. Cell *124*, 703–14.

Rivera, I.N.G., Chun, J., Huq, A., Sack, R.B., and Colwell, R.R. (2001). Genotypes associated with virulence in environmental isolates of *Vibrio cholerae*. Appl. Env. Microbiol. *67*, 2421–2429.

Waldor M.K., and Mekalanos, J.J. (1996). Lysogenic conversion by a filamentous phage encoding cholera toxin. Science. *272*, 1910–1914.

Coordinated Regulation of Gene Expression in *Vibrio cholerae*

9

Rupak K. Bhadra and Bhabatosh Das

Abstract

Vibrio cholerae, the causative agent of the severe diarrhoeal disease cholera, has evolved with intricate signal transduction and gene regulatory systems to survive and grow under various environmental conditions. The virulence regulon of *V. cholerae,* which involves multiple genes working in a coordinated manner, represents a regulatory paradigm for extracellular bacterial pathogens. Availability of the whole genome sequence has allowed microarray based transcriptome analyses of *V. cholerae* cells isolated directly from cholera patients. Such studies indicate that quite a large number of genes are involved in the disease process and their expression pattern changes as the infection progresses. Further understanding of the process came with the recent discoveries of small non-coding RNAs and intracellular signal molecule c-di-GMP as modulators of gene expression in *V. cholerae.* Transcriptome analysis has also shed light on synchronized gene expression related to chitin utilization and development of natural competence when the organism exists in the natural aquatic environment. Thus, the survival, evolution and pathogenesis of *V. cholerae* appear to be controlled by several intricate overlapping regulatory circuits.

Introduction

Vibrio cholerae, the causative agent of cholera, has two distinct lifestyles; one is the host-associated state during human intestinal infection and the other is the natural aquatic environment where it exists as part of the normal bacterial flora. Thus, the pathogen has evolved with a variety of intricate signal transduction and gene regulation systems to adapt to challenging environmental conditions. Signal transduction and coordinate regulation of genes and operons are fundamental mechanisms, which help microorganisms to survive and grow under different conditions (Miller *et al.,* 1989a). The coordinated expression of genes is particularly crucial for those microorganisms that occupy diverse environmental niches. For example, pathogenic *V. cholerae* must be able to express multiple virulence genes appropriately in a synchronized manner to cause the disease, and at the same time to survive and grow in the intestinal environment. Similarly, outside of the human host the coordinated expression of genes is required for adaptation of the pathogen to the aquatic environment.

Although there are multiple epidemic strains of *V. cholerae* including *V. cholerae* O1 of the El Tor and classical biotypes, and *V. cholerae* O139, (Kaper *et al.,* 1995), all these strains carry two critical clusters of virulence genes. These are the CTX prophage, and the TCP pathogenicity island (Faruque *et al.,* 1998). The expression of genes in these clusters are dependent on environmental signals as evident from various studies under laboratory conditions and in animal models.

Human intestinal infection due to pathogenic *V. cholerae* proceeds according to the following scenario. The bacterium enters its host through oral ingestion of contaminated water or food, passes through the gastric acidity barrier of the stomach and bile-rich initial part of the small intestine. Upon entering the ileum, motile *V.*

cholerae penetrates through the mucus lining and arrives at villous epithelial cell surface and colonizes there by expressing the colonization factor toxin-coregulated pilus (TCP). Colonization of the epithelial cell surface probably induces *V. cholerae* for the synthesis and secretion of the potent enterotoxin, cholera toxin (CT). CT is mainly responsible for the profuse watery diarrhoea in cholera patients, and the diarrhoea causes *V. cholerae* to spread into the environment (Faruque *et al.*, 2004; Kaper *et al.*, 1995). Thus, the 'environment-host-environment' cycle of pathogenic *V. cholerae* continues during an epidemic of cholera.

The coordinated expression of CT and TCP, in an appropriate way within the intestinal milieu appears to be crucial for the pathogenesis, and dissemination of *V. cholerae*. Thus, much efforts have been given to understand how the expression of these factors are regulated precisely under various environmental conditions. The *ctxAB* operon, which codes for CT is located in the genome of CTXϕ, a lysogenic filamentous phage (Waldor and Mekalanos, 1996), and the *tcpA* gene coding for TcpA protein, the major subunit of TCP (Taylor *et al.*, 1987), is located in the TCP pathogenicity island, alternatively referred to as the *Vibrio* pathogenicity island (VPI). This chromosomal island also harbours three important virulence regulatory genes, *tcpP*, *tcpH* and *toxT* (DiRita *et al.*, 1991; Häse and Mekalanos, 1998), all of which play important roles in cholera pathogenesis. Like *ctxAB* and *tcpA* genes, expression of each of these regulators is also dependent on similar kind of environmental signals. While *tcpPH* operon codes for two interacting transmembrane protein molecules, TcpP and TcpH, the *toxT* gene encodes ToxT which is an AraC-like transcriptional activator. The expression of ToxT is under the control of ToxR protein which is known as the master regulator that appears to sense environmental signals, and vary gene expression (Miller and Mekalanos, 1984). However, ToxT, is the most downstream positive regulator of the ToxR regulon and hence the subject of most intense research in recent years (Klose, 2001). Although the *ctxAB* and the TCP pathogenicity island genes are essential determinants of virulence, involvement of ancestral gene products, e.g. ToxR in the virulence regulatory network of *V. cholerae* emphasizes that *V. cholerae* requires functional cooperation from host genes to execute full-fledged virulence.

Our recent knowledge on environmental signal mediated virulence gene expression in *V. cholerae* has been rapidly enriched mainly by the availability of the whole genome sequence of *V. cholerae* O1 strain N16961 (Heidelberg *et al.*, 2000) and DNA microarray technology for transcriptional profiling (Bina *et al.*, 2003; LaRocque *et al.*, 2005; Merrell *et al.*, 2002; Xu *et al.*, 2003). Furthermore, the recent discovery of non-coding small RNAs (sRNAs) in controlling *V. cholerae* pathogenesis along with quorum sensing in modulation of gene expression in response to environmental stimuli (Lenz *et al.*, 2004), have shed light on intricate gene regulation networks in *V. cholerae*. Other recent findings that have added to our understanding include chitin-induced natural competence in *V. cholerae* (Meibom *et al.*, 2005) and involvement of intracellular signalling molecule c-di-GMP in modulation of various phenotypic processes including virulence gene expression (Lim *et al.*, 2006; Tischler and Camilli, 2005). In this chapter we intend to discuss the intricate mechanisms involved in regulating the expression of virulence and other genes in pathogenic *V. cholerae* in response to various environmental stimuli encountered by the bacterium inside and outside the host.

Virulence gene regulation studies under *in vitro* conditions

Currently available information regarding regulation of virulence gene expression in *V. cholerae* are derived from laboratory experiments using pure culture of either wild type pathogenic strains and/or their mutants. Among the various virulence-related regulators of *V. cholerae*, ToxR was discovered first based on its ability to transcriptionally activate the *ctxAB* operon in an *Escherichia coli* genetic background (Miller and Mekalanos, 1984). Thus, for historical reasons and because ToxR is a critical transcriptional activator, that acts early in the regulatory cascade to induce several virulence genes in a coordinated manner, this virulence gene regulatory network is widely known as the 'ToxR regulon' (Peterson and Mekalanos, 1988; Skorupski and Taylor,

1997a). In this model, as depicted in Fig. 9.1, under virulence inducing condition the ToxR protein encoded by the *toxR* gene activates transcription of the *toxT* gene, and ToxT activates expression of CT and TCP. Apart from activating *toxT*, ToxR also directly activates transcription of the gene encoding the outer membrane porin protein OmpU and represses transcription of *ompT* gene, which encodes another outer membrane porin. Further investigation has shown that various other transcriptional activators like AphA, AphB, TcpP, TcpH and additional factors may influence the ToxR regulon as shown in Fig. 9.1. Furthermore, recent transcriptional profiling experiments conducted on *V. cholerae* cells shed in stools by cholera patients showed activation of several genes including the known virulence genes of the ToxR regulon (Bina *et al.*, 2003) indicating that the function of ToxR is also important under *in vivo* conditions. Previously experiments conducted in a neonatal mouse model using a *toxR* mutant indicted that ToxR is essential for virulence under *in vivo* conditions (Lee *et al.*, 1999). However, the mechanisms by which environmental signals modulate the ToxR regulon are not yet fully understood.

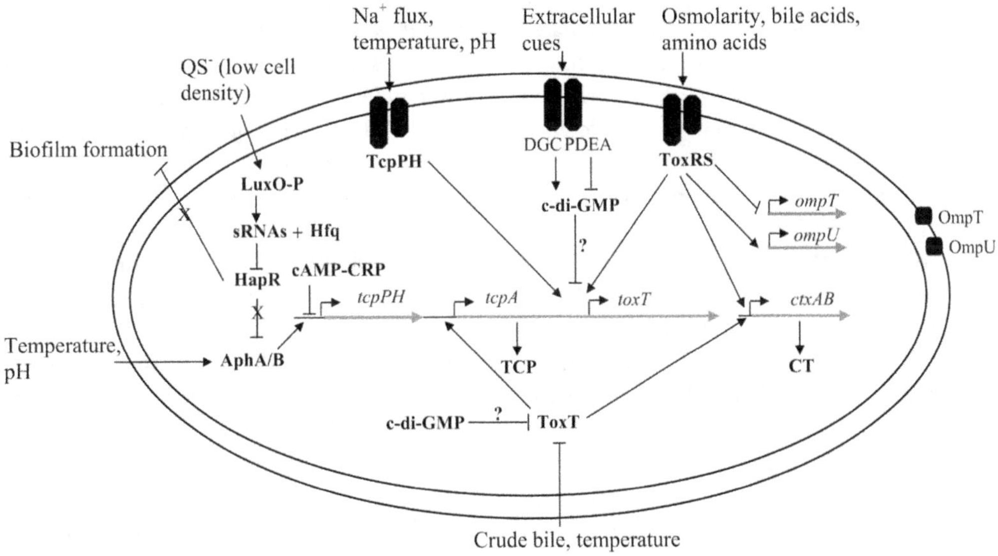

Figure 9.1 Model depicting the virulence regulatory cascade that controls CT and TCP expression in *V. cholerae* in response to various extracellular and intracellular signals. Thick arrows (gray in colour) indicate genes or operons. The *toxRS*, *hapR*, *aphA*, *aphB*, *luxO*, *ompT*, *ompU*, *cya* (coding for adenylate cyclase, which synthesizes cAMP), *crp* (coding for cAMP receptor protein) and c-di-GMP synthesizing (DGC) and degrading (PDEA) proteins coding genes reside in the ancestral genome of *V. cholerae*. Bent arrows indicate promoter regions of genes and operons. The transcription of *toxT* gene occurs from two promoters, one from upstream of *tcp* operon (shown as *tcpA* promoter) and the other one from its own promoter. Thin arrows indicate positive regulation while T-bars denote negative/ inhibitory effects. The question marks signify possible targets, which need further investigation. QS- indicates quorum sensing 'OFF mode' at low cell density when LuxO protein remains phosphorylated (LuxO-P) and inhibits the quorum sensing master regulator HapR and allowing AphA/B regulators to act positively on *tcpP/H* promoter (see text for details). In the absence of HapR, biofilm formation continues. The positive transcriptional activators ToxRS and TcpPH (both are localized in the inner membrane) act directly on *toxT* promoter region leading to expression of ToxT. ToxT, the key cytosolic transcriptional regulator of virulence, directly activates the tcpA and ctxAB genes. Apart from activating *toxT*, ToxRS positively and negatively regulate the expression of two outer membrane porin proteins, OmpU and OmpT, respectively. Environmental cues activate membrane bound or cytoplasmic (not shown) DGC (diguanylate cyclase, GGDEF/GGEEF domain containing proteins) or PDEA (phosphodiesterase A, EAL domain containing protein) enzyme [for example VieA in *V. cholerae*] which synthesizes or degrades c-di-GMP. The c-di-GMP may negatively regulate *toxT* transcription or it can bind and inhibit the function of the ToxT protein leading to down regulation of virulence genes. The intracellular cAMP-CRP complex binds at *tcpPH* promoter and negatively regulates TcpPH expression. Besides the other regulators, the chromosome architectural protein H-NS negatively regulates *tcpA*, *toxT* and *ctxAB* transcriptions, which are not shown in the diagram.

Environmental signals and the ToxR regulon

Although the structure of the ~32-kDa ToxR protein (Miller *et al.*, 1987) has been studied in detail, its exact mechanism of action is still not clearly understood. ToxR is an integral inner membrane protein with three distinct domains: an N-terminal cytoplasmic DNA-binding domain, a central transmembrane-domain and a C-terminal periplasmic domain (DiRita and Mekalanos, 1991). The cytoplasmic domain contains a predicted helix-turn-helix region and shows sequence homology to the DNA-binding domain of several prokaryotic transcriptional activator including OmpR (Miller *et al.*, 1987), and mutation of this region abolishes ToxR-mediated transcriptional activation of various genes it controls (Ottemann *et al.*, 1992; Skorupski and Taylor, 1997a). Thus, ToxR may be considered as a one-component derivative of a two-component signal transduction system of prokaryotes. It is noteworthy that a second regulatory gene *toxS*, which codes for a 19-kDa inner membrane protein, is present immediately downstream of *toxR* in the same transcriptional unit, and under ToxR inducing laboratory conditions ToxS is required concomitantly for activation of the *ctxAB* operon (Miller *et al.*, 1989b) as shown in Fig. 9.1. By mutational analysis it has been demonstrated that ToxR and ToxS proteins interact with each other and ToxS is essential for full ToxR regulon expression in *V. cholerae* (DiRita and Mekalanos, 1991). The finding that ToxR is involved in sensing a number of environmental signals to affect *ctxAB* expression helps explain its unusual cellular location. The environmental cues that affect the expression of CT include osmolarity, temperature, pH, and presence of certain amino acids e.g. asparagines, arginine, glutamate, and serine (Miller and Mekalanos, 1988) (Fig. 9.1). However, with the possible exception of fluctuation in osmolarity, evidence is lacking to support that ToxR acts alone in sensing all of these environmental parameters. Although the exact mechanism of activation of ToxR by environmental signals is still unknown, it was shown that the N-terminal portion of the activated ToxR binds to the 5'-TTTTGAT-3' heptamer repeats of the *ctxAB* promoter and this binding is modulated by the periplasmic C-terminal domain (Miller *et al.*, 1987). Under laboratory conditions, the ToxR regulon is induced when *V. cholerae* O1 classical biotype cells are grown in Luria broth (LB) at pH 6.5 and an osmolarity of 66 mmol/l at 30°C (virulence inducing conditions), but it is repressed at 37°C, pH 8.5 and an osmolarity of 300 mmol/l (virulence repressing conditions) (Gardel and Mekalanos, 1994). It has been proposed that the periplasmic domain senses osmolarity and is able to adopt a conformation that promotes transcriptional activation of *ctxAB* genes. Since in *V. cholerae*, ToxR can directly modulate the expression of two outer membrane porin proteins OmpT and OmpU in response to change in osmolarity (Fig. 9.1), it is believed that the primary function of ToxR is to sense the salinity levels in the natural environmental water, which is basically a physiological function of the organism. Paradoxically, during infection the ToxR protein remains functional even at high intestinal luminal osmolarity, which is close to 300 mM with respect to NaCl (Miller and Mekalanos, 1988). However, osmolarity throughout the gastrointestinal tract is not a constant parameter, and as a result signal transduction via the periplasmic domain of ToxR may fluctuate accordingly. At an optimal osmolarity ToxR may transfer the signal into multiple gene activation through its cytoplasmic transcriptional activator domain, but where osmolarity is high the periplasmic domain of ToxR adopts an inactive conformation leading to repression of the ToxR regulon. This model may be supported by the fact that ToxR is a constitutively expressed membrane localized transcriptional activator. However, further work is needed to fully understand the role of ToxR in *V. cholerae* gene expression.

AphA/AphB/TcpP/TcpH-mediated virulence regulatory cascade

The *tcpPH* operon located on the TCP pathogenicity island in *V. cholerae* is transcriptionally regulated by both temperature and pH (Thomas *et al.*, 1995). Thus, the function of TcpP/H appears to be very similar to that of ToxR/S. Importance of TcpP/H in the regulation of virulence gene expression in *V. cholerae* has been further examined by mutational analysis (Häse and Mekalanos, 1998). Like ToxR, TcpP is homologous to the OmpR family of transcriptional activators except that these proteins are of a special class containing DNA-binding motifs in their N-termini rather than in their C-termini as discussed above. Like ToxS, TcpH has been shown to enhance the activity of its cognate partner TcpP (Häse and Mekalanos, 1998). Recent studies indicate that transcription of *tcpP* is under the direct control of environmental cues like temperature, pH etc., and thus expression of *tcpP* is critical in regulation of expression of two principal virulence factors CT and TCP (Fig. 9.1). However, the activities of the TcpP protein is mediated by acting positively on the promoter region of the master virulence regulator gene *toxT* present within the same pathogenicity island (Fig. 9.1). Krukonis *et al.* (2000) showed by DNAse I foot-printing that TcpP binds to the *toxT* promoter (Fig. 9.1) between the ToxR and RNA polymerase-binding sites in order to activate transcription. It was shown that ToxR alone can activate the *ctxAB* promoter in the *E. coli* genetic background, but in homologous genetic background ToxR affect the *ctxAB* expression by activating the *toxT* gene; similarly TcpP indirectly activates the *ctxAB* promoter through activation of *toxT* (Fig. 9.1). Unlike ToxR, however, TcpP is not required to induce or repress the *ompU* or *ompT* gene (Fig. 9.1) since a TCP island deleted *V. cholerae* mutant strain is fully capable of modulating the expression of OmpU and OmpT proteins in a ToxR-dependent manner (Li *et al.*, 2000). Several lines of evidence suggest that the periplasmic domain of TcpP are critical for signalling across the inner membrane due to having a receptor-like topology. Further support in this regard comes from mutagenesis of *V. cholerae* cells with the transposon Tn*bla* (Häse and Mekalanos, 1998), which allows for identification of operon fusions to the periplasmic β-lactamase (responsible for resistance to the antibiotic ampicillin). A mutant expressing the Tn*bla* fused periplasmic domain of the TcpP protein was found to be constitutive in expression of virulence factors, e.g. expression of TCP through the activation of *toxT* gene. Thus, alteration of the periplasmic domain by fusion to β-lactamase led to a structural conformation, which constantly activated the *toxT* gene and thus, the signal detection mechanism is bypassed. Since TcpP is an inner membrane transcriptional activator like ToxR (Fig. 9.1), any perturbation in membrane topology of *V. cholerae* due to a change in environmental condition may lead to transcriptional activation of certain sets of genes to combat that situation.

Regulation of expression of *tcpP* appeared to be more intricate since the identification of two new regulators, AphA and AphB as activators of *tcpPH* (Kovacikova and Skorupski, 1999). These activators are chromosomally encoded proteins, which function synergistically at the *tcpP* promoter early in the virulence transcriptional cascade (Fig. 9.1). The AphA is a 20.4-kDa protein containing a predicted coiled-coil region, which is found in a number of transcriptional regulators. Mutation in the *aphA* gene prevents expression CT and TCP under virulence inducing conditions (Skorupski and Taylor, 1999). On the other hand, AphB is a 33.3-kDa protein belonging to the LysR family of transcriptional activators. AphB acts together with AphA to activate the transcription of *tcpPH* operon in response to environmental stimuli (Fig. 9.1). These signals have little effect on transcription of *aphA* and *aphB* themselves. Thus, AphA and AphB are most likely to be influenced by environmental conditions at the protein level. Kovacikova and Skorupski (2000) have shown that the base A at position –65 of the *tcpPH* promoter of a classical biotype strain is critical for transcriptional activation by AphB and the presence of G in this position in an El Tor promoter confers the classical and El Tor biotype specific *ctxAB* and *tcpA* expression in response to environmental stimuli.

Although much information is available regarding the activation of virulence cascade in *V. cholerae* under certain conditions, we know much less about the fate of these activated regulators when the pathogen is shifted from virulence inducing conditions to non-inducing conditions as happens when the organism is shed in the environment by cholera patient. A recent study has demonstrated YaeL protease-dependent degradation of membrane bound TcpP in *V. cholerae* TcpH$^+$ cells when shifted from virulence inducing condition to non-inducing condition (Matson and DiRita, 2005). The data indicate the role of another layer of signal processing mechanism that stimulates YaeL activity for rapidly shutting down virulence regulon under non-permissive conditions. However, YaeL mediated degradation mechanism was not detected in the case of ToxR mediated gene expression (Matson and DiRita, 2005).

ToxT, the linchpin virulence regulator

Although the transmembrane protein ToxR, is required for expression of virulence factors CT and TCP, the activation of *ctxAB* and *tcpA* is in fact carried out through the cytosolic transcriptional activator protein ToxT (Klose, 2001) (Fig. 9.1). The *toxT* gene was discovered in early 1990s (DiRita *et al.*, 1991) and like *tcpA*, *tcpP* and *tcpH* genes, *toxT* is also located in the TCP pathogenicity island of epidemic *V. cholerae* strains (Karaolis *et al.*, 1998). ToxT, the product of *toxT* gene, is a 32-kDa cytosolic protein with homology to the AraC/XlyS family of transcriptional activators, which have a conserved domain at their C-termini containing helix-turn-helix motifs typical of DNA binding proteins (Higgins *et al.*, 1992). The amino termini of this class of proteins are quite divergent and usually needed for effector binding and dimerization (Childers *et al.*, 2007). Transcription of *toxT* was greatly reduced in a *toxR* negative background suggesting a regulatory cascade in *V. cholerae* that controls expression of most virulence factors (DiRita *et al.*, 1991) (Fig. 9.1).

Under *in vitro* virulence inducing condition, ToxR along with TcpP directs the coordinated control of virulence gene expression by activating the *toxT* gene. ToxT expression leads to direct transcriptional activation of genes encoding the critical virulence factors CT and TCP (Krukonis *et al.*, 2000). In response to specific laboratory conditions the *toxT* gene expression is complex in that initially it is activated by ToxR and TcpP at a promoter situated just upstream of the gene (Higgins and DiRita, 1994; Krukonis *et al.*, 2000) (Fig. 9.1). Once activated, ToxT controls its own expression by an autoregulatory loop through transcriptional activation at the *tcpA* promoter, the first gene in the *tcp* operon (Fig. 9.1), leading to read-through transcription of the gene (Skorupski and Taylor, 1997a; Yu and DiRita, 1999). In addition, a third *toxT* transcript has been detected that initiates 105-bp upstream from the start site of the ToxR-regulated promoter (Higgins and DiRita, 1994). Activation of *toxT* transcription appears to be a critical event in expression of CT and TCP, because mutations that eliminate regulatory regions within the *toxT* promoter lead to loss of the virulence factors even under inducing conditions (Yu and DiRita, 1999).

Apart from ToxR inducing conditions, ToxT expression is also influenced by temperature and crude bile, a major constituent of digestive tract involved in lipid metabolism (Fig. 9.1). It was shown by Schuhmacher and Klose (1999) that ToxT expressed from an inducible promoter activated high-level production of CT and TCP in *V. cholerae* at 30°C, but expression of these virulence factors was substantially decreased or even abolished by the addition of 0.4% bile to the medium and/or an increase of the growth temperature to 37°C (Fig. 9.1). The results indicate additional level of complexity controlling the expression of virulence factors by environmental signals in *V. cholerae*. Thus, it appears that negative regulation of ToxT-dependent transcription by environmental stimuli, like temperature and bile, probably prevents the incorrect temporal and spatial expression of virulence factors during cholera pathogenesis. Furthermore, the amino acid residues that are critical for the function of the ToxT protein have recently been identified by scanning alanine mutagenesis (Childers *et al.*, 2007). Most of these mutated ToxT proteins caused reductions in CT and TCP expression in a *V. cholerae* Δ*toxT* background under virulence inducing conditions. From this study it was also apparent that amino-terminal residues are subject to environmental modulation and certain specific region of the protein contributes to promoter-specific activation (Childers *et al.*, 2007).

Bile mediated regulation of virulence genes

Bile aids in lipid metabolism in the intestinal tract. Several studies have attempted to address the role of bile as a potential signal in modulating the expression of virulence factors and motility of *V. cholerae*. Gupta and Chowdhury (1997) first reported that, in the sixth pandemic classical biotype *V. cholerae* strains, crude bovine bile drastically reduced expression of CT and TCP but significantly increased motility, apparently in a ToxR-independent manner (Fig. 9.1). Schuhmaker and Klose (1999) showed that bile prevents ToxT mediated transcriptional activation of virulence factor expression in *V. cholerae* (Fig. 9.1). It was recently shown that purified bile acids, the major components of crude bile, actually activate expression of CT in a classical strain under virulence non-inducing condition (LB with pH 8.5, 37°C), which is ToxRS-dependent but ToxT-independent (Hung *et al.*, 2005).

Response to low oxygen tension

Low oxygen tension is a characteristic feature of mammalian intestinal tissue and thus anaerobiosis may serve as a signal for gene regulation in enteric bacteria (Lynch and Lin, 1996). Recently, transcriptome analysis of *in vivo* (rabbit intestine) grown and *V. cholerae* cells shed by cholera patients have provided evidences that substantial metabolic shift occurs towards anaerobic metabolism of the *in vivo* grown bacterial cells (Bina *et al.*, 2003; Xu *et al.*, 2003). Both these studies have shown activation of the gene *arcA* the product of which is a global repressor of aerobic respiration genes and operons. Sengupta *et al.* (2003) showed that mutation in the *arcA* gene significantly decreases the expression of *V. cholerae* virulence gene activator protein ToxT with concomitant reduction in the expression of CT and TCP, although the mutation have no effect on expression of *toxR* and *tcpP* genes. Thus, it seems that ArcA positively modulates the expression of ToxT, which then influences expression of CT and TCP.

Additional modulators of the virulence regulatory cascade

cAMP- and CRP-mediated regulation

It was shown that mutation in either the adenylate cyclase gene (*cya*), which is responsible for cAMP synthesis or in the *crp* gene, which codes for cAMP receptor protein (CRP), results in expression of CT and TCP under ToxR repressing conditions (LB, pH 8.5, 37°C) suggesting that CRP negatively regulates *ctx* and *tcp* transcriptions (Skorupski and Taylor, 1997b). A cAMP-CRP binding site was found to be within the AphA/B binding regions of the *tcpPH* promoter, which raises the possibility that negative effect of cAMP-CRP on virulence gene expression is the result of its ability to influence AphA/B mediated transcriptional activation of *tcpPH* (Kovacikova and Skorupski, 2001).

H-NS and silencing of virulence regulon

Histone-like nucleoid structuring (H-NS) protein is a small, abundant bacterial nucleoid associated protein, which influences expression of various seemingly unrelated genes mainly by organizing promoter and related regulatory regions into nucleoprotein complexes in response to environmental cues like temperature, anaerobiosis, pH, osmolarity etc. Nye *et al.* (2000) first demonstrated a role for H-NS in negatively modulating the expression of several genes within the ToxR/TcpP/ToxT virulence regulon of *V. cholerae*. A *V. cholerae* Δ*hns* strain showed nearly constitutive levels of expression of the genes encoding CT, TCP and ToxT even under normally nonpermissive conditions. The results suggest that H-NS is a gene silencer, which permits activation of virulence regulators such as ToxT to induce virulence only under suitable environmental conditions. Yu and DiRita (2002) have shown that ToxT protein binds directly and specifically to the *ctxAB* and *tcpA* promoters by displacing the negative regulator H-NS leading to CT and TCP expression under virulence inducing condition. Recently, bile mediated repression of virulence gene expression was found to be relieved in an *hns* mutant strain of classical biotype *V. cholerae* (Ghosh *et al.*, 2006). The mutant also over-expressed CT under *in vivo* condition and showed decreased colonization efficiency.

Virulence gene expression by sensing *in vivo* signals

In vivo virulence gene expression in the mouse gut

V. cholerae, a paradigm of extracellular pathogens is an excellent model system for studying the regulations of several authentic virulence determinants like CT and TCP. Furthermore, using *V. cholerae* as a model, mutational analysis and gene fusion technology has allowed to understand the regulatory networks governing the expression of virulence factors in response to various environmental signals in the laboratory. In sharp contrast, our knowledge regarding how these regulatory responses are coordinated *in vivo* during pathogenesis is still very limited. Despite difficulties in addressing the problem, tremendous efforts have been made recently to understand CT and TCP expression during experimental infection in guts of neonatal mice. This effort used a novel genetic technology called recombinase-based *in vivo* expression technology (RIVET). The technique is based on the use of a resolvase reporter that when induced from a promoter, e.g. *ctxAB* or *tcpA* promoter, catalyzes the resolution of resolvase sites (*res* sites) that flank a tetracycline resistance gene cassette used as a marker (Lee *et al.*, 1999). The uniqueness of RIVET is that virulence gene activity can be monitored both temporally and spatially. Analysis using RIVET demonstrated that during intestinal infection activation of CT and TCP genes are temporally distinct with expression of TCP occurring before CT, and a second unknown signal for CT production might be encountered by *V. cholerae* cells after penetrating the mucous coat and colonization of intestinal epithelial cell surface. This could be due to expression of virulence factors by a particular subpopulation of cells, which experience the *in vivo* regulatory signals in a complete manner and respond accordingly. Thus, under *in vivo* condition expressions of CT and TCP genes are not coupled, although under laboratory conditions TCP synthesis is co-regulated with CT (Taylor *et al.*, 1987). RIVET experiments, however, established that ToxT is essential for expression of the *ctxA* and *tcpA* genes *in vivo* and the early burst of *tcpA* transcription requires ToxR and TcpP signalling proteins, since deletion of either of the genes results in delayed induction of *tcpA*. It has been suggested that both ToxR and TcpP sense signals in the intestinal lumen at an early stage of infection and tightly regulate the expression of *toxT*, and therefore *tcpA*. However, the late burst of *tcpA* transcription is maintained either by ToxR or TcpP since loss of ToxR or TcpP does not prevent this second induction from occurring, whereas loss of both does (Lee *et al.*, 1999).

Transcription profiling of *in vivo*-grown *Vibrio cholerae* cells

Availability of whole genome sequence information of the *V. cholerae* strain N16961 (Heidelberg, *et al.*, 2000) allowed direct genome-wide transcription profiling experiments using genomic microarrays, to identify *in vivo* induced virulence genes of *V. cholerae* recovered from cholera patients (Bina *et al.*, 2003; LaRocque *et al.*, 2005; Merrell *et al.*, 2002). Although these type of transcriptome analyses are extremely useful but each of them have drawbacks because of technical reasons which may lead to conflicting results (Bina *et al.*, 2003). Merrell *et al.* (2002) indicated that in human-passaged *V. cholerae* cells the genes needed for nutrient acquisition and motility were up-regulated, while the genes for chemotaxis were down-regulated. Human shed *V. cholerae* also showed up-regulation of a cold-induced protein (TIGR annotation number VCA0933), later named CspV (Datta and Bhadra, 2003). This gene was also activated in transcription profile of *V. cholerae* cells grown in rabbit ileal loops (Xu *et al.*, 2003) indicating that the CspV protein is needed in later stages of infection. Paradoxically, the transcriptome analysis of stool vibrios (Merrell *et al.*, 2002) did not detect induction of *ctxAB*, *tcpA*, *tcpP* or *toxT* transcripts relative to *in vitro* grown *V. cholerae* cells, presumably because the cells were taken from last phase of their pathogenic cycle and at this stage expression of virulence regulon must be in a declining phase. In contrast, Bina *et al.* (2003) showed activation of known virulence factors and regulators in *V. cholerae* cells recovered from cholera patients by using genomic DNA as a reference in their transcriptome analysis. Xu *et al.* (2003) also showed activation of virulence genes in their transcriptome profile, of bacterial cells recovered from rabbit ileal loops. Their *in vivo* expression data strongly suggested that low iron concentration in the host intestine is a potential environmental signal, which is sensed by *V. cholerae* cells and induces several genes involved in iron uptake (Bina *et al.*, 2003; Xu *et al.*, 2003). However, it is not clear how activation of low iron response is related with virulence regulon of *V. cholerae*.

LaRocque *et al.* (2005) compared the transcriptional profile of *V. cholerae* in the early phase of infection of the human upper intestine, represented by organisms in vomitus, with that of *V. cholerae* cells in stool, representing the last phase of human infection. The results of this microarray study revealed that the expression of *tcpA* was much more prominent in the early than the late phase of human infection. Similar conclusion was drawn from RIVET experiments carried out in neonatal mice by Lee *et al.* (1999). Interestingly, LaRocque *et al.* (2005) did not observe high levels of *ctxAB* expression in vibrios recovered from either human vomit or stool samples. This is not surprising, since CT has to be delivered on the epithelial cell surfaces of small intestine where the organism colonizes, the intestinal luminal *V. cholerae* cells recovered from vomitus or stool are thus expected to show down regulation of *ctxAB* expression. Thus, *V. cholerae* virulence gene expression within gastrointestinal environment differs substantially from patterns defined in laboratory-based virulence inducing conditions. This finding further demonstrates the complexity of the environmental signals experienced by *V. cholerae* cells during its passage through the human gut. Furthermore, the transcriptome studies indicate that the upper gastrointestinal tract of human is possibly an ideal environment for rapid replication of *V. cholerae*, since genes involved in DNA replication, ribosomal protein synthesis and energy production were highly expressed in early than in late human infection. The authors of this study also emphasized that human upper intestine is a most suitable niche for horizontal gene transfer events particularly that are important in the evolution of pathogenic strains since high levels of expression of *tcpA* (codes for the subunit protein TcpA of TCP pilus, a receptor of CTXΦ), *tolQRA* (codes for a membrane complex required for CTXΦ entry into *V. cholerae* cells) and *ruvA*, *ruvB*, *recA*, etc. (genes code for proteins involved in homologous recombination) occurred during early infection. The results of this genome-based study have provided a new knowledge about pathogen's differential gene expression profiles at two different phases of human intestinal infection and indicate further that the gene expression in *V. cholerae* rapidly changes as infection progress which again is intricately but in unknown ways governed by human gastrointestinal associated signals.

Recently, Nielsen *et al.* (2006) used rabbit ileal loop model, confocal microscopy, transcriptome analysis and mutants of *rpoS* (codes for RpoS protein, a stationary phase alternative sigma factor) and *hapR* (codes for HapR protein, a key regulator of the quorum sensing) to understand the molecular basis of detachment of colonized bacteria from intestinal epithelial cell surfaces followed by migration towards fluid-filled lumen of the bowel, the last step of the infection process, which has been designated by them the 'mucosal escape response' (Nielsen *et al.*, 2006). The study elegantly showed for the first time that the mucosal escape response phenotype is intricately coordinated by an RpoS-dependent genetic program that up-regulates chemotaxis and motility functions and in the same time down-regulates virulence gene expression of the colonized *V. cholerae* cells during the last intraintestinal stage of infection. HapR, the quorum sensing regulator, is also involved in the process, which enhances motility of *V. cholerae* cells to facilitate mucosal escape response. For colonization, *V. cholerae* cells should travel from luminal to epithelial surface, but mucosal escape response appears to be a reverse process where bacterial cells swim from epithelial cell surface to intestinal lumen. Although the probable *in vivo* signals evoking such a complex change of gene expression in *V. cholerae* during infection are currently not clear, it has been suggested that physiological state of the cholera pathogen at that particular niche may initiate and control the mucosal escape response. Taken together, *in vivo* gene expression studies indicate that both host microenvironment and growth state of the bacterium are important parameters for coordinate expression of virulence factors both spatially and temporally for cholera pathogenesis. Further mutational analyses of various new genes, particularly those coding for signalling proteins, identified by transcriptional profiling of *in vivo* grown cells will allow us to understand more clearly, the virulence regulatory cascade of *V. cholerae*.

Quorum sensing, small RNAs and coordinated regulation

Quorum sensing is a complex process of bacterial cell–cell communication to coordinate multiple gene expression in response to population density. Various gene products coordinately regulate quorum sensing in Gram-negative bacteria (Camilli and Bassler, 2006). For adaptation to different environmental conditions, *V. cholerae* has evolved with multiple parallel quorum sensing signalling systems (Lenz *et al.*, 2004; Miller *et al.*, 2002). Two of these systems are similar to those of *V. harveyi*, another species of the genus *Vibrio* and the model organism for studying quorum sensing, called CqsA/S and LuxS/PQ systems, which are mediated by the autoinducers CAI-1 and AI-2, respectively (Miller *et al.*, 2002). These systems converge at LuxO, a critical regulator of the quorum sensing cascade (Lenz *et al.*, 2004; Miller *et al.*, 2002) (Fig. 9.2). Like LuxR in *V. harveyi*, HapR is the key regulator of quorum sensing in *V. cholerae* (Jobling and Holmes, 1997) and how it regulates the signalling is shown in Fig. 9.2. At low cell density when autoinducers are not produced, the membrane bound sensors act as kinases and transfer the phosphate moiety via cytoplasmic LuxU protein to LuxO and thus activate LuxO (LuxO-P). Activated LuxO-P along with the sigma factor σ^{54} (a product of the *rpoN* gene) then negatively regulates the expression of HapR. In the absence of HapR, a negative regulator of *ctxAB* and *tcpA* expression (by affecting AphA/AphB/TcpP/ToxT cascade) and the polysaccharide genes (*vps*) involved in biofilm formation, *V. cholerae* expresses virulence factors CT and TCP and develop biofilm (Fig. 9.1). In contrast, at high cell density, LuxO remains no longer active and thus the *hapR* gene is derepressed leading to high cellular accumulation of HapR, which inhibits AphA/B function leading to down regulation of *tcpP* expression, hence no activation of *toxT*, and thus no CT and TCP productions. This is in contrast to other pathogens in which quorum sensing up-regulates virulence gene expression

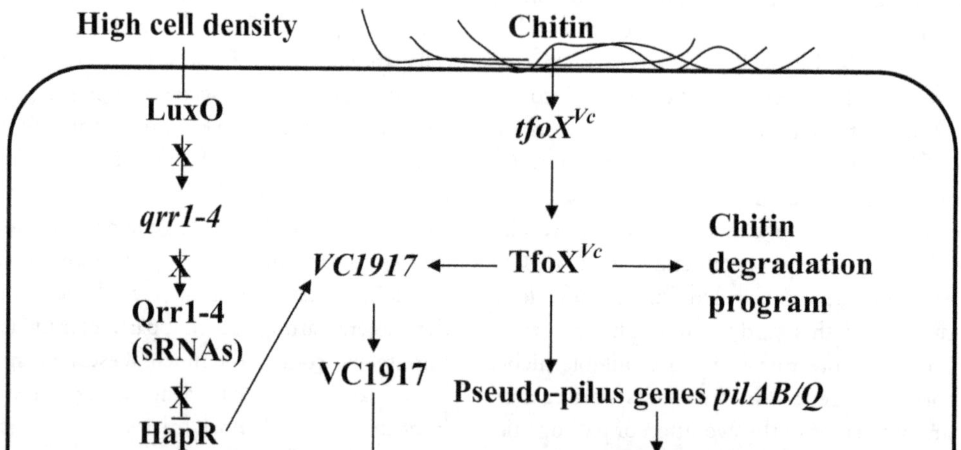

Figure 9.2 Model showing the proposed regulatory circuit governing natural competence in *V. cholerae* in response to environmental signals. Arrows indicate positive regulation while T-bars denote negative effects. Extracellular chitin positively regulates Tfox^{Vc} expression, which leads to transcriptional activation of several genes, the products of which are involved in extracytoplasmic chitin degradation. Tfox^{Vc}, also positively regulates expression of pseudo-pilus genes, the products (PilAB/Q) of which is crucial for competence development. In addition, Tfox^{Vc} has been shown to positively regulate a putative DNA binding protein coding gene VC1917. The VC1917 protein is required for the natural competence process. Thus, Tfox^{Vc} activates the chitin degradation program of *V. cholerae* to obtain nutrients and at the same time promotes natural competence development in *V. cholerae* cells. LuxO mediated repression of hapR is abolished during increasing cell density and hapR expression is further stimulated by stationary phase sigma factor, RpoS, due to lack of nutrient, decline in growth rate or possible other stresses. HapR and Tfox^{Vc} act positively at VC1917 promoter leading to expression of DNA binding protein VC1917 followed by competence development.

at high cell densities (Zhu *et al.*, 2002). Since HapR represses *vps* genes, biofilm formation is also inhibited by *V. cholerae* cells at high cell density (Zhu *et al.*, 2002).

It was shown that quorum-sensing regulators LuxO and HapR also control other cellular processes in *V. cholerae* like motility and protease production (Vance *et al.*, 2003; Zhu *et al.*, 2002). Thus, depending upon cell density, quorum sensing coordinately controls expression of several genes and operons involved in *V. cholerae* virulence, biofilm formation, motility, and protease production through the master regulator HapR (Hammer and Bassler, 2003; Kovacikova and Skorupski, 2002; Lenz *et al.*, 2004; Miller *et al.*, 2002; Vance *et al.*, 2003; Zhu and Mekalanos, 2003; Zhu *et al.*, 2002). However, recently it has been discovered that activated LuxO-P-σ^{54} complex does not directly inhibit the expression of *hapR* gene, rather it activates transcription of four homologous non-coding small regulatory RNAs (sRNAs) genes, called *qrr1*, *qrr2*, *qrr3* and *qrr4* (Lenz *et al.*, 2004). The expressed Qrr1–4 sRNAs (sizes in the range of 96–108 nucleotides) then with the help of small RNA chaperone protein Hfq bind to the *hapR* mRNA facilitating degradation of sRNA–mRNA complex and leading to a decreased cellular level of HapR followed by up regulation of virulence and biofilm formation (Lenz *et al.*, 2004) (Fig. 9.1). Lenz *et al.*, (2004) proposed that Hfq, together with four Qrr1–4 sRNAs, creates an ultra-sensitive regulatory switch that governs the critical transition of *V. cholerae* cells into the quorum-sensing mode. Recently, Lenz *et al.* (2005) showed that a third quorum sensing signal cascade in *V. cholerae* is mediated by a VarS–VarA phosphorelay system, although the environmental signal(s) of this pathway is still not known.

Cyclic di-GMP mediated intracellular signalling

The cyclic nucleotide bis-(3′–5′)-cyclic-diguanosine monophosphate (c-di-GMP) is a prokaryotic intracellular signalling molecule synthesized by the enzyme diguanylate cyclase (DGC) carrying Gly-Gly-Asp-Glu-Phe or Gly-Gly-Glu-Glu-Phe (GGDEF or GGEEF) protein domain and is degraded by the enzyme phosphodiesterase A (PDEA) containing Glu-Ala-Lys (EAL) protein motif (Romling *et al.*, 2005). Recent research indicates that c-di-GMP is widely used by bacteria as a second messenger and thus it serves as the pivotal point for managing several extracellular signals to adapt to different environmental niche. Presence of several dozens of GGDEF and EAL protein motifs coding genes in many bacterial genomes including in vibrios is perplexing. Recent research on c-di-GMP-mediated signal transduction pathways indicates that this small intracellular molecule regulates the transcription of *vps* genes, the products of which are responsible for exopolysaccharide synthesis and biofilm formation, an opposing activity of the quorum sensing master regulator HapR (Lim *et al.*, 2006). It has also been shown that c-di-GMP regulates *V. cholerae* virulence gene expression (Tischler and Camilli, 2005) as shown in Fig. 9.1. The response regulator VieA, an EAL domain containing protein, is needed to maintain a low cellular concentration of c-di-GMP. Low concentration of cellular c-di-GMP enhanced CT production under laboratory conditions, which was due to transcriptional activation of the *toxT* gene. Thus, it appears that the c-di-GMP is a negative regulator of ToxT virulence regulon and c-di-GMP signalling may be important for the transition of *V. cholerae* from the aquatic environment to the human host (Tischer and Camilli, 2005).

Chitin-induced gene regulation and competence development

V. cholerae being a normal flora of the aquatic environment, has to adapt and survive both inside and outside of the host. Our current understanding regarding molecular strategies for survival in natural aquatic environments and other behaviour of the organism have been advanced substantially due to the availability of the whole-genome sequence of the *V. cholerae* strain N16961 (Heidelberg *et al.*, 2000). *V. cholerae* can attach to and survive on the chitinous exoskeleton of zooplankton, particularly copepods, a subclass of crustacean zooplankton (Colwell, 1996), and certain fundamental questions like the expression and regulation of genes associated with chitin utilization and signals controlling attachment to chitin and colonization are beginning to be addressed. Chitin, an insoluble polymer of

N-acetylglucosamine (GlcNAc), is an abundant polysaccharide in the aquatic environment and a ready source of carbon and nitrogen for marine microorganisms like *V. cholerae*. Recently, microarray expression profiling and mutational studies of *V. cholerae* cells growing on a natural chitin surface allowed identification of the ChiS regulon (Meibom *et al.*, 2004). Chitin and chitin oligosaccharides were found to up-regulate several genes coding for proteins involved in the biogenesis of a type IV pilus named ChiRP, and PilA is a major component of this pilus. Interestingly, GlcNAc has been identified as a signal for coordinate regulation of genes involved with chitin chemotaxis, adherence and with the transport and assimilation of GlcNAc. Studies of Meibom *et al.* (2004) indicated a putative chitinase protein that is induced by GlcNAc. Altogether, it appears that chitin utilization in natural aquatic environments by *V. cholerae* is a complex multistage genetic program involving chitin sensing, attachment, degradation, and assimilation. This environmentally regulated genetic program allows survival and replication of *V. cholerae* cells in nature as well as helps the organism in early stages of human infection process.

Meibom *et al.* (2005) further extended their study on chitin signalling and discovered that chitin utilization program in *V. cholerae* helps the bacteria in development of genetic competence (Fig. 9.2), a transient physiological state of bacteria permitting direct uptake of DNA present in the natural environment. Meibom *et al.* (2005) showed that three convergent regulatory cascades in *V. cholerae* control natural transformation and that these genetic circuits are activated by chitin, nutritional deficiency or a decline in growth rate or other stresses and increase in bacterial cell density leading to quorum sensing (Fig. 9.2). The regulator of transformation competence in *V. cholerae* has also been identified. This regulator, called TfoX^{Vc} (Fig. 9.2), is an orthologue of TfoX domain containing competence regulator protein in *Haemophilus influenzae*. Transcriptional profiling indicated that a large number of genes (total 99 genes) including *pilA*, *pilB*, and *pilQ* along with other 28 chitin inducible genes are under the control of the competence regulator protein TfoX*Vc* of *V. cholerae* (Meibom *et al.*, 2005). Thus, chitin serves as a major signal in transcriptional activation of the key competence regulator protein TfoX*Vc*, which in turn controls the expression of genes encoding proteins with two quite different functions, one is chitin degradation and the other is chitin-induced natural competence phenotype (Fig. 9.2). Mutation in *hapR* or *rpoS* gene leads to competence negative phenotype indicating that quorum sensing regulatory network is also linked with chitin-induced competence development in *V. cholerae* (Fig. 9.2).

Concluding remarks

The availability of the whole genome sequence of the *V. cholerae* O1 El Tor strain N16961 has changed the scenario in every aspect of cholera research. This information has allowed developing the powerful microarray technology to analyse global gene expression in *V. cholerae* under any given condition whether it is recovered from clinical samples or grown in naturally occurring chitinous surfaces or in a specific mutant. Transcriptional profiling of *V. cholerae* genome revealed under a particular experimental condition, a large number of genes and operons are either expressed or repressed. Since in most cases these regulated genes are of hypothetical and of unknown functions, a future challenge would be to assign functions for each of them. It is only from bacterial whole-genome sequence analysis we came to know about the presence of abundant copies of GGDEF/GGEEF and EAL domain proteins coding genes. The products of these genes are involved in maintaining levels of intracellular small signalling molecule called c-di-GMP which appears to be crucial in bacteria including *V. cholerae* for controlling many vital biological processes including virulence gene expression. We have just begun to understand the c-di-GMP-mediated signalling pathways in *V. cholerae* and thus, much effort should be given to understand what are the targets of this novel second messenger and its mechanism of action. According to Galperin (2005) the total number of signalling proteins encoded in a given genome can be used as a measure of the adaptive potential of an organism, which may be called 'bacterial IQ'. Based on this bioinformatic analysis, *V. cholerae* has been positioned in the highest IQ class of environmental Gram-negative organisms, since it codes for 152 signal transducers. Presence of this large number of signalling proteins in *V. cholerae* further clarifies that the organism can adapt and survive in diverse ecological niches. Thus, it remains a challenge to characterize each of these signalling proteins to understand comprehensively about signal-dependent coordinated gene regulatory network, the bacterial sensory system.

Acknowledgement

Network Project (SMM003 and COR011) grants from Council of Scientific and Industrial Research, Government of India.

References

Bina, J., Zhu, J., Dziejman, M., Faruque, S.M., Calderwood, S., and Mekalanos, J.J. (2003). ToxR regulon of *Vibrio cholerae* and its expression in vibrios shed by cholera patients. Proc. Natl. Acad. Sci. USA *100*, 2801–2806.

Camilli, A., and Bassler, B. (2006). Bacterial small-molecule signaling pathways. Science *311*, 1113–1116.

Childers, B.M., Webera, G.G., Proutya, M.G., Castanedaa, M.M., Penga, F. and Klose, K.E. (2007). Identification of residues critical for the function of the *Vibrio cholerae* virulence regulator ToxT by scanning alanine mutagenesis. J. Mol. Biol. *367*, 1413–1430.

Colwell, R.R. (1996). Global climate and infectious disease: the cholera paradigm. Science *274*, 2025–2031.

Datta, P.P. and Bhadra, R.K. (2003). Cold shock response and major cold shock proteins of *Vibrio cholerae*. Appl. Environ. Microbiol. *69*, 6361–6369.

DiRita, V.J., Parsot, C., Jander, G. and Mekalanos, J.J. (1991). Regulatory cascade controls virulence in *Vibrio cholerae*. Proc. Natl. Acad. Sci. USA *88*, 5403–5407.

DiRita, V.J. and Mekalanos, J.J. (1991). Periplasmic interaction between two membrane regulatory proteins, ToxR and ToxS, results in signal transduction and transcriptional activation. Cell. *64*, 29–37.

Faruque, S.M., Albert, M.J. and Mekalanos, J.J. (1998). Epidemiology, genetics, and ecology of toxigenic *Vibrio cholerae*. Microbiol. Mol. Biol. Rev. *62*, 1301–1314.

Faruque, S.M., Nair, G.B. and Mekalanos, J.J. (2004). Genetics of stress adaptation and virulence in toxigenic *Vibrio cholerae*. DNA Cell Biol. *23*, 723–741.

Galperin, M.Y. (2005). A census of membrane-bound and intracellular signal transduction proteins in bacteria: Bacterial IQ, extroverts and introverts. BMC Microbiol. 5:35.

Gardel, C.L. and Mekalanos, J.J. (1994). Regulation of cholera toxin by temperature, pH and osmolarity. Methods Enzymol. *235*, 517–526.

Ghosh, A., Paul, K. and Chowdhury, R. (2006). Role of the histone-like nucleoid structuring protein in colonization, motility and bile-dependent repression of virulence gene expression in *Vibrio cholerae*. Infect. Immun. *74*, 3060–3064.

Gupta, S. and Chowdhury, R. (1997). Bile affects production of virulence factors and motility of *Vibrio cholerae*. *65*, 1131–1134.

Hammer, B.K. and Bassler, B.L. (2003). Quorum sensing controls biofilm formation in *V. cholerae*. Mol. Microbiol. *50*, 101–104.

Häse, C.C. and Mekalanos, J.J. (1998). TcpP protein is a positive regulator of virulence gene expression in *Vibrio cholerae*. Proc. Natl. Acad. Sci. USA *95*, 730–734.

Heidelberg, J.F., Eisen, J.A., Nelson, W.C., Clayton, R.A., Gwinn, M.L., Dodson, R.J., Haft, D.H., Hickey, E.K., Peterson, J.D., Umayam, L., Gill, S.R., Nelson, K.E., Read, T.D., Tettelin, H., Richardson, D., Ermolaeva, M.D., Vamathevan, J., Bass, S., Qin, H., Dragoi, I., Sellers, P., McDonald, L., Utterback, T., Fleishmann, R.D., Nierman, W.C., White, O., Salzberg, S.L., Smith, H.O., Colwell, R.R., Mekalanos, J.J., Venter, J.C. and Fraser, C.M. (2000). DNA sequence of both chromosomes of the cholera pathogen *Vibrio cholerae*. Nature *406*, 477–483.

Higgins, D.E., Nazareno, E. and DiRita, V.J. (1992). The virulence gene activator ToxT from *Vibrio cholerae* is a member of the AraC family of transcriptional activators. J. Bacteriol. *174*, 6974–6980.

Higgins, D.E. and DiRita, V.J. (1994). Transcriptional control of toxT, a regulatory gene in the ToxR regulon of *Vibrio cholerae*. Mol. Microbiol. *14*, 17–29.

Hung, D.T. and Mekalanos, J.J. (2005). Bile acids induce cholera toxin expression in *Vibrio cholerae* in a ToxT-independent manner. Proc. Natl. Acad. Sci. USA *102*, 3028–3033.

Jobling M.G. and Holmes, R.K. (1997). Characterization of *hapR*, a positive regulator of the *Vibrio cholerae* HA/protease gene *hap*, and its identification as a functional homologue of the Vibrio harveyi luxR gene. Mol. Microbiol. *26*, 1023–34.

Kaper, J.B., Morris Jr, J.G. and Levine, M.M. (1995). Cholera. Clin. Microbiol. Rev. *8*, 48–86.

Karaolis, D.K.R., Johnson, J.A., Bailey, C.C., Boedeker, E.C., Kaper, J.B. and Reeves, P.R. (1998). A *Vibrio cholerae* pathogenicity island associated with epidemic and pandemic strains. Proc. Natl. Acad. Sci. USA 95, 3134–3139.

Klose, K.E. (2001). Regulation of virulence in *Vibrio cholerae*. Int. J. Med. Microbiol. *291*, 81–88.

Kovacikova, G. and Skorupski, K. (1999). A *Vibrio cholerae* LysR homolog, AphB, cooperates with AphA at the tcpPH promoter to activate expression of the ToxR virulence cascade. J. Bacteriol. *181*, 4250–4256.

Kovacikova, G. and Skorupski, K. (2000). Differential activation of the tcpPH promoter by AphB determines biotype specificity of virulence gene expression in *Vibrio cholerae*. J. Bacteriol. *182*, 3228–3238.

Kovacikova, G. and Skorupski, K. (2001). Overlapping binding sites for the virulence gene regulator AphA, AphB and cAMP-CRP at the *Vibrio cholerae* tcpPH promoter. Mol. Microbiol. *41*, 393–407.

Kovacikova, G. and Skorupski, K. (2002). Regulation of virulence gene expression in *Vibrio cholerae* by quorum sensing: HapR functions at the aphA promoter. Mol. Microbiol. *46*, 1135–1147.

Krukonis, E.S., Yu, R.R. and DiRita, V.J. (2000). The *Vibrio cholerae* ToxR/TcpP/ToxT virulence cascade: distinct roles for two membrane-localized transcriptional activators on a single promoter. Mol. Microbiol. *38*, 67–84.

LaRocque, R.C., Harris, J.B., Dziejman, M., Li, X., Khan, A.I., Faruque, A.S.G., Faruque, S.M., Nair, G.B., Ryan, E.T., Qadri, F., Mekalanos, J.J. and Calderwood, S.B. (2005). Transcriptional profiling of *Vibrio cholerae* recovered directly from patient specimens during early and late stages of human infection. Infect. Immun. *73*, 4488–4493.

Lee, S.H., Hava, D.L., Waldor, M.K. and Camilli, A. (1999). Regulation and temporal expression patterns of *Vibrio cholerae* virulence genes during infection. Cell. *99*, 625–634.

Lenz, D.H., Mok, K.C., Lilley, B.N., Kulkarni, R.V., Wingreen, N.S., and Bassler, B.L. (2004). The small RNA chaperone Hfq and multiple small RNAs control quorum sensing in Vibrio harveyi and *Vibrio cholerae*. Cell. *118*, 69–82.

Lenz, D.H., Miller, M.B., Zhu, J., Kulkarni, R.V. and Bassler, B.L. (2005). CsrA and three redundant small RNAs regulate quorum sensing in *Vibrio cholerae*. *58*, 1186–1102.

Li, C.C., Crawford J.A., DiRita V.J. and Kaper, J.B. (2000). Molecular cloning and transcriptional regulation of ompT, a ToxR-repressed gene in *Vibrio cholerae*. Mol. Microbiol. *35*, 189–203.

Lim, B., Beyhan, S., Meir, J. and Yildiz, F.H. (2006). Cyclic-diGMP signal transduction systems in *Vibrio cholerae*: modulation of rugosity and biofilm formation. Mol. Microbiol. *60*, 331–348.

Lynch, A.S. and Lin, A.C. (1996). Responses to molecular oxygen, p. 1526–1538. In F. C. Neidhardt *et al.* (ed), *Escherichia coli* and Salmonella typhimurium: cellular and molecular biology, vol. 1, ASM Press, Washington D. C.

Matson, J.S. and DiRita, V.J. (2005). Degradation of the membrane-localized virulence activator TcpP by the YaeL protease in *Vibrio cholerae*. Proc. Natl. Acad. Sci. USA *102*, 16403–16408.

Meibom, K.L., Li, X.B., Nielsen, A.T., Wu, C.Y., Roseman, S. and Schoolnik, G.K. (2004). The *Vibrio cholerae* chitin utilization program. Proc. Natl. Acad. Sci. USA *101*, 2524–2529.

Meibom, K.L., Blokesch, M., Dolganov, N.A., Wu, C.Y. and Schoolnik, G.K. (2005). Chitin induces natural competence in *Vibrio cholerae*. Science. *310*, 1824–1827.

Merrell, D.S., Butler, S.M., Quadri, F., Dolganov, N.A., Alam, A., Cohen, M.B., Calderwood, S.B., Schoolnik, G.K. and Camilli, A. (2002). Host-induced epidemic spread of the cholera bacterium. Nature *417*, 642–645.

Miller, J.F., Mekalanos, J.J. and Falkow, S. (1989a). Coordinate regulation and sensory transduction in the control of bacterial virulence. Science. *243*, 916–922.

Miller, M.B., Skorupski, K., Lenz, D.H., Taylor, R.K. and Bassler, B.L. (2002). Parallel quorum sensing systems converge to regulate virulence in *Vibrio cholerae*. Cell *110*, 303–314.

Miller, V.L., DiRita, V.J. and Mekalanos, J.J. (1989b). Identification of toxS, a regulatory gene whose product enhances ToxR-mediated activation of the cholera toxin promoter. J. Bacteriol. *171*, 1288–1293.

Miller, V.L. and Mekalanos, J.J. (1984). Synthesis of cholera toxin is positively regulated at the transcriptional level by toxR. Proc. Natl. Acad. Sci. USA *81*, 3471–3475.

Miller, V.L. and Mekalanos, J.J. (1988). A novel suicide vector and its use in construction of insertion mutations: osmoregulation of outer membrane proteins and virulence determinants in *Vibrio cholerae* requires toxR. J. Bacteriol. *170*, 2575–2583.

Miller, V.L., Taylor, R.K. and Mekalanos, J.J. (1987). Cholera toxin transcriptional activator ToxR is a transmembrane DNA binding protein. Cell. *48*, 271–279.

Nielsen, A.T., Dolganov, N.A., Otto, G., Miller, M.C., Wu, C.Y. and Schoolnik, G.K. (2006). RpoS controls the *Vibrio cholerae* mucosal escape response. PLoS Pathogen. *2*, 0933–0948.

Nye, M.B., Pfau, J.D., Skorupski, K. and Taylor, R.K. (2000). *Vibrio cholerae* H-NS silences virulence gene expression at multiple steps in the ToxR regulatory cascade. J. Bacteriol. *182*, 526–528.

Ottemann, K.M., DiRita, V.J. and Mekalanos, J.J. (1992). ToxR proteins with substitution in residues conserved with OmpR fail to activate transcription from the cholera toxin promoter. J. bacterial. *174*, 6807–6814.

Peterson, K.M. and Mekalanos, J.J. (1988). Characterization of the *Vibrio cholerae* ToxR regulon: identification of novel genes involved in intestinal colonization. Infect. Immun. *56*, 2822–2829.

Romling, U., Gomelsky, M. and Galperin, M.Y. (2005). C-di-GMP: the dawning of a novel bacterial signalling system. Mol. Microbiol. *57*, 629–639.

Schuhmacher, D.A. and Klose, K.E. (1999). Environmental signals modulate ToxT-dependent virulence factor expression in *Vibrio cholerae*. J. Bacteriol. *181*, 1508–1514.

Sengupta, N., Paul, K. and Chowdhury, R. (2003). The global regulator ArcA modulates expression of virulence factors in *Vibrio cholerae*. Infect. Immun. *71*, 5583–5589.

Skorupski, K. and Taylor, R.K. (1997a). Control of the ToxR virulence regulon in *Vibrio cholerae* by environmental stimuli. Mol. Microbiol. *25*, 1003–1009.

Skorupski, K. and Taylor, R.K. (1997b). Cyclic AMP and its receptor protein negatively regulate the co-ordinate expression of cholera toxin and toxin co-regulated pilus in *Vibrio cholerae*. Proc. Natl. Acad. Sci. USA *94*, 265–270.

Skorupski, K. and Taylor, R.K. (1999). A new level in the *Vibrio cholerae* ToxR virulence cascade: AphA is required for transcriptional activation of the tcpPH operon. Mol. Microbiol. *31*, 763–771.

Taylor R.K., Miller, V.L., Furlong, D.B. and Mekalanos, J.J. (1987). Use of *phoA* gene fusions to identify a pilus colonization factor coordinately regulated with cholera toxin. Proc. Natl. Acad. Sci. USA *84*, 2833–2837.

Thomas, S., Willams, S.G. and Manning, P.A. (1995). Regulation of *tcp* genes in classical and El Tor strains of *Vibrio cholerae* O1. Gene. *166*, 43–48.

Tischler, A.D. and Camilli, A. (2005). Cyclic diguanylate regulates *Vibrio cholerae* virulence gene expression. Infect. Immun. *73*, 5873–5882.

Vance, R.E., Zhu, J. and Mekalanos, J.J. (2003). A constitutively active variant of the quorum-sensing regulator LuxO affects protease production and biofilm formation in *Vibrio cholerae*. Infect. Immun. *71*, 2571–2576.

Waldor, M.K. and Mekalanos, J.J. (1996). Lysogenic conversion by a filamentous phage encoding cholera toxin. Science. *272*, 1910–1914.

Xu, Q., Dziezman. M. and Mekalanos. J.J. (2003). Determination of the transcriptome of *Vibrio cholerae* during intraintestinal growth and midexponential phase in vitro. Proc. Natl. Acad. Sci. USA *100*, 1286–1291.

Yu, R.R. and DiRita, V.J. (1999). Analysis of an autoregulatory loop controlling ToxT, cholera toxin, and toxin-coregulated pilus production in *Vibrio cholerae*. J. Bacteriol. *181*, 2584–2592.

Yu, R.R. and DiRita, V.J. (2002). Regulation of gene expression in *Vibrio cholerae* by ToxT involves both antirepression and RNA polymerase stimulation. Mol. Microbiol. *43*, 119–134.

Zhu, J. and Mekalanos, J.J. (2003). Quorum sensing-dependent biofilms enhance colonization in *Vibrio cholerae*. Dev. Cell. *5*, 647–656.

Zhu, J., Miller, M.B., Vance, R.E., Dziejman, M., Bassler, B.L. and Mekalanos, J.J. (2002). Quorum-sensing regulators control virulence gene expression in *Vibrio cholerae*. Proc. Natl. Acad. Sci. USA 99, 3129–3134.

Evolutionary Relationships of Pathogenic Clones of *Vibrio cholerae* 10

O. Colin Stine

Abstract

Evolution refers to the differentiation of an ancestral genome into recognizably distinct genomes. Understanding the evolutionary history of an organism can provide insight into how it can be expected to evolve in the future and provide predictions that serve as the basis of where to best focus effort to prevent the emergence of new pathogenic variants. In order to accurately understand the evolutionary history, the methods used for interpreting the genetic variation need to reflect the mechanisms of genetic change. The critical mechanism for deciding how to interpret the genetic relatedness is the amount of recombination. If recombination is rare, then the traditional phylogenetic analysis based on bifurcating trees works well. If recombination is common, then a method that incorporates recombination must be used. Evolutionary relationships among pathogenic clones based on these assessments have been presented and discussed.

Introduction to evolutionary analysis

The starting point for the evolution of pathogenic *Vibrio cholerae* is the ancient ancestor of all *V. cholerae*. The *V. cholerae* genome, as is true of all bacterial genomes, has evolved by point mutations, insertions, deletions, recombination and selection. Each singular event represents the differentiation of the ancestor into new forms. In the absence of any genetic changes, two isolates are identical and referred to as being from the same clone. Two isolates that differ at specific loci, but that are derived from the same ancestor are referred to as being part of the same clade or lineage. In bacteria, there is no a priori reason for thinking that any particular form of genetic change is the most likely. In fact, different species or strains seem to evolve via different types of genetic change. For example, *Neisseria gonorrhoeae* has a high proportion of homologous recombination, while *Streptococcus pneunomiae* evolves predominately by point mutations in homologous genes (Spratt and Maiden, 1999) and it has been suggested that pathogenic clades of *Escherichia coli* have accelerated rates of mutation and recombination relative to non-pathogenic clades (Wirth *et al.*, 2006).

The evolutionary history of a species is often compared to a family tree. Trees are usually bifurcating structures. For genetic trees, each branch point occurs when there is a genetic change. In the absence of recombination, the changes, whether insertion/deletions or point mutations, occur sequentially. These sequential changes are consistent with the assumptions underlying a bifurcating tree. These trees represent the phylogeny of the samples. In the presence of recombination, the sequential nature of the changes is disrupted. Recombination moves a stretch of homologous DNA from one organism to another. In terms of the bifurcating tree a branch containing a series of changes is clipped off and grafted on to another branch of the tree. Thus, recombination disrupts the bifurcating nature of the tree. In addition, the potential to move multiple sequential mutations in a single recombination event disrupts the ability to use the number of mutations a metric to measure the distance between two isolates. In order to accurately understand the evolutionary history, the methods used for interpreting the genetic variation need to reflect the mechanisms of genetic change. Hence, the critical mechanism for deciding how to interpret the genetic relatedness is the amount of recombination. If recombination is rare, then the traditional phylogenetic analysis based on bifurcating trees works well. If recombination is common, then a method that incorporates recombination must be used.

As described above, in order to define genetic lineages, we need to define the extent of recombination, so that we know which are the appropriate types of analyses, and also to define our terminology. For example, an isolate may refer to the specimen in hand collected from a patient or the environment and possibly transferred between laboratories. A clone refers to identical individual isolates collected independently, whereas a clonal complex refers to isolates that are identical at many genetic traits, but distinct from each other at a few. While this term is standard in the MLST (multilocus sequence typing) literature referring to isolates that have identical alleles at many loci and alternative alleles at a few, it is reasonable to extend it to include isolates that share (a) most, but not all pulse gel field bands, (b) numerous but not all biochemical characteristics and (c) many but not all genes.

Assessing recombination in *Vibrio cholerae*

Vibrio cholerae is naturally competent in the presence of chitin and can be transformed by phage in an aquatic environment (Meibom *et al.*, 2005; Blokesch and Schoolnik, 2007). Both competence and transformation permit DNA to be transferred between cells. The key to knowing the appropriate methods for understanding evolutionary relationships in the presence of natural competence is to estimate the relative rate of point mutations versus recombination. There are 3 MLST based studies that have data that provide some insight into the question, although none of these studies was specifically designed to compare the rates of mutation and recombination. In each case only those alleles from isolates from the same clonal complex can be used to compare the rates of mutation and recombination because it is those cases in which a recent common ancestor can be inferred. The most recent study (Salim *et al.*, 2005) examined 26 loci in 12 cholera toxin positive (*ctx*+) isolates from a variety of places and years. Eleven of the loci were identical at every nucleotide in every isolate confirming the close genetic relatedness of the epidemic isolates. Among the variable alleles, there were seven in which only a single base changed. These are assumed to result from point mutations. There were 11 alleles for which there were many variable nucleotides (range 4–51). These 11 alleles were attributed to recombination of homologous genes. Thus, in this study, 61% (11 of 18) of the allelic changes were suggested to have occurred via homologous recombination. An earlier study had examined 96 *V. cholerae* O139 from a single location, Kolkata, India, over a 9-year period (Garg *et al.*, 2003). There were 63 allelic changes in the single clonal complex in this study. Thirty-four were the result of single base changes, while 29 were due to two or more changes. After allowing for the authors calculation that two of the alleles with two variable nucleotides might result from multiple point mutations, the proportion of allelic change due to recombination was shown to be 43% (27/63). A third study examined 6 loci from 29 isolates of *V. cholerae* O139 collected during the initial year of the epidemic (Farfan *et al.*, 2002). Applying the definition of a clonal complex, there were only two allelic changes that could be used to estimate the relative rate of point mutations and recombination. In this case, one of each or about 50% of the alleles were due to recombination. Thus, consistent with the finding that *V. cholerae* are naturally competent, it appears that about half the alleles are due to recombination events.

In the face of high levels of recombination, the use of bifurcating trees on sequence data to determine genetic relatedness is problematic and using the number of variable nucleotides between sequences to estimate the divergence is likely to be erroneous. Thus, the conservative approach to analyzing genetic relatedness between isolates using sequence data is to assign distinct alleles a number and define clonal complexes, a method that does not assume bifurcating structure (Spratt and Maiden, 1999). While clonal complexes accurately reflect the relatedness of closely related isolates, there is no good method for using sequence data of homologous genes to assess the relatedness of the distinct clonal complexes. However, it is possible to construct phylogenetic relationships based on the presence and absence of genes that are not found among all of the isolates. This is the traditional method of defining the phylogenetic relatedness among bacterial species using biochemical characteristics as phenotypic manifestations of the genes. However, the method can be updated to change from phenotypic characteristics (e.g. growth on media) to presence or absence of genes in either genomic sequences or by comparative hybridization to a microarray.

Pathogenic *Vibrio cholerae*

There are at least three and perhaps several other distinct pathogenic forms of *V. cholerae* each one with a distinct toxin and delivery system. The best known because of its propensity to cause seasonal outbreaks and world-wide epidemics is the *ctx*$^+$ form. The second well defined form carries a non-agglutinable heat-stable toxin (NAG-ST). The third form is currently characterized by the absence of both *ctx* and NAG-ST and has been reported from several locations, but the absence of identified genetic changes prevents an accurate classification of these variants. Nevertheless these three forms are paradigmatic of biochemical differences between isolates. The presence and absence of these virulence genes is then one method to differentiate among isolates.

ctx$^+$

The pathogenic form of *V. cholerae* that carries the cholera toxin gene has caused world wide epidemics seven times since 1817. Based on MLST and microarray data, the *ctx*$^+$ epidemic associated isolates are part of a single genetic clade or lineage. The group includes the isolates from various subtypes known as 'classical', 'El Tor', 'sixth pandemic', 'seventh pandemic', 'Australian clone', 'US Gulf Coast clone' O1, O139, O37 Sudan and O37 Czechoslovakia. Shown in Fig. 10.1 are representative isolates collected between 1921 and 1992 of most of the *ctx*$^+$ subtypes (Salim *et al.*, 2005). In order to be included in this figure, each isolate was identical at every nucleotide for eleven loci. The lines in the figure indicate differences between the isolates encircled at the nodes of the dendrogram. The loci listed next to the lines are those that have changed either by point mutation (m) or homologous recombination (r). Alleles differing by a single nucleotide among the over 850 nucleotides sequenced are considered to have diverged by a point mutation. Alleles that differ by 4 to 51 nucleotides are considered to have diverged based on a recombination event. The different nominal subtypes differ from each other at 3–13 loci, although there are a series of intermediates that connect the subtypes even more closely. The two isolates from each of the 'sixth pandemic clone' and the 'Australian clone' violate the definition of a clone, since genetic differences are observed between them and would be best discussed as clonal complex. Indeed the entire group of *ctx*$^+$ isolates can be considered to be members of a single clonal complex. The unitary nature of this clonal complex determined by the 11 identical loci and the relatedness of the other loci by definable genetic events contrasts with the observation that environmental isolates examined did not share any alleles with either themselves or the *ctx*$^+$ isolates. This is a remarkable observation; the environmental isolates did not share any alleles with the *ctx*$^+$ epidemic isolates. Thus, the *ctx*$^+$ isolates that share numerous alleles among themselves are a single genetic group.

Genetic variation among *ctx*$^+$ isolates has been observed on a smaller scale. *V. cholerae* O139 *ctx*$^+$ isolates are clonally related to the El Tor or seventh pandemic strains. In this branch of the *ctx*$^+$ clonal complex, variation was observed among isolates collected over a 9-year period in Kolkata (Garg *et al.*, 2003). In this MLST study using seven loci, although 95 isolates were in the same clonal complex, novel sequence types occurred at an average rate of about 0.5 per isolate.

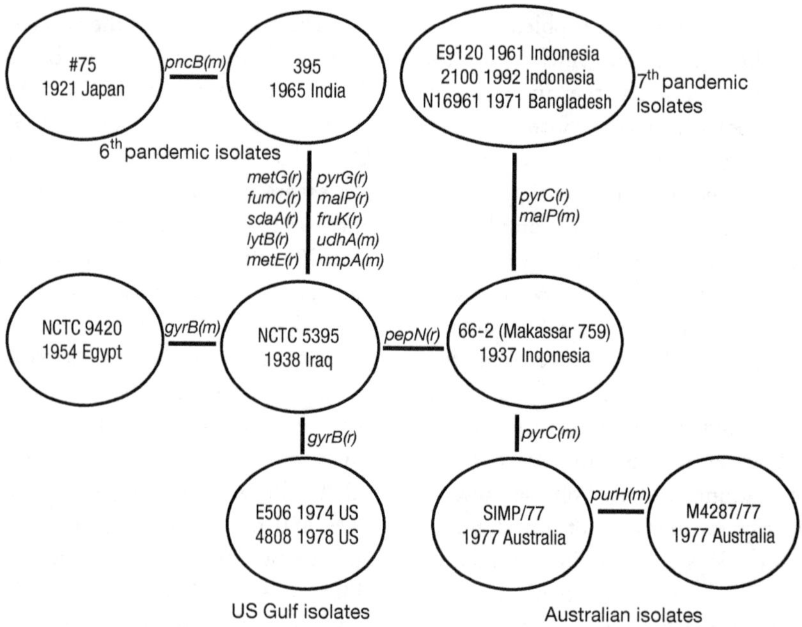

Figure 10.1 A dendrogram of the genetic relatedness between *ctx*+ isolates. Each node contains the names of the isolates and the date of collection. Each line represents the closest connection. The annotation next to each line indicates the variable locus and the genetic change either point mutation (m) or a recombination event (r). Taken from Salim *et al.* (2005) Emerging Infect. Dis. 11: 1758–1760.

In other words, every other isolate was genetically distinct. The high rate of genetic change was attributed to founder flush phenomena, in which a rapid increase in the population is accompanied by increased genetic variation (Garg *et al.*, 2003). It is important to note that this rate is similar to that from the study reported above (Salim *et al.*, 2005). Although this latter study used nearly four times the number of loci on isolates over 70 years, the average rate of novel sequence types was found to be 0.7. The observation of extensive variation in a short time and a similar amount over a much longer time scale implies that much of the short term variation must be lost. Thus the uniformity of the larger genetic group must be maintained by stabilizing selection (selection against most variants) during times when the population is decreasing or going through a bottleneck. The disappearance of *V. cholerae* from the human population in the inter-epidemic periods and into environmental reservoirs is a bottleneck in which the *V. cholerae* population and presumably the genetic variation is reduced.

The genetic relatedness of different isolates can be measured using microarrays. Microarrays constructed using a series of known genes based on one or more sequenced genomes can be used to identify the gene present in each tested genome. In 2007, two independent groups (Keymer *et al.*, 2007; Pang *et al.*, 2007) published data from microarrays based on the *V. cholerae* N16961 genome (Heidelberg *et al.*, 2000) that compared the genes present in various environmental isolates with those present in clinical isolates (e.g. O1 classical, O1 El Tor, O139). As illustrated in Fig. 10.2, the clinical isolates were distinct from the environmental isolates (Keymer *et al.*, 2007). However, considering the overall genome sequence, the clinical isolates and the environmental isolates are genetically related. The distinction rests on the approximately 13% of the genes whose presence or absence is variable. Among the environmental isolates, the presence or absence of specific genes correlates with selected environmental parameters consistent with the suggestion that these genes are subject to natural selection. In addition, the environmental isolates are likely to have some genes that are not in the sequenced genome. Of note, only 4% of the genes are found only in the clinical isolates and most of these are in the known pathogenesis islands (Keymer *et al.*, 2007).

The *ctx*+ clade is a paradigmatic example of an epidemic clone (Spratt and Maiden, 1999)

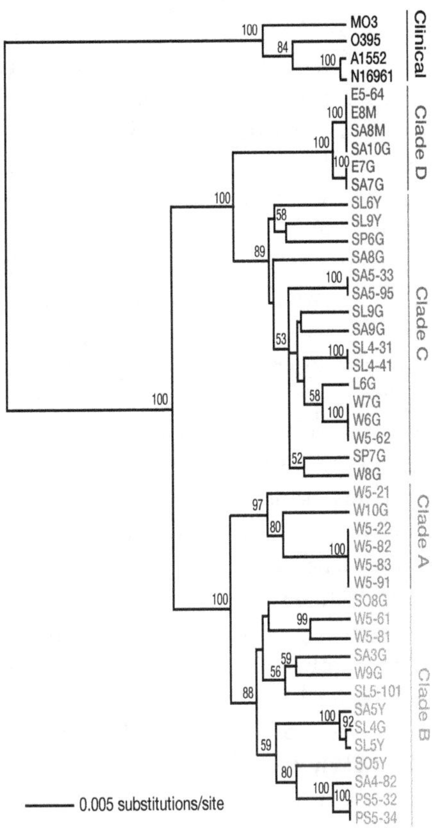

Figure 10.2 A cladogram of the genetic relatedness between clinical *ctx*+ isolates and environmental isolates of *Vibrio cholerae*. The differences are based on the presence and absence of specific genes on a microarray. The numbers at the nodes indicate the bootstrap values. From Keymer *et al.* (2007) Appl. Environ. Microbiol. *73:* 3695–3704.

that evolved by the acquisition of a few pathogenicity islands and has remained stable for decades and perhaps centuries. The small number of pathogenic islands means that environmental *V. cholerae* is capable of evolving into novel pathogenic forms. Many studies have examined the presence of *ctx* and TCP (toxin co-regulated pilius) in environmental isolates (Rivera *et al.*, 2001; Vital Brazil *et al.*, 2002; Li *et al.*, 2003). The TCP pilus is a receptor for the *ctx* phage (Waldor and Mekalanos, 1996). The other two major pathogenesis islands VSP-I and II have not been examined in environmental isolates.

A reflection of the clonal complex of the *ctx*+ isolates is the difficulty of separating these isolates into smaller genetic groups that can be used for epidemiological tracking of the population. Many techniques, biochemical tests, ribotyping, antibiograms, PFGE (pulse-field gel electrophoresis), RAPD (random amplified polymorphic DNA) (Chatterjee *et al.*, 2007), MLST (Farfan *et al.*, 2002; Garg *et al.*, 2003) have been utilized. In general, the most common genetic type is found in a large proportion of the isolates and there are a few rare variants from the common type. However, the use of variable number of tandem repeat (VNTR) loci has been successfully applied in *E. coli* O157 (Noller *et al.*, 2003a) where other methods (PFGE and MLST failed to distinguish among isolates (Noller *et al.*, 2003b). VNTR loci have been identified in *V. cholerae* (Danin-Poleg *et al.*, 2007) and distribution of the distinct genetic types is more uniform. When isolates from distinct locales in rural Bangladesh were examined with VNTR loci, they were found to be genetically distinct (Stine *et al.*, 2008). Thus, studies of the population dynamics of epidemic isolates are possible using VNTR loci.

Novel outbreak clones of the pathogenic type have been observed with novel serotypes. Both O139 (Stroeher *et al.*, 1992; Bik *et al.*,

1995; Comstock *et al.*, 1996) and O37 (Sarkar *et al.*, 2002) serogroups which appeared in 1992 and the 1960s, respectively, are examples of serotype switching. In each case, the genetic backbone of the *ctx*+ isolate remained the same, while the genes encoding the serogroup were replaced in an apparently single mutational event. Switching has been shown in the laboratory between isolates of *V. cholerae* growing in a biofilm on chitin. As in the natural cases, the genes encoding the serogroup were transferred en masse from one isolate to another. Switching was thought to be associated with the presence of IS1358 element present in O1 and O139 LPS encoding regions. However when this element was deleted from the strain, there was no effect on the rate of genetic exchange in the laboratory (Blokesch and Schoolnik, 2007). The JUMPstart (just upstream of many polysaccharide genes) signal has been postulated to be involved in O-antigen switching (Hobbs and Reeves 1994). In Argentina, the O1 serotype has been found in multiple genetic lineages and is likely to be another natural case of lateral gene transfer of the serotype encoding genes (Pichel *et al.*, 2003; González-Fraga *et al.*, 2007). The JUMPstart sequence has similarity and presumed homology to a DNA uptake signal sequence in *Haemophilus influenzae* (Redfield *et al.*, 2006). Regardless of whether there is a mechanism to increase the rate of transfer of serogroup encoding genes between isolates, it is clear that the members of the clinical clade can acquire serogroup encoding genes other than O1 and O139 that are currently the common wide spread causes of disease.

NAG-ST positive

The most studied NAG-ST-positive isolate is NRT36S that was isolated in Japan (Arita, *et al.*, 1986). Recently, an isolate with an identical MLST sequence type was found in Argentina (Gonzáles-Fraga *et al.*, 2007). When NRT36S was fed to volunteers, one patient produced a 5.3-litre purge, consistent with severe diarrhoea (Morris, *et al.*, 1990). The entire genome of this isolate has been sequenced and compared to *ctx*+ N16961 genome (Chen *et al.*, 2007). Eighty-nine per cent of the genes were similar between the two genomes. A third of the genome specific genes were involved in pathogenesis. NRT36S did not have the two largest pathogenesis related islands (CTX prophage, TCP pathogenicity island) nor two of the seventh pandemic smaller pathogenesis related islands (VSP-1 and VSP-2). NRT36S did have a mosaic structure for VPI 2 maintaining several genes in the middle of the island. It also had the gene encoding NAG-ST in the superintegron region of the genome. The superintegron is a region in which the genes are separated by a repeating element (~128 bp in *V. cholerae*) and thus having the same organizational structure as an integron. Also like integrons, this region is associated with extremely high, lateral gene transfer (Pang *et al.*, 2007). The NAG-ST gene was the first active gene identified in the superintegron of NRT36S. In addition, there were four pili systems, a type three secretion system and another toxin gene among the strain specific genes for NRT36S. Although the precise role of these genes in the pathogenic mechanism for this strain is unknown, the total number of insertions (6–10) to generate this pathogenic type from the core genome is small. The other genome specific genes included genes encoding proteins involved in synthesis of the surface polysaccharide, chemotaxis, transportation and metabolism, and transcriptional regulation as well as several hypothetical genes (Chen *et al.*, 2007).

Other variants (NAGST− and *ctx*−)

There are at least six independent reports of *ctx* and NAG-ST negative *V. cholerae* associated with local outbreaks. In Argentina, the Tucumán variant was first isolated in 1993 (Pichel *et al.*, 2003). Its name is derived from the province in which it was first found. It appears to be endemic having been isolated subsequently in 1994 and 1998. The Tucumán variant is distinct from the other pathogenic isolates. It is negative for both *ctx* and NAG-ST as well as for the secretion systems, and TCP. PFGE and RAPD analysis demonstrated that the Tucumán variant is distinct from the other pathogenic forms. However, there is one remarkable similarity between the Tucumán variant and many *ctx*+ isolates. They are serologically O1. This is one example in which the genetic history of the chromosome and the serotype are distinct, thus indicating that the genes encoding the serogroup are likely to have undergone lateral gene transfer (González-Fraga *et al.*, 2007).

A second *V. cholerae* O1 variant, the Amazonia variant was isolated first in the Amazon River basin near the intersection of the Brazilian, Columbian and Peruvian borders (Coelho, *et al.*, 1995). PFGE distinguishes it from isolates in the genes *ctx*+ clade. Although it has the O1 serotype, it does not have *ctx, sto, tcpA* and *zot*. It does have a nueromindase (nanH) gene that is known to be in VPI-2 in the *ctx*+ form. The *nan*H gene has an allele distinct from that found in NAG-ST and *ctx*+ isolates, further supporting the genetic uniqueness (Figueiredo *et al.*, 2005). A third O1 variant that does not have either *ctx* or NAG-ST was reported from an outbreak in Warangal in southern India (Saha *et al.*, 1996).

In addition to the O1 variants, O139 variants that do not have *ctx* have also been reported in India (Singh *et al.*, 2004), Mexico (Parveen *et al.*, 2003) and China (Qu *et al.*, 2003). The Mexican variants are distinct from the *ctx*+ clade in PFGE and ribotyping analyses. In contrast, the Chinese variants have a similar PFGE type, but distinct ribotype (Qu *et al.*, 2003) and are distinct based on presence and absence of genes in microarray analysis (Pang *et al.*, 2007). The array results also indicate that the variant isolates have the same set of serotype genes as the O139 *ctx*+ isolates and perhaps represent yet another example of lateral gene transfer of the serogroup encoding region. Although the pathogenic mechanism by which these variants cause diarrheal disease remains to be elucidated, it seems likely that the causative genes will be associated with lateral gene transfer events.

Unanswered questions and focus for future studies

The most important open question is how the epidemic *ctx*+ isolates spread so efficiently. What differentiates between localized outbreaks and pandemic spread of a pathogen? Is it a change in *V. cholerae* involving either mutation(s) or physiological variation? Is it rooted in the population dynamics of the pathogen or the host? Or is it derived from a confluence of environmental factors? It can also be due to a combined effect of multiple factors. Its also not known at this point why all pathogenic *V. cholerae* do not exhibit pandemic spread. It is important to understand the factors that cause explosive spread of an epidemic, in order to be able to devise strategies to control cholera.

Our understanding of *V. cholerae*, and cholera epidemics has expanded substantially but is not yet complete. *V. cholerae* is an organism whose core genome has undergone evolutionary changes to originate diverse pathogenic types. The existing pathogenic types can also mutate into new biotypes and serogroups and cause epidemic disease. Either of these evolutionary scenarios could give rise to the next world wide epidemic. Until further research defines how to stop these potentially devastating epidemics the best public health measure is constant vigilance tracking *V. cholerae* as it evolves.

References

Arita, M., Takeda, T., Honda, T. and Miwatani, T. (1986). Purification and characterization of *Vibrio cholerae* non-O1 heat-stable enterotoxin. Infect. Immun. 52, 45–9.

Bik, E.M., Bunschoten, A.E., Gouw R.D. and Mooi F.R. (1995). Genesis of the novel epidemic *Vibrio cholerae* O139 strain: evidence for horizontal transfer of genes involved in polysaccharide synthesis. Embo J. *14*, 209–16.

Blokesch, M. and Schoolnik, G.K. (2007). Serogroup Conversion of *Vibrio cholerae* in Aquatic Reservoirs. PLoS Pathog. 3, e81.

Chatterjee, S., Ghosh, K., Raychoudhuri, A., Pan, A., Bhattacharya, M.K., Mukhopadhyay, A.K., Ramamurthy, T., Bhattacharya, S.K. and Nandy, R.K. (2007). Phenotypic and genotypic traits and epidemiological implication of *Vibrio cholerae* O1 and O139 strains in India during 2003. J Med Microbiol. *56*, 824–32.

Chen, Y., Johnson, J.A., Pusch, G.D., Morris, Jr. J.G. and Stine, O.C. (2007). The genome of non-O1 *Vibrio cholerae* NRT36S demonstrates the presence of pathogenic mechanisms that are distinct from those of O1 *Vibrio cholerae*. Infect. Immun. 75, 2645–7.

Coelho, A., Andrade, J.R., Vicente, A.C. and Salles, C.A. (1995). New variant of *Vibrio cholerae* O1 from clinical isolates in Amazonia. J. Clin. Microbiol. 33, 114–8.

Comstock, L.E., Johnson, J.A., Michalski, J.M., Morris, J.G. Jr, and Kaper, J.B. (1996). Cloning and sequence of a region encoding a surface polysaccharide of *Vibrio cholerae* O139 and characterization of the insertion site in the chromosome of *Vibrio cholerae* O1. Mol. Microbiol. *19*, 815–26.

Danin-Poleg, Y, Cohen, L.A., Gancz, H., Broza, Y.Y., Goldshmidt, H., Malul, E., Valinsky, L., Lerner, L., Broza, M. and Kashi, Y. (2007). *Vibrio cholerae* strain typing and phylogeny study based on simple sequence repeats. J. Clin. Microbiol 45, 736–46.

Farfan, M., Minana-Galbis, D., Fuste, M.C. and Loren, J.G. (2002). Allelic diversity and population structure in *Vibrio cholerae* O139 Bengal based on nucleotide sequence analysis. J. Bacteriol. *184*, 1304–13.

Figueiredo, S.C., Neves-Borges, A.C. and Coelho, A. (2005). The neuraminidase gene is present in the non-toxigenic *Vibrio cholerae* Amazonia strain: a different allele in comparison to the pandemic strains. Mem. Inst. Oswaldo Cruz. *100*, 563–9.

Garg, P., Aydanian, A., Smith, D., J Glenn, M., Jr. Nair, G.B. and Stine, O.C. (2003). Molecular epidemiology of O139 *Vibrio cholerae*: mutation, lateral gene transfer, and founder flush. Emerg. Infect. Dis. *9*, 810–4.

González-Fraga, S., Pichel, M., Binsztein, N., Johnson, J.A., Morris, J.G., Stine, O.C. (2007). Lateral gene transfer of O1 serogroup encoding genes of *Vibrio cholerae*. U.S. - Japan Cooperative Medical Science Program, Austin, Texas.

Heidelberg, J.F., Eisen, J.A., Nelson, W.C., Clayton, R.A., Gwinn, M.L., Dodson, R.J., Haft, D.H., Hickey, E.K., Peterson, J.D., Umayam, L., Gill, S.R., Nelson, K.E., Read, T.D., Tettelin, H., Richardson, D., Ermolaeva, M.D., Vamathevan, J., Bass, S., Qin, H., Dragoi, I., Sellers, P., McDonald, L., Utterback, T., Fleishmann, R.D., Nierman, W.C., White, O., Salzberg, S.L., Smith, H.O., Colwell, R.R., Mekalanos, J.J., Venter, J.C. and Fraser, C.M. (2000). DNA sequence of both chromosomes of the cholera pathogen *Vibrio cholerae*. Nature *406*, 477–83.

Hobbs, M. and Reeves, P.R. (1994). The JUMPstart sequence: a 39 bp element common to several polysaccharide gene clusters. Mol. Microbiol. *12*, 855–6.

Keymer, D.P., Miller, M.C., Schoolnik, G.K. and Boehm, A.B. (2007). Genomic and phenotypic diversity of coastal *Vibrio cholerae* strains is linked to environmental factors. Appl. Environ. Microbiol. *73*, 3705–14.

Li, M., Kotetishvili, M., Chen, Y. Sozhamannan, S. (2003). Comparative genomic analyses of the Vibrio pathogenicity island and cholera toxin prophage regions in nonepidemic serogroup strains of *Vibrio cholerae*. Appl Environ Microbiol 69, 1728–38.

Meibom, K.L., Blokesch, M., Dolganov, N.A., Wu, C.Y. and Schoolnik, G.K. (2005). Chitin induces natural competence in *Vibrio cholerae*. Science *310*, 1824–7.

Miller, M.C., Keymer, D.P., Avelar, A., Boehm, A.B. and Schoolnik, G.K. (2007). Detection and transformation of genome segments that differ within a coastal population of *Vibrio cholerae* strains. Appl. Environ. Microbiol. 73, 3695–704.

Morris, J.G., Jr., Takeda, T., Tall, B.D., Losonsky, G.A. Bhattacharya, S.K., Forrest, B.D., Kay, B.A. and Nishibuchi, M. (1990). Experimental non-O group 1 *Vibrio cholerae* gastroenteritis in humans. J. Clin. Invest. 85, 697–705.

Noller, A.C., McEllistrem, M.C., Pacheco, A.G., Boxrub, D.J. and Harrison, L.H. (2003a). Multilocus variable-number tandem repeat analysis distinguishes outbreak and sporadic *Escherichia coli* O157:H7 isolates. J. Clin. Microbiol. *41*, 5389–97.

Noller, A.C., McEllistrem, M.C., Stine, O.C., Morris, J.G., Jr. Boxrub, D.J. Dixon, B. and Harrison, L.H. (2003b). Multilocus sequence typing reveals a lack of diversity among *Escherichia coli* O157:H7 isolates that are distinct by pulsed-field gel electrophoresis. J. Clin. Microbiol. *41*, 675–9.

Pang, B., Yan, M., Cui, Z., Ye, X., Diao, B., Ren, Y., Gao, S., Zhang, L. and Kan, B. (2007). Genetic diversity of toxigenic and nontoxigenic *Vibrio cholerae* serogroups O1 and O139 revealed by array-based comparative genomic hybridization. J. Bacteriol. *189*, 4837–49.

Parveen, S., Farrah, S.R., Gonzalez-Bonilla, C., Zamudio, A.V. and Tamplin, M.L. (2003). Characterization of a clinical *Vibrio cholerae* O139 isolate from Mexico. Can J Microbiol. *49*, 65–70.

Pichel, M., Rivas, M., Chinen, I., Martin, F. and Ibarra, C. (2003). Genetic diversity of *Vibrio cholerae* O1 in Argentina and emergence of a new variant. J. Clin. Microbiol. *41*, 124–34.

Qu, M., Xu, J., Ding, Y., Wang, R., Liu, P., Kan, B., Qi, G., Liu, Y. and Gao, S. (2003). Molecular epidemiology of *Vibrio cholerae* O139 in China: polymorphism of ribotypes and CTX elements. J. Clin. Microbiol. *41*, 2306–10.

Redfield, R.J., Findlay, W.A., Bosse, J., Kroll, J.S., Cameron, A.D. and Nash, J.H. (2006). Evolution of competence and DNA uptake specificity in the Pasteurellaceae. BMC Evol. Biol. *6*, 82.

Rivera, I.N., Chun, J., Huq, A., Sack, R.B. and Colwell, R.R. (2001). Genotypes associated with virulence in environmental isolates of *Vibrio cholerae*. Appl. Enviro. Microbiol. *67*, 2421–9.

Saha, P.K., Koley, H. *et al.* (1996). Nontoxigenic *Vibrio cholerae* 01 serotype Inaba biotype El Tor associated with a cluster of cases of cholera in southern India. J. Clin. Microbiol. *34*, 1114–7.

Salim, A., Lan, R. and Reeves P.R. (2005). *Vibrio cholerae* pathogenic clones. Emerg. Infect. Dis. *11*, 1758–60.

Sarkar, A., Nandy, R.K., Nair, G.B. and Ghose, A.C. (2002). Vibrio pathogenicity island and cholera toxin genetic element-associated virulence genes and their expression in non-O1 non-O139 strains of *Vibrio cholerae*. Infect. Immun. *70*, 4735–42.

Singh, D.V., Bhanumathi, R. and Colwell, R.R. (2004). Characterization of a cholera toxin gene-negative clinical strain of *Vibrio cholerae* O139 Bengal. J. Clin. Microbiol. *42*, 1381.

Spratt, B.G. and M.C. Maiden (1999). Bacterial population genetics, evolution and epidemiology. Phil. Trans. R.Soc. Lond. B Biol. Sci. *354*, 701–10.

Stine, O.C., Alam, M., Tang, L., Nair, G.B., Siddique, A.K., Faruque, S.M., Huq, A., Colwell, R.R., Sack, R.B., Morris, J.G. (2008). Seasonal cholera from multiple small outbreaks, rural Bangladesh. Emerg. Infect. Dis. *14*, May.

Stroeher, U.H., Karageorgos, L.E., Morona, R. and Manning, P.A. (1992). Serotype conversion in *Vibrio cholerae* O1. Proc. Natl. Acad. Sci. USA *89*, 2566–70.

Vital Brazil, J.M., Alves, R.M., Rivera, I.N., Rodrigues, D.P., Karaolis, D.K. and Campos, L.C. (2002). Prevalence of virulence-associated genes in clinical and environmental *Vibrio cholerae* strains isolated in Brazil between 1991 and 1999. FEMS Microbiol. Lett. *215*, 15–21.

Waldor, M.K. and Mekalanos, J.J. (1996). Lysogenic conversion by a filamentous phage encoding cholera toxin. Science *272*, 1910–4.

Wirth, T., D. Falush, Lan, R., Colles, F., Mensa, P., Wieler, L.H., Karch, H., Reeves, P.R., Maiden, M.C., Ochman, H. and Achtman, M. (2006). Sex and virulence in *Escherichia coli*: an evolutionary perspective. Mol. Microbiol. *60*, 1136–51.

Emerging Hybrid Variants of *Vibrio cholerae* O1

11

G. Balakrish Nair, Asish K. Mukhopadhyay, Ashrafus Safa and Yoshifumi Takeda

Abstract

Rapid emergence of genetic variants among toxigenic epidemic strains of *Vibrio cholerae*, contributes to the intricate epidemiological pattern of cholera. A remarkable event in recent years has been the emergence of strains of *V. cholerae* O1 which possess traits of both the classical and El Tor biotypes. The phenotypic, genetic, and epidemiological characteristics, of these new variants have been reviewed, with particular emphasis on the origin, significance and implications of these strains in the epidemiology of cholera.

Introduction

Vibrio cholerae strains are classified into serogroups on the basis of epitopic variations in the cell surface lipopolysaccharide, which identifies over 200 serogroups (Yamai *et al.*, 1997). However, the strains that have epidemic and pandemic potential belong to only two serogroups namely, O1 and O139. Until 1991, O1 was the only known serogroup of *V. cholerae* that caused endemic, epidemic and pandemic cholera. In 1992, serogroup O139 Bengal was recognized as the second epidemic serogroup of *V. cholerae* capable of causing epidemic and pandemic cholera (Albert *et al.*, 1993; Cholera Working Group, 1993; Ramamurthy *et al.*, 1993; Nair *et al.*, 1994, 1996). Extensive genetic analysis of *V. cholerae* O139 revealed that their O-antigen biosynthesis genes share maximum similarity with *V. cholerae* O22 serogroup (Dumontier and Berche, 1998), and homologous recombination would have replaced the O-antigen synthesis region of an ancestral El Tor strain with the O139-antigen synthesis genes (Comstock *et al.*, 1996; Ramamurthy *et al.*, 2003). The epidemiology of cholera due to both *V. cholerae* O1 and O139 has been extensively reviewed (Faruque *et al.*, 1993, 1998, 2003; Kaper *et al.*, 1995).

The O1 serogroup of *V. cholerae* is classified into two major biotypes, namely, classical and El Tor (Fig. 11.1). The classification into biotypes is based on a set of phenotypic traits that include susceptibility to polymixin B, Chicken cell (erythrocytes) agglutination (CCA), haemolysis of sheep erythrocytes, the Voges–Proskauer (VP) test, which measures the production of acetylmethylcarbinol, and phage susceptibilities (Kaper *et al.*, 1995, Faruque *et al.*, 1998). Biotype is a sub specific taxonomic classification of *V. cholerae* O1 but the bearing this classification has on the ecology, epidemiology and the disease is inadequately understood.

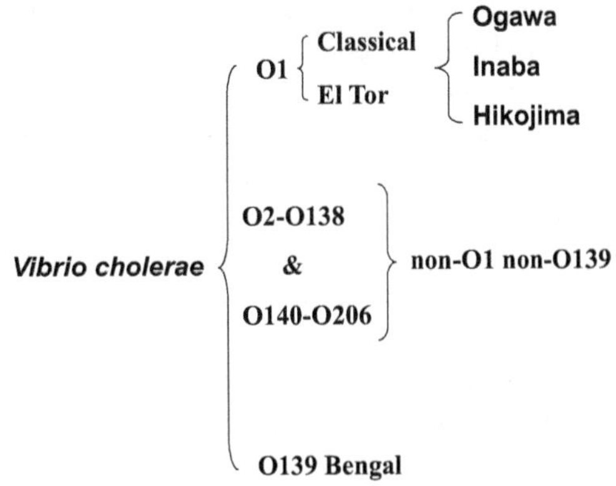

Figure 11.1 The revised classification of *Vibrio cholerae* (after the discovery of the O139 serogroup).

In recent years, cholera has become endemic in an increasing number of geographical areas, reflecting a collapse of socioeconomic infrastructure and difficulties in implementation of control measures in developing country settings. The annual cholera statistics released by WHO paints a gloomy picture with a dramatic rise in the number of cholera cases during 2006 reaching the level of the late 1990s (WHO, 2007). Cholera was notified from 52 countries in 2006 with the number of cases registering an overall increase of 79% compared with the number of cases reported in 2005 and the case fatality rate rising from 1.72% in 2005 to 2.66% in 2006 (WHO, 2007). The epidemiology of cholera continues to be perplexing and several extraordinary events, have been recorded in the recent past. Included among these enigmatic events are the bewildering Latin American episode of cholera after a 100-year absence in that continent (Levine 1991, Ries *et al.*, 1992), the unexplained reappearance of the classical biotype of *V. cholerae* O1 in southern Bangladesh until 1993 (Samadi *et al.*, 1983; Siddique *et al.*, 1991), the unprecedented and explosive emergence of *V. cholerae* O139 in 1992 (Albert *et al.*, 1993; Cholera Working Group 1993; Ramamurthy *et al.*, 1993), the reappearance in the Indian subcontinent of the O1 serogroup of *V. cholerae*, which apparently was a new clone different from those which existed earlier and shifts in the incidence of serotypes (Ogawa to Inaba and vice versa) of *V. cholerae* O1 in the Indian subcontinent (Dutta *et al.*, 2006; Raychowdhuri *et al.*, 2007). Another incomprehensible event in recent years has been the emergence of strains of *V. cholerae* O1 which possessed traits of both the classical and El Tor biotypes (Nair *et al.*, 2006; Safa *et al.*, 2006). In this chapter, we have reviewed the phenotypic genetic, and epidemiological characteristics, of these hybrid variants and have also speculated on the origin, significance and implications of these strains in the changing epidemiology of cholera.

Epidemiological relevance of the two biotypes of *Vibrio cholerae* O1

Seven times since 1817, cholera has spread into multiple countries in the form of pandemics. There is firm evidence that the fifth and sixth pandemics of cholera were caused by the classical biotype while the most extensive ongoing seventh pandemic which started in 1961 is caused by the El Tor biotype. Since 1961 classical strains were gradually replaced by the El Tor strains and within a decade or so the classical strains disappeared and were completely replaced by the El Tor strains (Bart *et al.*, 1970; Barua, 1992). However, the classical biotype re-emerged as the predominant epidemic strain in parts of Bangladesh in 1982 (Samadi *et al.*, 1983) and coexisted with the El Tor strains for about a decade. The classical biotype is now believed to be extinct with the last isolation in 1993 in southern Bangladesh.

Apart from the phenotypic differences based on bacteriological traits, there are dissimilarities in the infection pattern of the disease caused by the two biotypes. These include the occurrence of more asymptomatic carriers of El Tor strains that outnumber active cases by up to 50:1 (Sack *et al.*, 2004), better survival of El Tor strains in the environment and in the human host, as well as more efficient host-to-host transmission of El Tor strains as compared to classical strains. The level of asymptomatic *V. cholerae* infection is higher with El Tor strains than with classical strains, suggesting that the pathogenicity of El Tor strains is reduced compared with that of classical strains (Kaper *et al* 1995). The clinical manifestation of the disease caused by the classical biotype is more severe, while the El Tor biotype is considered to be more environmentally fit due to its capability to eliminate the classical biotype during the last pandemic. These observations led to the postulation that the epidemic nature of the two biotypes is different (Kaper *et al.*, 1995; Faruque *et al.*, 1998).

It has also been suggested that the displacement of the classical biotype by El Tor may be due to the ability of El Tor biotype strains to grow better in foods, such as cooked rice (Kolvin and Roberts, 1982). Because the infectious dose of *V. cholerae* is known to be quite high for humans with normal acidic stomach contents, transmission via contaminated food may be more important than transmission through contaminated water alone. Recently, the results of mathematical modelling of phage-mediated modulation of cholera epidemics have provided compelling evidence that amplification of *V. cholerae* in food sources may be an important rate-limiting step in cholera epidemics (Jensen *et al.*, 2006).

Differential regulation and expression of virulence factors

The seminal work that led to the discovery of cholera toxin was made by SN De in 1959 (De, 1959). The pathogenesis of cholera is a complex process and involves a number of factors, which helps the pathogen to reach and colonize the epithelium of the small intestine and produce cholera toxin (CT) that disrupts ion transport by intestinal epithelial cells. Thus, the major virulence factors, CT and the colonization factor toxin-coregulated pilus (TCP) are required for infection by both biotypes (Kaper *et al.*, 1995; Taylor *et al.*, 1987). There are significant differences in the regulation of virulence factors between the classical and El Tor biotype strains (DiRita *et al.*, 1996). Production of CT and TcpA (the major subunit of TCP) is readily achieved in classical strains of *V. cholerae* by growth in Luria–Bertani (LB) medium (pH 6.5–7.0) at 25–30°C. El Tor strains do not produce these virulence factors when grown under the same conditions but requires a complex growth medium termed AKI, a bicarbonate-containing medium, and specific unusual biphasic growth conditions that include the incubation of cultures at 37°C under static conditions for 4 h, followed by overnight incubation with shaking at 37°C to induce virulence gene expression. (Iwanaga *et al.*, 1986; DiRita *et al.*, 1996). The classical strains produced high levels of CT and TcpA when grown in either LB or AKI, although LB seemed the more favourable medium but in contrast, the El Tor strains produced detectable CT and TcpA only under AKI conditions (DiRita *et al.*, 1996)

The expression of the genes encoding TCP and CT is positively activated by the regulator ToxT (Champion *et al.*, 1997; DiRita *et al.*, 1991; DiRita, 1994; Higgins *et al.*, 1992), which is encoded within the TCP-ACF pathogenicity element, and its expression is dependent upon two transcriptional activator pairs, ToxRS and TcpPH, as well as by stimuli from the environment, including pH, temperature, osmolarity, and amino acids (DiRita and Mekalanos 1991; Häse and Mekanalos, 1998). The expression of *tcpPH* is also under the control of LysR regulator AphB along with a second protein, AphA. (Kovacikova and Skorupski, 1999; Skorupski and Taylor, 1999). Each of these activator proteins appears to play a similar role in both biotypes of *V. cholerae* O1. However, the expression of *tcpA*, *ctx*, *toxT*, and *tcpPH* is significantly lower in the El Tor biotype than in the classical biotype under most *in vitro* culture conditions (DiRita *et al.*, 1996; Murley *et al.*, 1999). The overproduction of AphB from a plasmid in the El Tor biotype has been observed to dramatically increase *tcpPH* and *tcpA* expression to levels close to classical biotype *in vitro* (Kovacikova and Skorupski 1999). This result, together with the finding that the expression of *aphB* is similar in both biotypes, raised the possibility that the biotype-specific expression of *tcpPH* might be related to the ability of AphB to activate transcription at each of these promoters. The molecular basis for differential regulation of virulence factor production has been extensively studied by different groups. Kovacikova and Skorupski (2000) have shown that the biotype specificity of virulence gene expression is due to a single base pair difference (65 and 66 bp) between the classical and El Tor *tcpPH* promoters, respectively, upstream of the transcriptional start, which alters the ability of the LysR regulator AphB to activate transcription under various conditions. To understand the influence of these base-pair changes on the production of TCP and CT in the two biotypes, they introduced a nucleotide (G at position −65) in the *tcpPH* promoter of a classical strain, and an A at position −66 in the *tcpPH* promoter of an El Tor strain. The presence of a G at position −65 in the classical *tcpPH* promoter reduced the production of TCP in LB medium (pH 6.5) at 30°C to a level undetectable by Western blotting and decreased the toxin production from 1500 ng/ml to less than 10 ng/ml. Conversely, an A at position −66 in the El Tor *tcpPH* promoter increased TCP production to a detectable level in LB medium (pH 6.5) at 30°C and increased toxin production from less than 10 ng/ml to almost 1400 ng/ml. Using this elegant experiment, they established that it is the ability of AphB to activate *tcpPH* expression which largely determines the biotype-specific pattern of virulence gene expression observed in *V. cholerae*.

The important question is how this single base pair change influences biotype-specific virulence gene expression. The same group showed the differential affinity of AphB for its recognition site in the classical and El Tor *tcpPH* promoters due to the single base pair alteration. By gel retardation experiments, it has been shown that AphB binds greater than 10-folds more efficiently to a wild type classical *tcpPH* promoter fragment with an A at -65 relative to a wild type El Tor fragment that has a G at this position. It has also been shown that the timing of the transcription of *tcpPH* is different between the classical and El Tor biotypes (Murley *et al.*, 2000).

Recently, Beyhan *et al.*, (2006) used transcriptional profiling to gain insight into variations in global gene expression between the classical and El Tor biotypes of *V. cholerae* O1 and determined that 13.5% of the total genes are differentially expressed between the two biotypes under conditions that induce virulence gene expression in the classical biotype. In particular, messages of *vieS*, *vieA*, and *vieB* genes, which encode regulatory proteins that control *ctx* transcription through their modulation of intracellular cyclic diguanylate (c-di-GMP) concentration (Tischler *et al.*, 2002), were increased in the classical biotype. The *vieA* gene encodes a response regulator possessing hydrolase activity against c-di-GMP (Sambrook and Russell, 2001; Tamayo *et al.*, 2005) and performs the critical role of lowering c-di-GMP concentration to allow for virulence gene expression. To determine the contribution of VieA to the transcriptome under the classical biotype virulence gene-inducing condition, they compared the whole-genome expression profile of the wild type classical strain to that of a *vieA* deletion mutant. The results revealed that VieA is a major regulator of gene expression under classical virulence gene-inducing conditions, and regulates the expression of 10.3% of the genes. In contrast, VieA does not significantly contribute to the regulation of gene expression in the El Tor biotype.

There are other differences between the biotypes beside the way they regulate toxin expression. El Tor strains produce a fimbrial structure termed the mannose-sensitive haemagglutinin (MSHA) that is not expressed by the majority of the classical strains (Jonson *et al.*, 1992). Studies in animal models suggested that MSHA pili may be important in the pathogenesis of El Tor cholera (Osek *et al.*, 1992, 1994). However, a more recent study (Attridge *et al.*, 1996) on El Tor strains of *V. cholerae* with deletions in the *mshA* gene did not show any significant attenuation or loss of colonization potential in an infant-mouse model. Antibodies to the pilin subunit were also not sufficient to mediate significant protection against a challenge with the wild-type El Tor strain (Attridge *et al.*, 1996). In addition to MSHA, El Tor strains also produce a haemolysin, which is not expressed by classical strains. The role of this factor in virulence has not been explained, but animal model experiments suggest that factors coordinately regulated with haemolysin may be important for successful colonization by El Tor strains (Williams and Manning, 1991; Alm and Manning, 1990). Production of virulence-associated factors like MSHA and haemolysin by El Tor strains, which distinguishes them from classical strains, may play a role in the differences observed in the epidemic nature of the two biotypes (DiRita *et al.*, 1996). Recently, two genes in the tRNA operon 1 were reported to be absent in the classical biotype strain as compared to that of the El Tor biotype (Ghatak *et al.*, 2005).

Differences in sugar metabolism

Yoon and Mekalanos (2006) showed a dramatic difference in carbohydrate metabolism between strains of the classical and El Tor biotypes. This difference leads classical biotype strains to produce organic acids when metabolizing carbohydrates, with a concomitant drop in pH of the growth media. Because *V. cholerae* is a profoundly acid-sensitive organism, this drop in pH rapidly inhibits further growth of strains of classical biotype on carbohydrate-rich media. El Tor strains do not accumulate such acids but instead produce 2,3-butanediol as their fermentation end product and thus grow to much higher densities in media containing carbohydrates. A mutant of an El Tor biotype strain deficient in the 2,3-butanediol fermentation showed the same growth defect displayed by classical biotype strains under carbohydrate-rich conditions. Thus, El Tor biotype strains which produce 2,3-butanediol are predicted to have a profound selective advantage over classical biotype strains in any environment that is rich in metabolizable carbohydrates. Such a metabolic selection can explain the dramatically improved fitness of toxigenic El Tor compared with the earlier classical biotype strains (Yoon and Mekalanos, 2006).

It is interesting to note that oral rehydration solution (ORS) containing 20 g/l glucose (Guerrant *et al.*, 2003; Ramakrishna *et al.*, 2000) is frequently used to treat the dehydration caused by *V. cholerae* infection (Schultz and Curran 1970; Serebro *et al.*, 1968). ORS has been used since the early 1960s, and this period happens to correspond to the same decade in which the seventh pandemic clone emerged and replaced the classical biotype. It has been suggested that the glucose present in ORS may have contributed to an *in vivo* selection of El Tor strains, at least in those patients that were not simultaneously treated with antibiotics. Thus, ORS may have been antibacterial for classical biotype strains that would be sensitive to added glucose but stimulatory to El Tor biotype strains that actually could utilize this sugar with no deleterious effects (Yoon and Mekalanos, 2006).

Cholera toxin of the two biotypes

CT is the principal toxin produced by *V. cholerae* O1 and O139 and is responsible for most of the manifestations of the disease cholera. Based on the B subunit of CT, two immunologically related but not identical epitypes and three genotypes of CT are currently recognized (Fig. 11.2). Epitype CT1 is the prototype elaborated by classical biotype strains and by the US Gulf Coast strains while CT2 is produced by the El Tor biotype and O139 strains (Finkelstein *et al.*, 1987). Another classification identifies three types of *ctxB* genes based on three non-random base changes resulting in a change in the deduced amino acid sequence. Genotype 1 is found in strains of classical biotype worldwide and the US Gulf Coast, genotype 2 is found in El Tor biotype strains from Australia, and genotype 3 is found in El Tor biotype strains from the seventh pandemic and the Latin American epidemic (Olsvik *et al.*, 1993). Thus, *V. cholerae* O1 El Tor biotype of the ongoing seventh pandemic produces CT of the CT2 epitype and genotype 3 while the classical biotype CT belongs to the CT1 epitype and genotype 1. The B subunits of El Tor and classical CT share major epitopes but also have one or more weaker biotype-specific epitopes (Dubey *et al.*, 1990). DiRita *et al.*, (1996) suggested that if the history of interaction between *V. cholerae* and human populations indeed resulted in the decreased severity of the El Tor biotype compared with the classical biotype, it is logical to consider that the target of this natural selection might have been the control of toxin production.

Hybrids of *Vibrio cholerae* O1 biotype

In the conventional tests used for biotyping of *V. cholerae* O1 cited earlier in this chapter, classical strains of *V. cholerae* O1 are negative for CCA, haemolysis of sheep erythrocytes, VP test and

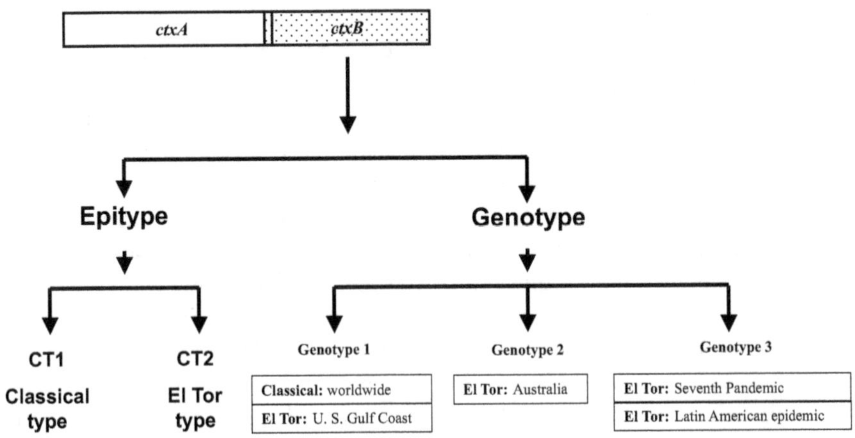

Figure 11.2 Currently recognized epitypes (immunological types) and genotypes (genetic types) of cholera toxin produced by *Vibrio cholerae* O1

are sensitive to polymyxin B and to phage IV but resistant to phage 5. Conversely, El Tor strains are positive for haemolysis of sheep erythrocytes, CCA, and VP test but resistant to polymyxin B and phage IV and sensitive to phage 5. However, over period of time, most of these tests have proven to be inadequate in classifying strains of *V. cholerae* O1 into their biotype. Currently, haemolysis of sheep erythrocytes is of limited use since, apart from isolates in the early 1960s, most isolates of *V. cholerae* O1 El Tor are negative for haemolysis of sheep erythrocytes and therefore non-haemolytic (Barrett and Blake, 1981). More recently, Nair *et al.*, (2002) reported polymyxin B sensitive El Tor strains of *V. cholerae* O1 from southern parts of Bangladesh. The phage sensitivity test has also shown variable results as reported with *V. cholerae* O1 strains that are either sensitive or resistant to both phages (Nair *et al.*, 2002, Safa *et al.*, 2006). On the other hand, the CCA and VP tests are more consistent in distinguishing the two biotypes, till date. With the advent of molecular techniques and with a better understanding of the genome of *V. cholerae* O1, biotype specific alleles of some critical virulence genes of *V. cholerae* were detected. Although these alleles encode identical function, they differ partly in their DNA sequence. Two major virulence genes of *V. cholerae* O1 strains, *tcpA* and *hlyA* possess classical- and El Tor-specific alleles (Keasler and Hall, 1993). These and other genes have been found to be useful genetic markers to distinguish between the classical and El Tor biotypes. It was clear that the current biotyping scheme based on phenotypic traits could be augmented by genotypic tests. For example, *ctxB* genotyping largely matches with the conventional biotyping scheme, except for the US Gulf Coast El Tor strains (Olsvik *et al.*, 1993). Since typically classical strains possess classical CT genes and El Tor strains possess El Tor CT genes with sequence slightly different from the classical, possession of classical CT gene by the US Gulf Coast strains despite being El Tor is an anomaly.

Since the current biotyping scheme using phenotypic and genotypic traits are inadequate to definitely categorize strains of *V. cholerae* O1 into classical or El Tor biotype, an alternate designation for such strains seemed necessary. Therefore, to designate strains, which could not be biotyped, we chose the term 'hybrid' to distinguish *V. cholerae* O1 strains, which possessed a mixture of phenotypic and genetic traits of the classical and El Tor biotypes from the typical classical and El Tor biotypes. In the recent past, different hybrids of *V. cholerae* O1 strains have been identified and these are described below.

The Matlab variants

The first description of naturally occurring hybrids of the classical and El Tor biotype associated with clinical cholera came from Bangladesh (Nair *et al.*, 2002). A retrospective analysis of *V. cholerae* O1 strains isolated from hospitalized cases of diarrhoea in Matlab, a rural area about 50 km south of Dhaka, revealed that some strains could not be biotyped as El Tor or classical using the conventional biotyping tests, and these were designated as Matlab variants (Nair *et al.*, 2002). However, responses of Matlab variants to the phenotypic traits examined were not uniform amongst the strains and thereby sub classified into three types namely Matlab I, II and III. Genotypically, the Matlab variants carried the set of typical virulence genes. However, in the different Matlab subtypes, the classical and El Tor alleles of *tcpA* and the CTX prophage repressor gene (*rstR*) were found in different combination (Nair *et al.*, 2002). Since CT plays the principal role in the pathogenesis of cholera, CT genes were also examined. Interestingly, the nucleotide-sequence-derived CtxB amino acid sequence of the Matlab variants was of the classical biotype (Safa *et al.*, 2006).

Ribotyping was used as a phylogenetic tool for assessing similarity between the hybrid strains versus standard strains of classical and El Tor biotypes. Ribotyping demonstrated that the Matlab I, II and III strains were similar to the El Tor reference strain and were completely different from the classical reference strain suggesting that the hybrids originated from an El Tor progenitor. Therefore, overall the Matlab strains were of the El Tor lineage displaying some traits of the classical biotype. Dendrogram constructed using the unweighted-pair group method using average linkages generated from *Not*I restriction patterns of whole-chromosomal DNA grouped the Matlab variants into two major clusters that were found to be similar but not identical to those of either of the biotypes (Safa *et al.*, 2005).

The Mozambique variants

In 1970, the seventh cholera pandemic reached the continent of Africa with devastating consequences and since then cholera caused by the El Tor biotype has affected most parts of Africa including the sub-Saharan region. In Mozambique, cholera continues to be endemic, mainly concentrated in Beira and Maputo, two of the largest cities. The most recent cholera outbreak in Beira occurred in 2004 (Ansaruzzaman *et al.*, 2004) and the outbreak was primarily believed to be caused by typical toxigenic El Tor strains but subsequent laboratory investigation revealed that the epidemic was associated with an unusual clone of *V. cholerae* O1 El Tor biotype. Biotyping of Mozambique strains of *V. cholerae* O1, isolated in the cholera treatment centre in Beira, showed the El Tor phenotype profile but the genotyping test for biotype detection, targeting *tcpA* and *rstR* genetic markers, generated incongruous results as the *tcpA* gene was of the El Tor type while the *rstR* gene was of the classical type (Ansaruzzaman *et al.*, 2004). Interestingly, DNA sequence analysis of the *ctxB* gene showed that the Mozambique strains harboured classical CT genes, despite belonging to the El Tor biotype (Ansaruzzaman *et al.*, 2004). Extensive analysis showed that the Mozambique strain carries two copies of CTX-^{class} prophage located on the small chromosome in a tandem array that allows excision of the prophage but the excised phage genome was deficient in replication and did not produce CTX^{class} virion (Kimsey and Waldor 1998; Faruque *et al.*, 2007; Das *et al.*, 2007; Lee *et al.*, 2006). Typically, two copies of CTX prophages are present in the classical strains, one in each of the chromosomes and are dysfunctional with regard to their phage virion production. In contrast, El Tor genome accommodates single or multiple copies of CTX phages, which are tandem in arrangement in the large chromosome and are capable of producing phage virion (Figures 11.3 and 11.4). Comparison of Mozambique strains and typical classical strains suggests that multiple recombination events might have taken place in the origination of the Mozambique strain which has a typical El Tor genome but its CTX prophage is of the typical classical type (Faruque *et al.*, 2007)

El Tor biotype strain producing classical cholera toxin

A collection of *V. cholerae* O1 strains isolated in Bangladesh during the past four and a half decades were recently examined by specific ELISA using monoclonal antibodies produced against classical and El Tor CT. This study revealed that the epitype and genotype of CT of the El Tor strains since 2001 associated with cholera in Bangladesh has shifted from epitype CT2 to epitype CT1 and from genotype 3 to genotype 1. The presence of classical CT by El Tor biotype *per se* is not novel and has been reported (Nair *et al.*, 2002; Tamplin *et al.*, 1990; Ansaruzzaman *et al.*, 2004). In fact, the US Gulf Coast clone of *V. cholerae* O1 is El Tor strains that possess the classical CT have been described several years ago (Olsvik *et al.*, 1993). Interestingly most of these strains harbour a hybrid CTX prophage which has an El Tor type *rstR* gene but the *ctxB* gene is of classical type. The fact that the El Tor strain producing classical CT has completely replaced the prototype seventh pandemic El Tor strains that produced the El Tor CT in Bangladesh is significant. It is clear that a cryptic change has taken place and the impact of this

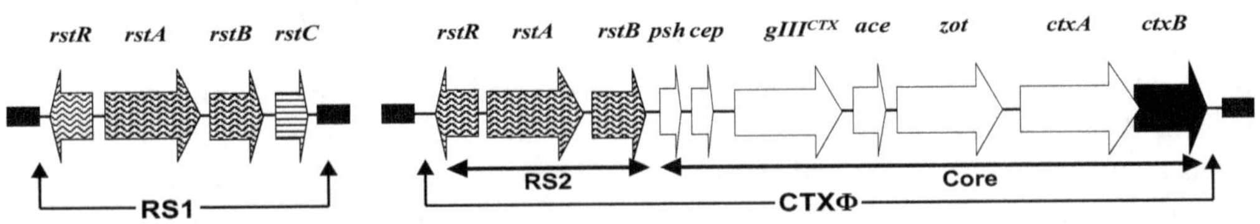

Figure 11.3 Typical genetic structure of the CTX and RS1 filamentous phages has been shown. The almost identitcal genes of the RS1 and RS2, except the addition of *rstC* (which has been shown in different shade) in RS1, are shown. The six of the seven core genes of the CTXΦ shown in white boxes while the remaining *ctxB* shown in Black. ■ boxes represent repeated sequences that flank the integrated phage DNA in *V. cholerae*.

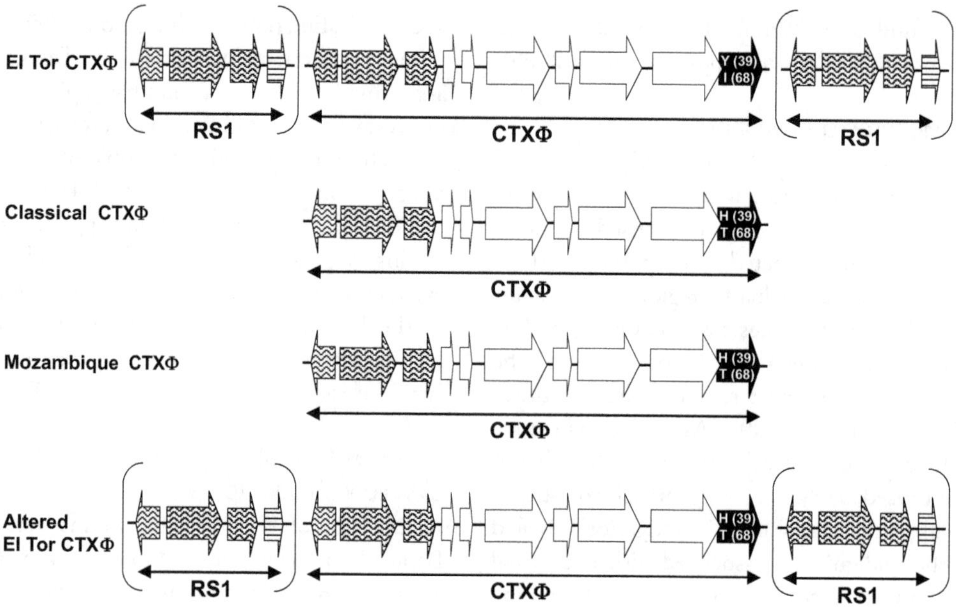

Figure 11.4 Arrangement of CTX and RS1 filamentous phages in different *V. cholerae* O1 variants have been shown. The double bracket indicates that the RS1 may flank both the upstream and downstream of the integrated CTXΦ genome. As shown in Figure 11.2, in the El Tor and altered El Tor strains CTXΦ usually flanked by RS1 whereas in the classical and recently identified Mozambique strains RS1 is absent. The number inside the black-coloured arrow (*ctxB*) represent the amino acid and the corresponding position, respectively, of *ctxB* mutation. H, T, Y and I stand for Histidine, Threonine, Tyrosine and Isoleucine respectively. Comparison shown in Figure 11.2 ascertains that classical, Mozambique and altered El Tor CTX phages possess identical amino acids at position 39 and 68 which differ from that of the El Tor type.

change is still being assessed from the point of view of the clinical presentation of the disease. Continuing studies have shown that the El Tor strains producing classical CT are also becoming the dominant type in India and Zambia.

Evolution of the two biotypes of *Vibrio cholerae* O1

Classical and El Tor strains of *V. cholerae* are closely related but are not directly derived from each other (Karaolis *et al.*, 2001) and are believed to have evolved from separate lineages (Kaper *et al.*, 1982; 1995). The two biotypes are closely related in their O-antigen biosynthetic genes (Manning *et al.*, 1994; Yamasaki *et al.*, 1999) although these two biotypes differ in other regions of their genomes (Beltran *et al.*, 1999; Karaolis *et al.*, 1995, 2001; Voss and Attridge, 1993; Wachsmuth *et al.*, 1994). Comparative genomic analysis using microarrays based on the whole genome sequenced O1 El Tor strain N16961 (Heidelberg *et al.*, 2000) has identified genes unique to the classical strains, pre-seventh pandemic El Tor strains and genes specific to seventh pandemic El Tor strains (Dziejman *et al.*, 2002).

Although the classical and El Tor biotypes have different lineages, hybrids between the classical and El Tor biotypes resulting from genetic exchange between different bacterial lineages also exist in nature (Chakraborty *et al.*, 2000, Mukhopadhyay *et al.*, 2001). The classical and El Tor biotypes had coexisted for at least over a decade following the emergence of the El Tor biotype in 1961 and the slow exit of the classical biotype in the 1970s (Samadi *et al.*, 1983, Siddique *et al.*, 1991). During this time, the two biotypes are likely to have interacted and exchanged genetic material either in the human intestinal milieu or in the aquatic environment. The biotype hybrids that are now being isolated from cholera cases possibly resulted from such genetic exchanges (Nair *et al.*, 2002). It is well recognized that genetic exchange between divergent bacterial lineages can contribute importantly to the success of a species in complex and inconstant environments, such as those in which *V. cholerae* may reside. Several studies have also pointed to such exchanges as an important factor in *V. cholerae* population genetics and evolution (Beltran *et al.*, 1999; Bik *et al.*, 1995; Faruque *et al.*, 1998). Using a *V. cholerae* genomic microarray that displayed over 93% of the predicted genes of the whole genome of the El Tor strain N16961, Dziejman *et al.* (2002) showed that only seven genes were absent solely in classical strains but present in other strains leading them to speculate that classical biotype strains may be derived from a primordial environmental strain that was more El Tor-like than previously thought. IS*1004* fingerprinting has determined that an O37 strain of *V. cholerae*, which was responsible for a large outbreak of cholera in Sudan in 1968 (Bik *et al.*, 1995) is closely related to classical O1 strains (Bik *et al.*, 1996). This indicates that horizontal exchange of genes have occurred not only between O1 biotypes but also between classical biotype and non-O1 strains, and the Sudan strain is a typical example of how an improved genotype can cause a large outbreak.

Conclusion

The prevalence of hybrid strains of *V. cholerae* may be more common than currently recognized because phenotypic methods are inadequate to precisely distinguish between the two biotypes, and genotypic techniques are not routinely used in clinical microbiology laboratories. By currently used bacteriological methods, the El Tor biotype strains that produce classical CT cannot be identified and therefore a revision in methods is needed to track the spread of such strains. The recognition of such hybrid strains and tracking their global spread is important because each of these types possess all the traits necessary to initiate a pandemic.

Strains of the El Tor biotype that produces classical CT are now dominant among strains currently associated with cholera in Bangladesh, India and in several other countries. At this time, it is not certain whether the change in CT subtype in the El Tor strains will enhance their epidemic potential. However, given that there are differences between the classical and El Tor biotypes, the selection of the El Tor biotype which produces classical CT would seem to indicate an evolutionary optimization of the El Tor biotype and could represent a new more efficient emerging form of this biotype (Nair *et al.*, 2006).

The implications in the heterogeneity in the B subunit as seen in the CT of the El Tor strains that produce classical CT also needs to be assessed from an epidemiological, ecological and clinical perspective. It has been shown that various polyclonal antisera raised against CT1 antigens neutralized CT2 considerably less effectively than they neutralized the homologous toxin (Marchlewicz and Finkelstein, 1983). Vaccines partly based on the B subunit from classical strains have been shown to protect less efficiently against El Tor strains than against classical strains in field trials (Clemens *et al.*, 1990). How the El Tor biotype strains that produce classical CT will influence the global dynamics of cholera remains to be observed.

References

Albert, M.J., Siddique, A.K., Islam, M.S., Faruque, A.S.G., Ansaruzzaman, M., Faruque, S.M. and Sack, R.B. 1993. Large outbreak of clinical cholera due to *V. cholerae* non-O1 in Bangladesh. Lancet *341*, 704.

Alm, R.A. and Manning, P.A. 1990. Characterization of the hlyB gene and its role in the production of the El Tor haemolysin of *Vibrio cholerae* O1. Mol. Microbiol. *4*, 413–425.

Ansaruzzaman, M., Bhuiyan, N.A., Nair, G.B., Sack, D.A., Lucas, M., Deen, J.L., Ampuero, J., Chaignat, C.L. and the Mozambique Cholera Vaccine Demonstration Project Coordination Group. 2004. Cholera in Mozambique, variant of *Vibrio cholerae*. Emerg. Infect. Dis. *10*, 2057–2059.

Attridge, S.R., Manning, P.A., Holmgren, J. and Jonson, G. 1996. Relative significance of mannose-sensitive hemagglutinin and toxin-coregulated pili in colonization of infant mice by *Vibrio cholerae* El Tor. Infect. Immun. *64*, 3369–3373.

Barrett, T.J. and Blake, P.A. 1981. Epidemiological usefulness of changes in hemolytic activity of *Vibrio cholerae* biotype El Tor during the seventh pandemic. J. Clin. Microbiol. *13*, 126–129.

Bart, K.J., Huq, Z., Khan, M. and Mosley, W.H. 1970. Seroepidemiologic studies during a simultaneous epidemic of infection with El Tor Ogawa and classical Inaba *Vibrio cholerae*. J. Infect. Dis. *121* (Suppl.), S17-S4.

Barua, D. 1992. Cholera. In: Barua D and Greenough WB, eds. 3rd ed. New York: Plenum Medical Book Co., pp.1–36.

Beltran, P., Delgado, G., Davarro, A., Trujillo, F., Selander, R.K. and Cravioto, A. 1999. Genetic diversity and population structure of *Vibrio cholerae*. J. Clin. Microbiol. *37*, 581–590.

Beyhan, S., Tischler, A.D., Camilli, A. and Yildiz, F.H. 2006. Differences in gene expression between the classical and El Tor biotypes of *Vibrio cholerae* O1. Infect. Immun. *74*, 3633–3642.

Bik, E.M., Bunschoten, A.E. Gouw, R.D. and Mooi, F.R. 1995. Genesis of the novel epidemic *Vibrio cholerae* O139 strain: evidence for horizontal transfer of genes involved in polysaccharide synthesis. EMBO J. *14*, 209–216.

Bik, E.M., Gouw, R.D. and Mooi, F.R. 1996. DNA fingerprinting of *Vibrio cholerae* strains with a novel insertion sequence element: a tool to identify epidemic strains. J. Clin. Microbiol. *34*, 1453–1461.

Chakraborty, S, Mukhopadhyay, A.K., Bhadra, R.K., Ghosh, A.N., Mitra, R., Shimada, T., Yamasaki, S., Faruque, S.M., Takeda, Y., Colwell, R.R. and Nair, G.B. 2000. Virulence genes in environmental strains of *Vibrio cholerae*. Appl. Environ. Microbiol. *66*, 4022–4028.

Champion, G.A., Neely, M.N., Brennan, M.A. and DiRita, V.J. 1997. A branch in the ToxR regulatory cascade of *Vibrio cholerae* revealed by characterization of *toxT* mutant strains. Mol. Microbiol. 23, 323–331.

Cholera Working Group. 1993. Large epidemic of cholera-like disease in Bangladesh caused by *Vibrio cholerae* O139 synonym Bengal. Lancet *342*, 387–390.

Clemens, J.D., Sack, D.A., Harris, J.R., van Loon, F., Chakraborty, J., Ahmed, F., Rao, M.R., Khan, M.R., Yunus, M.D., Huda, N., Stanton, B.F., Kay, B.A., Walter, S., Eeckels, R., Svennerholm, A.-M and Holmgren, J. 1990. Field trial of oral cholera vaccines in Bangladesh: results from a three-year follow-up. Lancet *335*, 270–273.

Comstock, L.E., Johnson, J.A., Michalski, J.M., Morris Jr., J.G. and Kaper, J.B. 1996. Cloning and sequence of a region encoding a surface polysaccharide of *Vibrio cholerae* O139 and characterization of the insertion site in the chromosome of *Vibrio cholerae* O1. Mol. Microbiol. *19*, 815–826.

Das, B., Haldar, K., Pal, P. and Bhadra, R.K. 2007. Small chromosomal integration site of classical CTX prophage in Mozambique *Vibrio cholerae* O1 biotype El Tor strain. Arch. Microbiol. *188*, 677–683.

De, S.N. 1959. Enterotoxicity of bacteria-free culture-filtrate of *Vibrio cholerae*. Nature *183*, 1533–1534.

DiRita, V.J. and Mekalanos, J.J. 1991. Periplasmic interaction between two membrane regulatory proteins, ToxR and ToxS, results in signal transduction and transcriptional activation. Cell *64*, 29–37.

DiRita, V.J. 1994. Multiple regulatory systems in *Vibrio cholerae* pathogenesis. Trends Microbiol. 2, 37–38.

DiRita, V.J., Neely, M., Taylor, R.K. and Bruss, P.M. 1996. Differential expression of the ToxR regulon in classical and El Tor biotypes of *Vibrio cholerae* is due to biotype-specific control over *toxT* expression. Proc. Natl. Acad. Sci. USA *93*, 7991–7995.

Dubey, R.S., Lindblad, M. and Holmgren, J. 1990. Purification of El Tor cholera enterotoxins and comparisons with classical toxin. J. Gen. Microbiol. *136*, 1839–1847.

Dumontier, S. and Berche, P. 1998. *Vibrio cholerae* O22 might be a putative source of exogenous DNA resulting in the emergence of the new strain of *Vibrio cholerae* O139. FEMS Microbiol. Lett. *164*, 91–98.

Dutta, B., Ghosh, R., Sharma, N.C., Pazhani, G.P., Taneja, N., Raychowdhuri, A., Sarkar, B.L., Mondal, S.K., Mukhopadhyay, A.K., Nandy, R.K., Bhattacharya, M.K., Bhattacharya, S.K. and Ramamurthy, T. 2006. Spread of cholera with newer clones of *Vibrio cholerae* O1 El Tor, serotype inaba, in India. J. Clin. Microbiol. *44*, 3391–3393.

Dziejman, M., Balon, E., Boyd, D., Fraser, C.M., Heidelberg, J.F. and Mekalanos, J.J. 2002. Comparative genomic analysis of *Vibrio cholerae*: genes that correlate with cholera endemic and pandemic disease. Proc. Natl. Acad. Sci. USA *99*, 1556–1561.

Faruque, S.M., Albert, M.J. and Mekalanos, J.J. 1998. Epidemiology, genetics and ecology of toxigenic *Vibrio cholerae*. Microbiol. Mol. Biol. Rev. *62*, 1301–1314.

Faruque, S.M., Alim, A.R.M.A., Rahman, M.M., Siddique, A.K., Sack, R.B. and Albert, M.J. 1993. Clonal relationships among classical *Vibrio cholerae* O1 strains isolated between 1961 and 1992 in Bangladesh. J. Clin. Microbiol. *31*, 2513–2516.

Faruque, S.M., Sack, D.A., Colwell, R.R., Sack, R.B. and Nair, G.B. 2003. Emergence and Evolution of *Vibrio cholerae* O139 Proc. Natl. Acad. Sci, USA *100*, 1304–1309.

Faruque, S.M., Tam, V.C., Chowdhury, N., Diraphat, P., Dziejman, M., Heidelberg, J.F., Clemens, J.D., Mekalanos, J.J. and Nair, G.B. 2007. Genomic analysis of the Mozambique strain of *Vibrio cholerae* O1 reveals the origin of El Tor strains carrying classical CTX prophage. Proc. Natl. Acad. Sci. USA *104*, 5151–5156.

Finkelstein, R.A., Burks, F., Zupan, A., Dallas, W.S., Jacob, C.O. and Ludwig, D.S. 1987. Epitopes of the cholera family of enterotoxins. Rev. Infect. Dis. *9*, 544–561.

Ghatak, A., Majumdar, A. and Ghosh, R.K. 2005. Structural organization of the transfer RNA operon I of *Vibrio cholerae*: differences between classical and El Tor strains. J. Biosci. *30*, 469–474.

Guerrant, R.L., Carneiro-Filho, B.A. and Dillingham, R.A. 2003. Cholera, diarrhea, and oral rehydration therapy: triumph and indictment. Clin. Infect. Dis. *37*, 398–405.

Häse, C.C. and Mekalanos, J.J. 1998. TcpP protein is a positive regulator of virulence gene expression in *Vibrio cholerae*. Proc. Natl. Acad. Sci. USA *95*, 730–734.

Heidelberg, J.F., Eisen, J.A., Nelson, W.C., Clayton, R.A., Gwinn, M.L., Dodson, R.J., Haft, D.H., Hickey, E.K., Peterson, J.D., Umayam, L., Gill, S.R., Nelson, K.E., Read, T.D., Tettelin, H., Richardson, D., Ermolaeva, M.D., Vamathevan, J., Bass, S., Qin, H., Dragoi, I., Sellers, P., McDonald, L., Utterback, T., Fleishmann, R.D., Nierman, W.C., White, O., Salzberg, S.L., Smith, H.O., Colwell, R.R., Mekalanos, J.J., Venter, J.C. and Fraser, C.M. 2000. DNA sequence of both chromosomes of the cholera pathogen *Vibrio cholerae*. Nature. *406*, 477–483.

Higgins, D.E., Nazareno, E. and DiRita, V.J. 1992. The virulence gene activator ToxT from *Vibrio cholerae* is a member of the AraC family of transcriptional activators. J. Bacteriol. *174*, 6974–6980.

Iwanaga, M., Yamamoto, K., Higa, N., Ichinose, Y., Nakasone, N. and Tanabe, M. 1986. Culture conditions for stimulating cholera toxin production by *Vibrio cholerae* O1 El Tor. Microbiol. Immunol. *30*, 1075–1083.

Jensen, M.A., Faruque, S.M., Mekalanos, J.J. and Levin, B.R. 2006. Modeling the role of bacteriophage in the control of cholera outbreaks. Proc. Natl. Acad. Sci. USA *103*, 4652–4657.

Jonson, G., Holmgren, J. and Svennerholm, A.M. 1992. Analysis of expression of toxin-coregulated pili in classical and El Tor *Vibrio cholerae* O1 in vitro and in vivo. Infect. Immun. *60*, 4278–4284.

Kaper, J.B., Bradford, H.B., Roberts, N.C. and Falkow, S. 1982. Molecular epidemiology of *Vibrio cholerae* in the US Gulf Coast. J. Clin. Microbiol. *16*, 129–134.

Kaper, J.B., Morris Jr., J.G. and Levine, M.M. 1995. Cholera. Clin. Microbiol. Rev. *8*, 48–86.

Karaolis, D.K.R., Lan, R. and Reeves, P.R. 1995. The sixth and seventh cholera pandemics are due to independent clones separately derived from environmental, nontoxigenic, non-O1 *Vibrio cholerae*. J. Bacteriol. *177*, 3191–3198.

Karaolis, D.K.R., Lan, R., Kaper, J.B. and Reeves, P.R. 2001. Comparison of *Vibrio cholerae* pathogenicity islands in sixth and seventh pandemic strains. Infect. Immun. *69*, 1947–1952.

Keasler, S.P. and Hall, R.R. 1993. Detection and biotyping *Vibrio cholerae* O1 with multiplex polymerase chain reaction. Lancet *341*, 1661.

Kimsey, H.H. and Waldor, M.K. 1998. CTXϕ immunity: Application in the development of cholera vaccines. Proc. Natl. Acad. Sci. USA *95*, 7035–7039.

Kolvin, J.L. and Roberts, D. 1982. Studies on the growth of *Vibrio cholerae* biotype eltor and biotype classical in foods. J. Hyg. (London) *89*, 243–252.

Kovacikova, G. and Skorupski, K.A. 1999. *Vibrio cholerae* LysR homolog, AphB, cooperates with AphA at the tcpPH promoter to activate expression of the ToxR virulence cascade. J. Bacteriol. *181*, 4250–4256.

Kovacikova, G. and Skorupski, K. 2000. Differential activation of the tcpPH promoter by AphB determines biotype specificity of virulence gene expression in *Vibrio cholerae*. J. Bacteriol. *182*, 3228–3238.

Lee, J.H., Han, K.H., Choi, S.Y., Lucas, M.E., Mondlane, C., Ansaruzzaman, M., Nair, G.B., Sack, D.A., von Seidlein, L., Clemens, J.D., Song, M., Chun, J., Kim, D.W. and Mozambique Cholera Vaccine Demonstration Project Coordination Group. 2006. Multilocus sequence typing (MLST) analysis of *Vibrio cholerae* O1 El Tor isolates from Mozambique that harbour the classical CTX prophage. J. Med. Microbiol. *55*, 165–170.

Levine, M.M. 1991. South America: the nature of cholera. Lancet *338*, 45–46.

Manning, P.A., Stroeher, U.H. and Morona, R. 1994. Molecular basis for O-antigen biosynthesis in *Vibrio cholerae*: Ogawa-Inaba switching. In: Wachsmuth IK, Blake PA, Olsvik O, eds. *Vibrio cholerae* and Cholera: Molecular to Global Perspectives. Washington DC: ASM Press, pp. 77–94.

Marchlewicz, B.A. and Finkelstein, R.A. 1983. Immunological differences among the cholera/coli family of enterotoxins. Diagn. Microbiol. Infect. Dis. *1*, 129–138.

Mukhopadhyay, A.K., Chakraborty, S., Takeda, Y., Nair, G.B. and Berg, D.E. 2001. Characterization of VPI Pathogenicity Island and CTXΦ prophage in environmental strains of *Vibrio cholerae*. J. Bacteriol. *183*, 4737–4746.

Murley, Y.M., Behari, J., Griffin, R. and Calderwood, S.B. 2000. Classical and El Tor biotypes of *Vibrio cholerae* differ in timing of transcription of tcpPH during growth in inducing conditions. Infect. Immun. *68*, 3010–3014.

Murley, Y.M., Carroll, P.A., Skorupski, K., Taylor, R.K. and Calderwood, S.B. 1999. Differential transcription of the tcpPH operon confers biotype-specific control of the *Vibrio cholerae* ToxR virulence regulon. Infect. Immun. *67*, 5117–5123.

Nair, G.B., Albert, M.J., Shimada, T. and Takeda, Y. 1996. *Vibrio cholerae* O139 Bengal: the new serogroup causing cholera. Rev. Med. Microbiol. *7*, 43–51.

Nair, G.B., Faruque, S.M., Bhuiyan, N.A., Kamruzzaman, M., Siddique, A.K. and Sack, D.A. 2002. New variants of *Vibrio cholerae* O1 biotype El Tor with attributes of the classical biotype from hospitalized patients with acute diarrhea in Bangladesh. J. Clin. Microbiol. *40*, 3296–3299.

Nair, G.B., Qadri, F., Holmgren, J., Svennerholm, A.M., Safa, A., Bhuiyan, N.A., Ahmad, Q.S., Faruque, S.M., Faruque, A.S., Takeda, Y. and Sack, D.A. 2006. Cholera due to altered El Tor strains of *Vibrio cholerae* O1 in Bangladesh. J. Clin. Microbiol. *44*, 4211–4213.

Nair, G.B., Ramamurthy, T., Bhattacharya, S.K., Mukhopadhyay, A.K., Garg, S., Bhattacharya, M.K., Takeda, T., Shimada, T., Takeda, Y. and Deb, B.C. 1994. Spread of *Vibrio cholerae* O139 Bengal in India. J. Infect. Dis. *169*, 1029–1034.

Olsvik, O., Wahlberg, J., Petterson, B., Uhlen, M., Popovic, T., Wachsmuth, I.K. and Fields, P.I. 1993. Use of automated sequencing of polymerase chain reaction-generated amplicons to identify three types of cholera toxin subunit B in *Vibrio cholerae* O1 strains. J. Clin. Microbiol. *31*, 22–25.

Osek, J., Jonson, G., Svennerholm, A.M. and Holmgren, J. 1994. Role of antibodies against biotype-specific *Vibrio cholerae* pili in protection against experimental classical and El Tor cholera. Infect. Immun. *62*, 2901–2907.

Osek, J., Svennerholm, A.M. and Holmgren, J. 1992. Protection against *Vibrio cholerae* El Tor infection by specific antibodies against mannose-binding hemagglutinin pili. Infect. Immun. *60*, 4961–4964.

Ramakrishna, B.S., Venkataraman, S., Srinivasan, P., Dash, P., Young, G.P. and Binder, H.J. 2000. Amylase-resistant starch plus oral rehydration solution for cholera. N. Engl. J. Med. *342*, 308–313.

Ramamurthy, T., Garg, S., Sharma, R., Bhattacharya, S.K., Nair, G.B., Shimada, T., Takeda, T., Karasawa, T., Kurazano, H., Pal, A. and Takeda, Y. 1993. Emergence of a novel strain of *V. cholerae* with epidemic potential in southern and eastern India. Lancet *341*, 703–704.

Ramamurthy, T., Yamasaki, S., Takeda, Y. and Nair, G.B. 2003. *Vibrio cholerae* O139 Bengal: odyssey of a fortuitous variant. Microbes Infect. 5, 329–344.

Raychoudhuri, A, Chatterjee, S., Pazhani, G.P., Nandy, R.K., Bhattacharya, M.K., Bhattacharya, S.K., Ramamurthy, T. and Mukhopadhyay, A.K. 2007. Molecular characterization of recent *Vibrio cholerae* O1, El Tor, Inaba strains isolated from hospitalized patients in Kolkata, India. J. Infect. *55*, 431–438.

Ries, A.A., Vugia, D.J., Beingolea, L., Palacios, A.M., Vasquez, E., Wells, J.G., Baca, N.G., Swerdlow, D.L., Pollack, M., Bean, N.H., Seminario, L. and Tauxe, R.V. 1992. Cholera in Piura, Peru: a modern urban epidemic. J. Infect. Dis. *166*, 1429–1433.

Sack, D.A., Sack, R.B., Nair, G.B. and Siddique, A.K. 2004. Cholera. Lancet *363*, 223–233.

Safa, A., Bhuiyan, N.A., Alam, M., Sack, D.A. and Nair, G.B. 2005. Genomic relatedness of the new Matlab variants of *Vibrio cholerae* O1 to the classical and El Tor biotypes as determined by pulsed-field gel electrophoresis. J. Clin. Microbiol. *43*, 1401–1404.

Safa, A., Bhuyian, N.A., Nusrin, S., Ansaruzzaman, M., Alam, M., Hamabata, T., Takeda, Y., Sack, D.A. and Nair, G.B. 2006. Genetic characteristics of Matlab variants of *Vibrio cholerae* O1 that are hybrids between classical and El Tor biotypes. J. Med. Microbiol. *55*, 1563–1569.

Samadi, A.R., Shahid, N., Eusuf, A., Yunus, M., Huq, M.I., Khan, M.U., Rahman, A.S.M.M. and Faruque, A.S.G. 1983. Classical *Vibrio cholerae* biotype displaces El Tor in Bangladesh. Lancet, *i*, 805–807.

Sambrook, J. and Russell, D.W. 2001. Molecular cloning: a laboratory manual. Cold Spring Harbor Laboratory Press, Cold Spring Harbor, N.Y.

Schultz, S.G. and Curran, P.F. 1970. Stimulation of intestinal sodium absorption by sugars. Am. J. Clin. Nutr. *23*, 437–440.

Serebro, H.A., Bayless, T.M., Hendrix, T.R., Iber, F.L. and McGonagle, T. 1968. Absorption of d-glucose by the rabbit jejunum during cholera toxin-induced diarrhoea. Nature *217*, 1272–1273.

Siddique, A.K., Baqui, A.H., Eusof, A., Haider, K., Hossain, M.A., Bashir, I. and Zaman, K. 1991. Survival of classic cholera in Bangladesh. Lancet *337*, 1125–1127.

Skorupski, K. and Taylor, R.K. 1999. A new level in the *Vibrio cholerae* ToxR virulence cascade: AphA is required for transcriptional activation of the tcpPH operon. Mol. Microbiol. *31*, 763–771.

Tamayo, R., Tischler, A.D. and Camilli, A. 2005. The EAL domain protein VieA is a cyclic di-guanylate phosphodiesterase. J. Biol. Chem. *280*, 33324–33330.

Tamplin, M.L., Jalali, R., Ahmed, M.K. and Colwell, R.R. 1990. Variation in epitopes of the B subunit of *Vibrio cholerae* non-O1 and *Vibrio mimicus* cholera toxins. Can. J. Microbiol. *36*, 409–413.

Taylor, R.K., Miller, V.L., Furlong, D.B. and Mekalanos, J.J. 1987. Use of *phoA* gene fusions to identify a pilus colonization factor coordinately regulated with cholera toxin. Proc. Natl. Acad. Sci. USA *84*, 2833–2837.

Tischler, A.D., Lee, S.H. and Camilli, A. 2002. The *Vibrio cholerae* vieSAB locus encodes a pathway contributing to cholera toxin production. J. Bacteriol. *184*, 4104–4113.

Voss, E. and Attridge, S.R. 1993. In vitro production of toxin-coregulated pili by *Vibrio cholerae* El Tor. Microb. Pathog. *15*, 255–268.

Wachsmuth, K., Olsvik, O., Evins, G.M. and Popovic, T. 1994. Molecular epidemiology of cholera. In: Wachsmuth IK, Blake PA, Olsvik O, eds. *Vibrio cholerae* and cholera: molecular to global perspectives, Washington DC: ASM Press, pp. 357–70.

Williams, S.G. and Manning, P.A. 1991. Transcription of the *Vibrio cholerae* haemolysin gene, hlyA, and cloning of a positive regulatory locus, hlyU. Mol. Microbiol. *5*, 2031–2038.

World Health Organization. 2007. Cholera 2006. Wkly. Epidemiol. Rec. *82*, 273–284.

Yamai, S., Okitsu, T., Shimada, T. and Katsube, Y. 1997. Distribution of serogroups of *Vibrio cholerae* non-O1 non-O139 with specific reference to their ability to produce cholera toxin, and addition of novel serogroups. Kansenshogaku Zasshi. *71*, 1037–1045.

Yamasaki, S., Garg, S., Nair, G.B. and Takeda, Y. 1999. Distribution of *Vibrio cholerae* O1 antigen biosynthesis genes among O139 and other non-O1 serogroups of *Vibrio cholerae*. FEMS Microbiol. Lett. *179*, 115–121.

Yoon, S.S. and Mekalanos, J.J. 2006. 2,3-Butanediol synthesis and the emergence of the *Vibrio cholerae* El Tor Biotype. Infect. Immun. *74*, 6547–6556.

Antibiotic Resistance in *Vibrio cholerae* 12

T. Ramamurthy

Abstract

Antimicrobial resistance has become a major medical and public health problem as it has direct link with the disease management. *Vibrio cholerae*, the cholera-causing pathogen is increasingly developing resistance towards many antimicrobials used for the treatment of diarrhoea. However, the pattern of resistance differs from country to country. The well-known factor responsible for development and spread of resistance is injudicious use of antimicrobial agents, which is directly related to the stimulation of several mechanisms of resistance. In *V. cholerae*, several resistance mechanisms such as plasmid encoded resistance, mutation in the quinolones resistance determining regions, integrons, efflux pumps and SXT constins have been established. Considering the importance of drug resistance, quick diagnostic assay methods are available for the identification of multidrug resistant (MDR) *V. cholerae*. Many new-generation antimicrobials were discovered, which are effective against *V. cholerae* in the *in vitro* studies. The resistance pattern of *V. cholerae* to several antimicrobials are not always uniform as it depends on the source of isolation. In some of the recent findings it was established that vibrios can act as reservoirs of antimicrobial resistance as cross-spread is common in *in vitro* studies. Promotion of indigenous drugs should be considered in the future and studied in detail for their efficacy.

Introduction

From the early 1920s, more than 25 potent antimicrobials were discovered and widely used in the clinical practice. Due to excessive use, most of these antimicrobials became less significant as many human and animal pathogens developed resistance using several mechanisms. Cholera is an endemic disease caused by toxigenic strains of *Vibrio cholerae* belonging to the O1 and O139 serogroups. In addition to these, more than 200 serogroups of *V. cholerae* are involved in causing sporadic cholera-like diarrhoea in many parts of the world. Wide circulation of multidrug resistant (MDR) *V. cholerae* strains complicates the problem in etiotrophic therapy. A study conducted in Kenya showed that the empiric treatment of diarrhoea with antimicrobials against different enteric bacteria was not successful as 51% of the pathogens were resistant to one or more antimicrobials (Shapiro *et al.*, 2001). This result may be true in other regions, where indiscriminate use of antimicrobials exists. This practice needs reevaluation to improve the clinical management of diarrhoea. The studies on prevalence of MDR strains and their mechanisms of resistance are evolved from carefully planed epidemiological investigations. This chapter describes the epidemiology and important mechanisms involved in MDR *V. cholerae*.

Clinical Investigations

Many clinical studies proved that the antimicrobial therapy for cholera effectively reduces the duration of diarrhoea, volume of the stool, and the bacterial load in the stool. Considering the emergence of MDR *V. cholerae* O1, animal model studies/clinical trials were conducted using alternative antimicrobials. In 1950s, furoxone (furazolidone) was widely used in the treatment of diarrhoea for its low toxicity and less chances for the development of resistant organisms (Phillips and Hailey, 1986). The efficacy of furazolidone for the treatment of cholera in children was tested either with a single dose of 7 mg/kg/day or in divided doses for 3 days (Rabbani *et al.*, 1991). In this double-blind placebo-controlled trial, stool culture was negative for *V. cholerae* on days 2–4 after the start of the therapy with cessation of diarrhoea within 72 hrs. In terms of clinical outcome, there was no significant difference between responses after 3-days of administration and single dose regime of furazolidone. Colonization of furazolidone-resistant strains of *V. cholerae* O1 and O139 in the rabbit intestine was prevented effectively with erythromycin and azithromycin (Chakrabarti *et al.*, 1998). Infection caused by mutants of *V. cholerae* O1 resistant to fluoroquinolones (FQs) in experimental animals could not be treated with the same drug (Ryzhko *et al.*, 2000). An Egyptian study showed that the mean number of *V. cholerae* O1 in stool reduced from 5.2×10^8 per ml to nil within 36 h after the administration of either erythromycin or cotrimoxazole (Burans *et al.*, 1989). Studies conducted in Lima, Peru, have shown that efficacy of cotrimoxazole (160 mg of trimethoprim and 800 mg of sulfamethoxazole every 12 h for 3 days) and tetracycline (500 mg every 6 h for 3 days) was comparable in cholera patients (Grados *et al.*, 1996). In the case of doxycycline, difference in antimicrobial susceptibility was reported in both *in vitro* and *in vivo* conditions. Treatment was clinically failed in 52% of doxycycline-treated cholera patients infected with tetracycline-resistant *V. cholerae* O1 compared with 8% of patients infected by tetracycline susceptible *V. cholerae* (Khan *et al.*, 1996). In a Peruvian study, it was shown that efficacy of ciprofloxacin is comparable with tetracycline in moderate to severe cholera patients (Gotuzzo *et al.*, 1995).

Studies conducted in Calcutta, India, have shown the efficacy of norfloxacin in the treatment of cholera caused by O1 and O139 serogroups (Bhattacharya *et al.*, 1990; Dutta *et al.*, 1996). In these findings it was established that compared to doxycycline and cotrimoxazole, patients treated with norfloxacin had reduced stool output, duration of diarrhoea, and oral rehydration requirement. Compared to single-dose doxycycline, ciprofloxacin was also very effective in the treatment of cholera patients infected with *V. cholerae* O1 or O139 (Khan *et al.*, 1996). Considering its potential use, single-dose ciprofloxacin is preferred for the treatment of cholera in areas where tetracycline-resistance *V. cholerae* is common. Studies on treatment of cholera patients and carriers of vibrios with different antimicrobials showed that bacterial relapses had occurred in 29.7% of the subjects with chloramphenicol, 16.5% with tetracycline, 15% with combination of chloramphenicol and tetracycline, 4.3% with furazolidone and 2.8% with ciprofloxacin (Karbyshev *et al.*, 1995). Based on this study, it was proposed to use ciprofloxacin for the treatment of cholera patients and furazolidone for *Vibrio* carriers. In a clinical trial with different antimicrobials, it was shown that the bacterial clearance within 24 h was 100%, 20% and 8.3% for ciprofloxacin, erythromycin and tetracycline, respectively (Khan *et al.*, 1995).

In Pakistan, the *in vitro* susceptibility of *V. cholerae* isolated from children was high to the third generation cephalosporins, nalidixic acid and chloramphenicol (Memon and Murtaza, 2002). Based on this study, it was proposed that the children could be treated with these antimicrobials. Progressive increase in the resistance towards nalidixic acid and FQs were reported in many findings. This trend might lead to treatment failure of cholera if the patients continue to receive either nalidixic acid or FQs.

Epidemiology of multiple drug-resistant (MDR) *Vibrio cholerae*

Comparison of *V. cholerae* infection status showed that the secondary infection rate was high (57%) with MDR strain than the susceptible ones (29%). Many investigations conducted in different geographical areas have demonstrated increase in the antimicrobial resistance spectrum among epidemic *V. cholerae*. Analysis of the antibiograms of *V. cholerae* O1 from various regions between 1938 and 1993 showed resistance to 1–3 antimicrobials, whereas the strains isolated from 1994 to 2005 possessed 3–8 resistance markers (Ryzhko *et al.*, 2005).

Tetracycline resistant strains of *V. cholerae* were isolated from pilgrims arrived from Mecca (Ng and Taha, 1994). *V. cholerae* O1 and O139 strains isolated in Bangladesh were consistently resistant to trimethoprim-sulfamethoxazole (cotrimoxazole) and furazolidone for very long time (Dhar *et al.*, 1996). In Indonesia, most of the enteric pathogens including *V. cholerae* O1 were susceptible to quinolones (Oyofo *et al.*, 2002) and resistance to ampicillin, cotrimoxazole and chloramphenicol was comparatively less (Tjaniadi *et al.*, 2003). Studies conducted in India during early 1990s showed that most of the *V. cholerae* strains were susceptible to many antimicrobials including nalidixic acid, kanamycin, cotrimoxazole, and chloramphenicol (Saraswathi and Deodhar, 1990), but furazolidone and cotrimoxazole resistant strains of *V. cholerae* were more common (Ramamurthy *et al.*, 1992a; Singh *et al.*, 1998), and an increasing trend to furazolidone resistance was reported from many parts of India (Khanna *et al.*, 1990; Niyogi *et al.*, 1994, 1995). In Calcutta, the *V. cholerae* O1 and O139 strains isolated from epidemic and endemic cholera cases were resistant to furazolidone and ampicillin (Sur *et al.*, 2002), whereas nalidixic acid and trimethoprim resistant *V. cholerae* strains were isolated in Chandigarah, North India (Taneja *et al.*, 2005). In Madras, most of the *V. cholerae* O1 and O139 strains were susceptible to chloramphenicol (Sundaram *et al.*, 2002). *V. cholerae* O1 isolated in Delhi from 1992–94 were resistant to polymyxin B and furazolidone but susceptible to tetracycline, chloramphenicol and cotrimoxazole (Rudra *et al.*, 1997). *V. cholerae* belonging to different serogroups in Calcutta, had about 119 resistance profiles between 1992 and 1997 with a peak during 1993–94 among O1 and O139 serogroups (Ramamurthy *et al.*, 2000). In the same study, cluster analysis demonstrated the dissemination of some of the common resistance patterns from non-O1, non-O139 serogroups to the O139 serogroup during subsequent years. Clinical *V. cholerae* strains resistant to O/129 compound and cotrimoxazole were identified in many countries (Ramamurthy *et al.*, 1992b; Bougoudogo *et al.*, 1992). Though both *V. cholerae* O1 and O139 serogroups were isolated during the same period and in the same region, their resistance profiles were different, especially to nalidixic acid (Chakraborty *et al.*, 2001a; Chhotray *et al.*, 2002). The general prevalence of tetracycline resistance in *V. cholerae* O1 and O139 are low in India (Jain *et al.*, 1993, Yamamoto *et al.*, 1995; Jesudason, 2006).

Interestingly, cholamphenicol, cotrimoxazole and nalidixic acid resistance among *V. cholerae* O139 strains has not always been constant. In many cholera endemic regions, after initial increase in resistance to cholramphenicol and cotrimoxazole during 1993–94, all the *V. cholerae* O139 strains became susceptible to these two antimicrobials during 2000- 2001 (Garg *et al.*, 2000; Jabeen and Hasan, 2003; Falklind-Jerkerus *et al.*, 2003; Das and Gupta, 2005). *V. cholerae* O139 isolated in 1996 were mostly susceptible to nalidixic acid but resistant to cotrimoxazole (91.5%) (Vijayalakshmi *et al.*, 1997). However, the remerged strains during 2000 were nalidixic acid resistant (Chakraborty *et al.*, 2001a). The molecular typing studies conducted with O139 *V. cholerae* collected during 1992–1998 from Bangladesh and India revealed prevalence of discrete antimicrobial resistance patterns with respect to their ribotypes and CTX genotypes (Faruque *et al.*, 2000). This investigation has shown that the changes in the antimicrobial resistance patterns might contribute to the selection of different genetic variants. Comparative study of *V. cholerae* O1 and O139 strains isolated from cholera patients showed difference in antimicrobial susceptibility, especially for tetracycline (58% vs. 100%) and furazolidone (27% vs. 93%) (Dhar *et al.*, 1996).

In Africa, the prevalence of MDR *V. cholerae* O1 including resistance to tetracycline was very high. During 1977–78 cholera epidemic in Tanzania, 76% of the *V. cholerae* O1 strains were resistant to tetracycline (Mhalu *et al.*, 1979). Though the strains isolated in November 1977 were susceptible to tetracycline, sudden upsurge in resistance was noticed during the subsequent months due to extensive therapeutic and prophylactic use of this drug. However, in Cameroon, routine use of antimicrobial therapy for the treatment of cholera and large-scale chemoprophylaxis of close contacts did not compromise the stability of susceptible *V. cholerae* (Noeske *et al.*, 2006). But, such reports are rare and lack explanation for the delay or no development of resistance due to therapeutic and prophylactic use of antimicrobials. Strains isolated during 1982–85 in Kenya were resistant to tetracycline, ampicillin and cotrimoxazole (Finch *et al.*, 1988). About 75% of the O1 strains isolated from cholera patients in Kenya (Nyanza Province) during 1983 were resistant to tetracycline, streptomycin and ampicillin (Ichinose *et al.*, 1986). *V. cholerae* O1 strains collected from cholera epidemics during 1985–1986 in Somali were resistant to ampicillin, kanamycin, streptomycin, sulfonamide, and tetracycline (Coppo *et al.*, 1995). It was assumed that these strains would have an origin from drug-susceptible Ethiopian strain. Majority of the *V. cholerae* strains isolated in Angola were resistant to ampicillin, kanamycin, streptomycin, cotrimoxazole, and chloramphenicol (Ferreira *et al.*, 1992). During 2000, spread of *V. cholerae* O1 resistant to cotrimoxazole, streptomycin, chloramphenicol, ampicillin and tetracycline was detected in Madagascar (Rakoto Alson *et al.*, 2001).

In Zambia, the tetracycline resistance was low (2–3%) during early 1990s, but increased to about 95% in 1996 (Mwansa *et al.*, 2006). Antimicrobial susceptibility pattern of *V. cholerae* O1 isolated from cholera outbreaks during 1994–96 showed interesting results. Strains from Tanzania and Rwanda were resistant to tetracycline, chloramphenicol and cotrimoxazole whereas, strains from Kenya, Sudan, and Somalia were susceptible to tetracycline (Materu *et al.*, 1997). Resistance to chloramphenicol and cotrimoxazole increased in Kenya and Somalia from 15% in 1994 to 90% in 1996 (Materu *et al.*, 1997). The number of cholera cases in Guinea-Bissau rose considerably during 1996–97 due to the emergence of MDR *V. cholerae* O1, with the increase in fatality rate from 1.0% to 5. 3% (Dalsgarred *et al.*, 2000a). These strains were susceptible (75%) to nalidixic acid and erythromycin.

V. cholerae isolated during the largest cholera outbreak in Kenya during 1998–99 were mostly resistant to chloramphenicol, streptomycin, and cotrimoxazole (Scrascia *et al.*, 2006). In Tanzania, *V. cholerae* strains were resistant to sulphamethoxazole, ampicillin, streptomycin, chloramphenicol, tetracycline and trimethoprim (Young and Amyes, 1986). The *dfrA1* gene encoding dihydrofolate reductase was discovered for the first time in *V. cholerae* with the Tanzanian strains isolated during early 1980s (Young and Amyes, 1986). Significant increase in the proportion of resistance among *V. cholerae* O1 was detected mostly to tetracycline, ampicillin, nalidixic acid, and erythromycin during 1997–99 (Urassa *et al.*, 2000). Almost during the same period, the O1 strains isolated in a cholera outbreak in Mozambique were resistant to tetracycline and cotrimoxazole (52% and 100%, respectively) (Folgosa *et al.*, 2001). Incidence of tetracycline resistant *V. cholerae* was also reported from Angola during 1987–90 (Ferreira *et al.*, 1992), Malaysia in 1992, and 1998 (Hassan and Teh, 1993; Ranjit and Nurahan, 2000), Pakistan in 1994 (Nizami and Farooqui, 1998), Italy in 1994 (Maggi *et al.*, 1997), Madagascar in 2000 (Dromigny *et al.*, 2002) and Thailand during 1997–98 (Kondo *et al.*, 2001).

The O1 strains isolated from a cholera outbreak in Albania and Italy during 1994 were resistant to tetracycline, streptomycin, trimethoprim, spectinomycin, sulfathiazole and the O/129 compound (Falbo *et al.*, 1999). In Europe, though the incidence of cholera is very less, the O1 strains isolated from sporadic cases were resistant to cotrimoxazole and tetracycline (Maggi *et al.*, 1997). Studies conducted by the CDC, USA based on reported cholera cases showed the increasing proportion of MDR *V. cholerae* O1 from 3% in 1992 to 93% during 1994 (Mahon *et al.*, 1996). However, in *V. cholerae* O139, tetracycline resistance is rarely reported (Yamamoto *et al.*, 1995). The extent of travel-associated *V. cholerae* O1 infections with strains resistant to cotrimoxazole, streptomycin, and furazolidone increased in the USA from 8% during 1990–94 to 31% during 1995–2000 (Steinberg *et al.*, 2001). The O1 strains isolated from a cholera outbreak in Senegal and Guiena-Bissau were resistant to cotrimoxazole and chloramphenicol (Sow *et al.*, 1997; Aidara *et al.*, 1998). In Ecuador, the outbreak strain of *V. cholerae* O1 in 1998 was susceptible to most of the antimicrobials (tetracycline and quinolones) used in the treatment of cholera (Gabastou *et al.*, 2002). Studies conducted in Brazil between 1991 and 1999 showed that the clinical and environmental isolates of *V. cholerae* O1 were resistant to several antimicrobials compared to the non-O1 strains (Campos *et al.*, 2004).

Generally, *V. cholerae* of the non-O1 and non-O139 serogroups are more resistant to various antimicrobials than O1 and O139 serogroups. In early 1980s, the Indian non-O1 vibrios were susceptible to cotrimoxazole, nalidixic acid, and furazolidone but resistant to chloramphenicol, streptomycin, tetracycline and ampicillin (Sundaram and Murthy, 1984). However, the rough strains of *V. cholerae* were additionally resistant to aminoglycosides, kanamycin, gentamicin, and neomycin. In Thailand, 41% of the non-O1, non-O139 serogroups isolated from children with diarrhoea during 1993–95 were multidrug resistant, especially to streptomycin and sulphisoxazole (Dalsgarrd *et al.*, 1999a). From 1995, *V. cholerae* non-O1, non-139 strains isolated from diarrhoeal patients in Calcutta were resistant to more than 8 antimicrobials (Garg *et al.*, 2000). In a recent cholera outbreak in South India (Hubli), 56% and 47% of the non-O1, non-O139 strains were resistant to ciprofloxacin and norfloxacin, respectively, as opposed to 12% for the O1 serogroup (Krishna *et al.*, 2005). *In vitro* studies with *V. cholerae* non-O1, non-O139 in Taiwan have shown the *in vitro* inhibition by cefazolin, cefotaxime and minocycline alone (Su *et al.*, 2005). However, combination of cefazolin or cefotaxime with minocycline exhibited synergistic inhibitory effect on these vibrios.

Methods of identification

Automation in the identification of pathogens and their susceptibility towards different antimicrobials are useful in clinical practice. Considering this, many assay systems are being developed. The Vitek system was evaluated with the reported trend in antimicrobial resistance and good correlation was established in the resistance patterns of different serogroups of *V. cholerae* (Sciortino *et al.*, 1996). Isolation and identification of toxigenic *V. cholerae* from environmental samples is always difficult. The antibiotic selection technique (AST) was reported to be useful for the isolation of *V. cholerae* O1 from environmental samples (Faruque *et al.*, 2006). In addition to the identification of *V. cholerae* and its virulence genes, other genes encoding resistance to antimicrobials such as the class 1 integron, and constins can be detected through a microarray-based detection system (Vora *et al.*, 2005). PCR based methods are also available for the identification *V. cholerae* strains harbouring sulfamethoxazole-trimethoprim (SXT) resistance genetic element and other resistance encoding genes (Mantri *et al.*, 2006; Ramachandran *et al.*, 2007).

Mechanisms of antimicrobial resistance

Several mechanisms are involved in *V. cholerae* in order to survive the prevailing antimicrobial pressure, and these involve either the modification of existing genetic material or the acquisition of new genetic material. The common mechanisms reported for the antimicrobial resistance includes, spontaneous mutations (primary resistance) and transfer of such change(s) to the progeny (vertical gene transfer). The widespread development of MDR in many serogroups simultaneously led us to believe the involvement of other mechanisms in *V. cholerae*. These mechanisms can either chemically modify the antibiotic or alternatively render it inactive through physical removal from the cell, or through modification of its target site. Presence of efflux system, lateral or horizontal gene transfer through mobile genetic elements is the well-established mechanisms for the antimicrobial resistance in *V. cholerae*.

Mutations in the quinolone resistance-determining regions (QRDRs)

Quinolones are antimicrobial agents that can act by forming ternary complexes with DNA gyrase (GyrA) and DNA topoisomerase IV (ParC) on bacterial chromosomal DNA. In Gram-negative bacteria, the primary targets of quinolones are topoisomerase II enzyme, the A and B subunits of DNA gyrase and topoisomerase, which is also composed of two subunits (ParC and ParE). Alternations in DNA gyrase have shown to reduce the affinity of the enzyme for quinolones and decrease the susceptibility of the organisms. Resistance is generally associated with amino acid substitutions in GyrA and ParC proteins located in the quinolone resistance-determining regions (QRDR).

Most of the pathogenic vibrios are susceptible *in vitro* to norfloxacin (Morris *et al.*, 1985). Since its first clinical use in Turkey during 1994

(Doganci *et al.*, 1995), ciprofloxacin is considered for the treatment of cholera in adults. The early isolates of *V. cholerae* are highly susceptible to fluoroquinolones (FQs), but the emergence of resistant strains was reported in India (Garg *et al.*, 2001). Most of the *V. cholerae* O1 and non-O1, non-O139 strains resistant to FQs had mutation at codon 83 in *gyrA* resulting in substitution of serine by isoleucine and at codon 85 with substitution of serine by leucine in GyrA (Baranwal *et al.*, 2002). However, there was no correlation between mutation in the QRDRs of GyrA and ParC and the minimal inhibitory concentration of FQs (Baranwal *et al.*, 2002). This study suggests that mutation in the QRDRs alone is not the prime factor for resistance to FQs in *V. cholerae.*

Efflux system

The efflux system pumps out a variety of structurally and functionally distinct antimicrobials from bacterial cells. In vibrios, a family of multidrug efflux proteins such as VmrA, NorM, VcrM. VcmA have been identified, which exploit the Na^+ transport across membranes as driving force (Morita *et al.*, 1998; Nozmul Huda *et al.*, 2001, 2003; Chen *et al.*, 2002; Singh *et al.*, 2006). VceR, a member of tetracycline resistance (Tet^R) family of transcriptional regulators is a repressor of the *vceCAB* operon that encodes a multidrug efflux pump in *V. cholerae*. The VceR binds within the *vceR–vceC* intergenic region and is dissociated from this site with CCCP, a pump substrate (Borges-Walmsley *et al.*, 2005). The co-operative transition between the free VceR and VceR-CCCP allows the repressor to respond to relatively small changes in the drug concentration in the cytoplasm of the bacterium (Borges-Walmsley *et al.*, 2005).

Recently, two bile-regulated RND family of efflux systems known as VexAB and VexCD were identified in *V. cholerae,* which are involved mainly in bile resistance (Bina *et al.*, 2006). The vexAB system was associated with resistance for several antimicrobials as determined by mutational analysis (Bina *et al.*, 2006). In non-O1 strains, about six putative genes for multidrug and toxic compound extrusion (MATE) family of efflux pumps have been reported (Begum *et al.*, 2005). Involvement of genes such as *vcmB, vcmD, vcmH* and *vcmN* with energy-dependent efflux towards FQs and aminoglycosides has been demonstrated in the *E. coli* background. The efflux activity of VcmN is sodium ion-independent, whereas in rest of the genes the efflux activity was sodium ion dependent (Begum *et al.*, 2005). A proton motive force-dependent efflux involving quinolone resistance has been identified in clinical strains of *V. cholerae* (Baranval *et al.*, 2002). Recently, it was shown that disruption of the *norM* gene of *V. cholerae* confers susceptibility for FQs (Singh *et al.*, 2006).

Horizontal transfer of antibiotic resistance determinants

The ability of bacteria to acquire and disseminate various antimicrobial resistance genes has been a major factor in the development of MDR. In many bacteria, resistance (R) plasmids, and transposons support acquisition and dissemination of genetic determinants of MDR (Recchia and Hall, 1995). The newly described integrating conjugative elements (ICEs) carry resistance for SXT, which is located in a specific DNA fragment called SXT element. In most of the clinical isolates, the frequency of the sulfonamides suggests the involvement of other mechanism linked to a novel group of naturally occurring mobile genetic elements, integrons. Integrons may be found as a part of the Tn21 transposon family or located on broad-host-range plasmids (Falbo *et al.*, 1999). In clinical isolates, at least five integrons (Intl) denoted Intl1 through Intl5 has been described, of which Intl1 and Intl4 are common among *V. cholerae.*

Plasmids

Generally, the wild strains of *V. cholerae* are a poor host for plasmids carrying genes that encode antimicrobial resistance (Davey and Pittard, 1975). The frequency of plasmid carriage in clinical and environmental strains of *V. cholerae* O1 is less (2%) compared with clinical and environmental non-O1 strains (25%) (Newland *et al.*, 1984). Most of the clinical strains of *V. cholerae* O1 isolated from different geographical regions lacked the plasmids (Bag *et al.*, 1998; Clark *et al.*, 1998), and when present, the size and functions of plasmids were not uniform. In Malaysia, *V. cholerae* O1 and non-O1 strains isolated during 1993–1997 from cholera patients and environmental sources harboured several small plasmids ranging from 1.3 to 4.6 MDa (Radu *et al.*, 1999). The Thai non-O1, non-O139 strains had one or more small plasmids but did not encode resistance to any antimicrobials (Dalsgaard *et al.*, 1999a). Similarly, the non-O1 strains from Asia and Peru though harboured low molecular weight plasmids, did not correlate with MDR (Dalsgaard *et al.*, 1995). The *V. cholerae* O1 strains from Iran harboured a large plasmid (Pourshafie *et al.*, 2002). In some studies it was shown that MDR environmental non-O1 *V. cholerae* strains harboured a large plasmid (74.8 MDa) carrying resistance markers for ampicillin, amoxicillin and sulfanilamide which were transferable by conjugation (Amaro *et al.*, 1988).

Transfer of MDR from *V. cholerae* to *Escherichia coli* (K12) has also been demonstrated *in vitro* with conjugative plasmids. Strains isolated in Angola co-transferred tetracycline and chloramphenicol phenotype to *E. coli* (Ferreira *et al.*, 1992). Inc C and Inc J groups of plasmids were detected in *V. cholerae* (Kuwahara, 1978; Rahal *et al.*, 1978), and the stability of plasmids carrying MDR with different Inc groups showed that except Inc C and J groups, rest of the plasmids were lost during subsequent culture in antibiotic-free medium (Rahal *et al.*, 1978). Further studies showed that Inc C plasmid was highly stable in El Tor vibrios than Inc J or F1 (Rahal *et al.*, 1978; Rahal, 1982). In Bangladesh, the 1979 cholera outbreak was caused by the MDR *V. cholerae* O1 El Tor strains harbouring Inc C type of plasmid (Glass *et al.*, 1983). MDR *V. cholerae* isolated in Kenya harboured a 100 MDa plasmid of group Inc C (Finch *et al.*, 1988). RFLP of this plasmid was the same with strains isolated before and after the emergence of MDR but differed with *V. cholerae* strains isolated from Tanzania, Nigeria and Bangladesh (Finch *et al.*, 1988). MDR strains isolated from Somali cholera epidemic during 1985–1986 harboured two Inc C group plasmids, which carried resistance for ampicillin, kanamycin, streptomycin, sulfonamide, and tetracycline (Coppo *et al.*, 1995). In *V. cholerae* O139 strains isolated during 1994 in India, a 200-kb self-transmissible plasmid (STP) belonging to the Inc C group was identified with several antimicrobial resistance genes (Yamamoto *et al.*, 1995).

Extended-spectrum beta-lactamases (ESBL) are very common among members of *Enterobacteriaceae*. In Tanzania, the El Tor strains harbouring the Inc C group of plasmids encoded production of SAR-1 type beta-lactamase, which hydrolyse carbenicillin and penicillin G (Reid and Amyes, 1986). SAR-1 is distinct from other plasmid-born beta-lactamase by virtue of its low isoelectric point, size (33.7 kDa), substrate profile and inhibition properties. In some of the *V. cholerae* O1 strains isolated during cholera outbreaks in Argentina, the ESBL activity was detected by isoelectric focusing with CTX-M or PER-2 and TEM-a like enzymes (Petroni *et al.*, 2002). The genes that encode beta-lactamases were found in a 150-kb plasmid and transferred the resistance to *E. coli*. The transconjugants also received other non-beta-lactam markers such as resistance for gentamicin, kanamycin, and sulfamethoxazole, or amikacin and streptomycin or trimethoprim, tetracycline or chloramphenicol (Petroni *et al.*, 2002).

Co-transfer of trimethoprim and O/129 compound resistance-encoding genes were demonstrated from *V. cholerae* to other members of enteric bacteria through Inc 6-C plasmid (Goldstein *et al.*, 1986). Subsequent to this transfer, the target genes were detected in a 14 kb transposon, Tn1527, emphasizing the importance of this transposon in the transmission of antimicrobial resistance through plasmids. In a clinical study, it was shown that the Inc 6-C plasmid was responsible for cotrimoxazole resistance (Dupont *et al.*, 1985). The size of the Inc 6-C group plasmid was 100 MDa (Ferreira *et al.*, 1992). In Uganda, the *V. cholerae* O1 strains possessed Inc 6-C plasmid with a size of 130-MDa and had several antimicrobial resistance genes including *dhfrI*, *sulII*, *tet*C, and *catI* encoding resistance for trimethoprim, sulphonamides, tetracycline and chloramphenicol, respectively (Kruse *et al.*, 1995). This plasmid (pRVC1) was transferred to other enteric bacteria with a very high frequency. Some of the *V. cholerae* O1 strains isolated during 1991–1993 in Ceara, Brazil harboured a 147 kb STP, which also had the marker gene for O/129 compound resistance (Hofer *et al.*, 1999). The 60 MDa conjugative plasmid of

V. cholerae O1 strains isolated from Albania and Italy harboured several genes encoding resistance to tetracycline, streptomycin, trimethoprim, sulfathiazole and O/129 compound. *dfrA1* gene was identified in these strains, which was responsible for resistance to trimethoprim and O/129 compound (Falbo *et al.*, 1999). A novel (Inc J) group of plasmid was detected for the first time in *V. cholerae* isolated from cholera outbreaks in Algeria during 1994 and carried resistance marker genes for chloramphenicol, tetracycline, streptomycin, and cotrimoxazole (Korichi *et al.*, 1997).

V. cholerae O1 strains isolated during a cholera outbreak in Nigeria harboured one or more plasmids with size ranging from 4.5 to 150 kb (Olukoya *et al.*, 1995). Most of these plasmids were transferred to *E. coli* K12 strain with specific resistance to ampicillin, tetracycline, and trimethoprim. The MDR strains of *V. cholerae* O1 from Madagascar harboured a 26-kb STP (Rakoto Alson *et al.*, 2001). A 150 kb STP that carried the *aadB* gene cassette was identified in Thai O1 strains isolated during 1982 (Dalsgarrd *et al.*, 2000b). *V. cholerae* O1 strains isolated in Zambia (1996–1997) showed high level of resistance to tetracycline due to the presence of a 14-MDa conjugative plasmid (Mwansa *et al.*, 2006). The O1 strains isolated between 1991 and 1996 from Angola harboured a large plasmid (p3iANG) with class 1 integron containing the resistance gene cassettes *dfrA15*, *blaP1* and *aadA8* genes (Ceccarelli *et al.*, 2006a).

The SXT element

The SXT element is an integrative and conjugative element (ICE) of *V. cholerae*, which also has been known as conjugative transposon. The SXT element generally carries genes encoding resistance for sulfamethoxazole, trimethoprim, chloramphenicol, and streptomycin (Hochhut *et al.*, 2001; Iwanaga *et al.*, 2004). Before the emergence of *V. cholerae* O139, SXT was rarely detected in O1 serogroup (Amita *et al.*, 2003). The ICEs are often found in Asian *V. cholerae* strains, which were isolated after the emergence of the serogroup O139 (Amita *et al.*, 2003; Burrus *et al.*, 2006a). In *V. cholerae*, ICEs devoid of resistance genes have also been reported in many countries (Ehara *et al.*, 2004; Iwanaga *et al.*, 2004; Ceccarelli *et al.*, 2006b; Bani *et al.*, 2007).

The Inc J elements such as R391 are closely related to SXT and more than 25 members of the SXT/R391 family of ICEs have been identified in many bacterial species (Burrus *et al.*, 2006a). R391 is an archetype of Inc J class of mobile genetic element detected in *Providencia rettgeri* consisting of genes originating from bacteriophages, plasmids, and transposable elements (Boltner *et al.*, 2002). Sequence analysis suggests a homology of large section of R391 with *Salmonella* Inc HI1 plasmid R27 (Boltner *et al.*, 2002). The 62-kb SXT element excises spontaneously from the host chromosome and forms a circular, non-replicative extrachromosomal molecule and transfer into a new host by conjugation, where it integrates into the *prfC* gene that encodes peptide chain release factor-3 (Hochhut and Waldor, 1999). In the absence of *prfC*, the SXT element integrates into several secondary sites at the 5′ end of the *pntB* (Burrus and Waldor, 2003). The mechanisms of integration and excision of SXT element is almost similar to the site-specific recombination reported in lambdoid phages. Excision and self-transfer of the SXT element require an element encoded integrase (Int), which is a SXT element-encoded recombinase, and the *int* expression in recipient cells in turn requires the SXT activators SetC and SetD (Burrus and Waldor, 2003). The *setR* locus carrying the function similar to lambdoid phage CI repressors also play a role in the regulation of *sxt* expression (Beaber *et al.*, 2002). The *xis*-like gene was not reported in the SXT genome, however, a novel SXT Xis that function as a recombination directionality factor (RDF) was identified in *V. cholerae*, facilitating excision and inhibiting the SXT integration in the same cells (Burrus and Waldor, 2003). Considering several properties, Hochhut and Waldor (1999) classified SXT element as a constin, an acronym for a conjugative, self-transmissible, integrating element.

Freshly isolated SXT exoconjugates harbour tandem SXT arrays *in vitro*, but these were not stable in subsequent cultures and might promote formation of novel ICEs (Burrus and Waldor, 2004). Functional hybrid elements were detected in these exoconjugates, which arose from *recA*-independent recombination between the two ICEs. The specific requirement of host factors in the transmission of ICEs has not been studied in detail. Nucleoid proteins such as integration host factor (IHF), Fis and lambda integrase play crucial role in the transfer and integration of SXT element. The importance of IHF for *V. cholerae* to act as a SXT element donor has been demonstrated with IHF mutants of *V. cholerae* and *E. coli*. These mutants were defective in conjugation and lacked the capacity to act as donors or recipients, respectively (McLeod *et al.*, 2006). This analysis indicated that IHF is a *V. cholerae*-specific conjugative factor.

Co-transfer of the SXT element genes was demonstrated in mating experiments (Dalsgaard *et al.*, 2001). Considering the recent O139 strains harbouring SXT element without antimicrobial resistance genes, it appears that selection for increasing resistance may not be the primary factor for the spread of SXT-related ICEs in vibrios (Burrus *et al.*, 2006b). Interestingly, the SXT constin of *V. cholerae* O1 El Tor and O139 strains are not the same, as the location of gene encoding resistance to trimethoprim varies in the two different serogroups (Hochhut *et al.*, 2001).

The nomenclature for SXT/R391 elements was proposed recently (Burrus *et al.*, 2006a). The ICEs were designated with genera, species, name of the country and number of resistance gene cluster. For example, the Indian *V. cholerae* carrying the ICE are designated as ICE*Vch*Ind1. The SXT-related ICE, ICEVchMex1 was first reported in a Mexican environmental strain of *V. cholerae* (Burrus *et al.*, 2006a). Significant difference between the genomes of SXT from the Mexican and Asian strains suggests their independent emergence (Burrus *et al.*, 2006a). Strains isolated in Vietnam during 1990 contained ICEVchVie0 without any resistance gene cluster (Bani *et al.*, 2007). A variant type of *V. cholerae* SXT element was identified in *V. fluvialis*, which has a 99% identity to that of *V. cholerae* (Ahmed *et al.*, 2005). The hot spot regions of SXT might carry genomic islands associated with multidrug resistance in other species (Iwanaga *et al.*, 2004). Thus, there is constant flux of genes among *V. cholerae* SXT constins.

Integrons

Integrons represent the primary mechanism for antimicrobial resistance gene capture and dissemination among most of the Gram-negative bacteria. The integron consists of an integrase gene (*intI*) belonging to the tyrosine recombinase family (Nunes-Duby *et al.*, 1998), *attI* and *attC* sites for recombination. The *attC* site is generally associated with an ORF in a circularized structure termed as gene cassette. Based on the divergence of their integrase genes, about five classes of integrons have been reported (Rowe-Magnus and Mazel, 2002). As seen with other enteric bacteria, the class 1 integron is widely prevalent in *V. cholerae*. Class 4 integron is specific for *V. cholerae*, while the class 2 integron is very rare in this species (Ahmed *et al.*, 2006).

The class 4 integron (*IntI4*) is associated with a *V. cholerae* repeated sequences (VCR) that carries several antimicrobial resistance gene cassettes or multiple resistance integrons (MRIs). The similarity of *IntI4* was identified by recombination of a VCR cassette into class 1 integron (Mazel *et al.*, 1998). *Intl4* a member of the integron family of integrases is widely found in the chromosome of *V. cholerae* at the 3′ end of the *attI* site (Clark *et al.*, 2000). The superintegron (SI), which was first discovered in the *V. cholerae* genome, is an archetype component of the genomes in many bacterial species (Mazel *et al.*, 1998; Rowe-Magnus *et al.*, 2002). The SI in *V. cholerae* harbours hundreds of gene cassettes encoding functions such as antimicrobial resistance, adaptation to the environment and pathogenicity (Rowe-Magnus *et al.*, 1999). It was also demonstrated that MRI could randomly select genes through integron-mediated recombination events from the SI cassettes and act as a genetic source for phenotypic conversion of sensitive strains to resistant ones (Rowe-Magnus *et al.*, 2002).

Considering its sedentary and homogeneous nature of the *attC*, the SI is structurally different from other integrons. Assays on resistance gene cassette deletion and integration of cognate *attI* sites demonstrated that the range of *attC* sites recombined by the SI were narrower than the *intI1* (Biskri *et al.*, 2005). This indicates that the substrate recognition and recombination reactions mediated by SI differ from the *intI1*. In some of the *E. coli* strains the resistance gene cassette arrangement at the *attC* region was similar to that of *V. cholerae* superintegron (Mazel *et al.*, 2000). This supports the view that the resistance gene cassettes are chosen from the SI pools and independently assembled at the *attC* site.

The constitution of the resistance gene cassettes in class 1 integrons is not the same. The resistance gene cassettes reported in the class 1 integrons of *V. cholerae* are shown in Table 12.1. The epidemic strains of *V. cholerae* O1 from Guinea-Bissau harboured a 150-kb conjugative multiple-antimicrobial resistance plasmid with gene cassettes *dhfrIII* and *ant (3')-1a*, encoding resistance for trimethoprim and aminoglycosides, respectively (Dalsgaard *et al.*, 2000a). The *aadB*, *aadA2*, *blaP1*, *dfrA1* and *dfrA15* gene cassettes were detected in *V. cholerae* O1 strains from Thailand in a single cassette array (Dalsgaard *et al.*, 2000b). The *aadA2* gene cassette was found in *V. cholerae* O1 strains isolated from Mozambique (Dalsgarrd *et al.*, 2001). In African strains of *V. cholerae* O1, the *intI1* carried *aadA1* and *dfrA15* resistance gene cassettes (Ceccarelli *et al.*, 2006b). A new gene cassette *ant(3')-1a* encoding resistance for streptomycin and spectinomycin was detected in El Tor vibrios isolated from Vietnam (Dalsgaard *et al.*, 1999b). In Kolkata, India, *V. cholerae* non-O1, non-O139 strains isolated during 1997 to 1998 harboured *dfrA5*, *dfrA15*, *aac(6')-Ib* and *ereA2* cassettes (Thungapathra *et al.*, 2002). However, the clinical isolates of non-O1, non-O139 *V. cholerae* were found to harbour different resistance gene cassettes such as *dfrA1*, *dfrA15*, *aadA1*, *aadA2*, *blaP1*, *aar-3*, *aacA4* (Shi *et al.*, 2006).

Role of environmental vibrios

The environmental factors that regulate transfer of antimicrobial resistance genes among *V. cholerae* are not fully studied. Prevalence of MDR *V. cholerae* in the environment is very dynamic as the antimicrobial selection pressure is low compared to the human host. Some of the acquired resistance characteristics by *V. cholerae* strains are long lasting and transferable to other bacteria. In the cholera-affected areas, the MDR *V. cholerae* were also recovered from the water sources and seafood (Khan *et al.*, 1986; Weber *et al.*, 1994). These studies indicate that MDR *V. cholerae* may have additional features for survival and secondary infection from the environment. Conjugation experiments suggested that the antimicrobial resistance might be carried on a self-transmissible genetic element (Clark *et al.*, 1998). In environmental non-O1 strains, transmissible R factor was detected for resistance to many antimicrobials by intraspecific and intergeneric mating experiments (Amaro *et al.*, 1988). Presence of therapeutic drugs in the environment itself promotes the spread of antibiotic resistance (Beaber *et al.*, 2004). As seen with clinical vibrios, the environmental strains were also resistant to

Table 12.1 Resistance gene cassettes reported in class 1 integrons of *Vibrio cholerae*

Resistance gene/ gene allele	Target antimicrobials	Reference
aadA1, aadA2, aadB	Gentamicin	Dalsgarrd *et al.*, 2000a,b, 2001
aadA2	Kanamycin, tobramycin streptomycin, spectinomycin	Ceccarelli *et al.*, 2006b
blaP1, blaCTX-M-2	Beta-lactams	Shi *et al.*, 2006
dhfrIII, dfrA1, dfrA5, dfrA12, dfrA15,	Trimethoprim	Thugapathra *et al.*, 2002
ant(3')-1a	Aminoglycosides	Dalsgarrd *et al.*, 1999b, 2000a
aacA4, aac(6')-1b,	Kanamycin and gentamicin	Shi *et al.* 2006
aar-3	Rifampicin	Shi *et al.*, 2006
aac(6')-Ib	Amikacin and tobramycin	Thugapathra *et al.*, 2002
ereA2	Erythromycin	Thugapathra *et al.*, 2002

beta-lactams, SXT, ampicillin, chloramphenicol, tetracycline, streptomycin, ciprofloxacin etc., (Sathiyamurthy *et al.*, 1997; Thomson *et al.*, 1998; Burrus *et al.*, 2006b; Wang *et al.*, 2006). Thus, the similar antibiotic resistance patterns of the vibrios may indicate that they share a common mode of developing resistance.

In many regions the antimicrobial resistance patterns of clinical O1 and environmental non-O1 strains of *V. cholerae* are not uniform (Sinha *et al.*, 2001; Chakraborty *et al.*, 2001b; Israil *et al.*, 2003). In Brazil, the clinical and environmental O1 strains showed greater drug resistance than the non-O1 strains (Campos *et al.*, 2004). The susceptibility pattern of *V. cholerae* O1 isolated from clinical and environmental strains are different in Russia (KhaÄtovich *et al.*, 1996). Overall, it can be assumed that the antimicrobial resistance is governed by many factors and the environmental O1 and non-O1 strains can act as a reservoir for antimicrobial resistance.

Conclusion

The cholera treatment regimes in some countries are based on the surveillance information on antimicrobial susceptibility of *V. cholerae* (Gabastou *et al.*, 2002; Memon and Murtaza, 2002). Since emergence of MDR *V. cholerae* may significantly influence the strategies for controlling cholera, epidemiological studies should give importance to this aspect. Insights into these areas should lead to better strategies for managing MDR development and limiting the spread of resistance. Newer antimicrobials such as meropenem (Clark, 1992), fleroxacin (Imagawa *et al.*, 1994), rifaximin (Scrascia *et al.*, 2003) and sitafloxacin (Okuda *et al.*, 2007) appear to be promising drugs for the treatment of cholera. There are few compounds, which can act as alternative to the existing antimicrobials, for example virstatin, which blocks the production of virulence factors of *V. cholerae* and also inhibits the infection (Hung *et al.*, 2005). In recent years, promotion of indigenous medical system in primary health care has become more of a concern. Knowledge on pharmacology is slowly expanding to understand the importance of naturally derived products (Thakurta *et al.*, 2007).

The best methods for control of emergence and spread of antimicrobial resistance are (i) rational use of these agents including in aquaculture and animal husbandry, (ii) change to newer antimicrobials, and (iii) constant surveillance of emerging bacterial resistance. Further understanding of the complex ecological, biochemical and molecular aspects of antimicrobial resistance is needed, to develop new antimicrobials and to protect the potency of the existing ones.

References

Ahmed, A.M., Shinoda, S. and Shimamoto, T. (2005). A variant type of *Vibrio cholerae* SXT element in a multidrug-resistant strain of *Vibrio fluvialis*. FEMS Microbiol. Lett. *242*, 241–247.

Ahmed, A.M., Kawaguchi, F. and Shimamoto, T. (2006). Class 2 integrons in *Vibrio cholerae*. J. Med. Microbiol. *55*, 643–644.

Aidara, A., Koblavi, S., Boye, C.S., Raphenon, G., Gassama, A., Grimont, F. and Grimont, P.A. (1998). Phenotypic and genotypic characterization of *Vibrio cholerae* isolates from a recent cholera outbreak in Senegal: comparison with isolates from Guinea-Bissau. Am. J. Trop. Med. Hyg. *58*, 163–167.

Amaro, C., Aznar, R., Garay, E. and Alcaide, E. (1988). R plasmids in environmental *Vibrio cholerae* non-O1 strains. Appl. Environ. Microbiol. *54*, 2771–2776.

Amita, Chowdhury, S.R., Thungapathra, M., Ramamurthy, T., Nair, G.B. and Ghosh, A. (2003). Class 1 integrons and SXT elements in El Tor strains isolated before and after 1992 *Vibrio cholerae* O139 outbreak, Calcutta, India. Emerg. Infect. Dis. *9*, 500–502.

Bag, P.K., Maiti, S., Sharma, C., Ghosh, A., Basu, A., Mitra, R., Bhattacharya, S.K., Nakamura, S., Yamasaki, S., Takeda, Y. and Nair, G.B. (1998). Rapid spread of the new clone of *Vibrio cholerae* O1 biotype El Tor in cholera endemic areas in India. Epidemiol. Infect. *121*, 245–251.

Bani, S., Mastromarino, P.N., Ceccarelli, D., Le Van, A., Salvia, A.M., Ngo Viet, Q.T., Hai, D.H., Bacciu, D., Cappuccinelli, P. and Colombo, M.M. (2007). Molecular characterization of ICEVchVie0 and its disappearance in *Vibrio cholerae* O1 strains isolated in 2003 in Vietnam. FEMS Microbiol. Lett. *266*, 42–48.

Baranwal, S., Dey, K., Ramamurthy, T., Nair, G.B. and Kundu, M. (2002). Role of active efflux in association with target gene mutations in fluroquinolone resistance in clinical isolates of *Vibrio cholerae*. *46*, 2676–2678.

Beaber, J.W., Hochhut, B. and Waldor, M.K. (2002). Genomic and functional analyses of SXT, an integrating antibiotic resistance gene transfer element derived from *Vibrio cholerae*. J. Bacteriol. *184*, 4259–4269.

Beaber, J.W., Hochhut, B. and Waldor, M.K. (2004). SOS response promotes horizontal dissemination of antibiotic resistance genes. Nature. *427*, 72–74.

Begum, A., Rahman, M.M., Ogawa, W., Mizushima, T., Kuroda, T. and Tsuchiya, T. (2005). Gene cloning and characterization of four MATE family multidrug efflux pumps from *Vibrio cholerae* non-O1. Microbiol. Immunol. *49*, 949–957.

Bhattacharya, S.K., Bhattacharya, M.K., Dutta, P., Dutta, D., De, S.P., Sikdar, S.N., Maitra, A., Dutta, A. and Pal, S.C. (1990). Double-blind, randomized controlled clinical trial of norfloxacin for cholera. Antimicrob. Agents Chemother. *34*, 939–940.

Bina, J.E., Provenzano, D., Wang, C., Bina, X.R. and Mekalanos, J.J. (2006). Characterization of the *Vibrio cholerae* VexAB and VexCD efflux system. Arch. Microbiol. *186*, 171–181.

Biskri, L., Bouvier, M. Guerout, A.M. Boisnard, S. and Mazel, D. (2005). Comparative study of class 1 integron and *Vibrio cholerae* superintegron integrase activities. J. Bacteriol. *187*, 1740–1750.

Boltner, D., MacMahon, C., Pembroke, J.T., Strike, P. and Osborn, A.M. (2002). R391: a conjugative integrating mosaic comprised of phage, plasmid, and transposon elements. J. Bacteriol. *184*, 5158–5169.

Borges-Walmsley, M.I., Du, D., McKeegan, K.S., Sharples, G.J. and Walmsley, A.R. (2005). VceR regulates the VceCAB drug efflux pump operon of *Vibrio cholerae* by alternating between mutually exclusive conformations that bind either drugs or promoter DNA. J. Mol. Biol. *349*, 387–400.

Bougoudogo, F., Fournier, J.M., Gastellu-Etchegorry, M., Guillou, M. and Dodin, A. (1992). Resistance of *Vibrio cholerae* O:1 to the O/129 compound and multiresistance to antibiotics. Bull. Soc. Pathol. Exot. *85*, 136–141. [French].

Burans, J.P., Podgore, J., Mansour, M.M., Farah, A.H., Abbas, S., Abu-Elyazeed, R. and Woody, J.N. (1989). Comparative trial of erythromycin and sulphatrimethoprim in the treatment of tetracycline-resistant *Vibrio cholerae* O1. Trans. R. Soc. Trop. Med. Hyg. *83*, 836–838.

Burrus, V. and Waldor, M.K. (2003). Control of SXT integration and excision. J. Bacteriol. *185*, 5045–5054.

Burrus, V. and Waldor, M.K. (2004). Formation of SXT tandem arrays and SXT-R391 hybrids. J. Bacteriol. *186*, 2636–2645.

Burrus, V., Marrero, J. and Waldor, M.K. (2006a). The current ICE age: biology and evolution of SXT-related integrating conjucative elements. Plasmid. 55, 173–183.

Burrus, V., Quezada-Calvillo, R., Marrero, J. and Waldor, M.K. (2006b). SXT-related integrating conjucative element in New World *Vibrio cholerae*. Appl. Environ. Microbiol. 72, 3054–3057.

Campos, L.C., Zahner, V., Avelar, K.E., Alves, R.M., Pereira, D.S., Vital, B.J., Freitas, F.S., Salles, C.A. and Karaolis, D.K. (2004). Genetic diversity and antibiotic resistance of clinical and environmental *Vibrio cholerae* suggest that many serogroups are reservoirs of resistnace. Epidemiol. Infect. *132*, 985–992.

Ceccarelli, D., Salvia, A.M., Sami, J., Cappuccinelli, P. and Colombo, M.M. (2006a). A new cluster of plasmid-located class 1 integrons in *Vibrio cholerae* O1 and *Vibrio parahaemolyticus* isolated in Angola. Antimicrob. Agents Chemther. *50*, 2493–2499.

Ceccarelli, D., Bani, S., Cappuccinelli, P. and Colombo, M.M. (2006b). Prevalence of *aadA1* and *dfrA15* class 1 integron cassettes and SXT circulation in *Vibrio cholerae* O1 isolates from Africa. J. Antimicrob Chemother. *58*, 1095–1097.

Chakrabarti, M.K., Bhattacharya, M.K., Bhattacharya, J., Bhattacharya, S.K. and Mahalanabis, D. (1998). Evaluation of different antibiotics in inhibiting colonization of *Vibrio cholerae* O1 and O139 in the rabbit. Chemotherapy. *44*, 108–111.

Chakraborty, S., Deokule, J.S., Garg, P., Bhattacharya, S.K., Nandy, R.K., Nair, G.B., Yamasaki, S., Takeda, Y. and Ramamurthy, T. (2001a). Concomitant infection of enterotoxigenic *Escherichia coli* in an outbreak of cholera caused by *Vibrio cholerae* O1 and O139 in Ahmedabad, India. J. Clin. Microbiol. *39*, 3241–3246.

Chakraborty, S., Garg, P., Ramamurthy, T., Thungapathra, M., Gautam, J.K., Kumar, C., Maiti, S., Yamasaki, S., Shimada, T., Takeda, Y., Ghosh, A. and Nair, G.B. (2001b). Comparison of antibiograms, virulence genes, ribotypes and DNA fingerprints of *Vibrio cholerae* of matching serogroups isolated from hospitalized diarrhoea cases and from the environment during 1997–1998 in Calcutta, India. J. Med. Micriobiol. *50*, 879–888.

Chhotray, G.P., Pal, B.B., Khuntia, H.K., Chowdhury, N.R., Chakraborty, S., Yamasaki, S., Ramamurthy, T., Takeda, Y., Bhattacharya, S.K. and Nair, G.B. (2002). Incidence and molecular analysis of *Vibrio cholerae* associated with cholera outbreak subsequent to the super cyclone in Orissa, India. Epidemiol. Infect. *128*, 131–138.

Chen, J., Morita, Y., Nazmula Huda, M., Mizushima, T. and Tsuchiya, T. (2002). VmrA, a member of a novel class of Na^+ -coupled multidrug efflux pumps from *Vibrio parahaemolyticus*. J. Bacteriol. *184*, 572–576.

Clark, R.B. (1992). Antibiotic susceptibilities of the Vibrionaceae to meropenem and other antimicrobial agents. Diagn. Microbiol. Infect. Dis. *16*, 453–455.

Clark, C.G., Kravetz, A.N., Alekseenko, V.V., Krendelev, Yu.D. and Johnson, W.M. (1998). Microbiological and epidemiological investigations of cholera epidemic in Ukraine during 1994 and 1995. Epidemiol. Infect. *121*, 1–13.

Clark, C.A., Purins, L, Kaewrakon, P., Focareta, T. and Manning, P.A. (2000). The *Vibrio cholerae* O1 chromosomal integron. Microbiology. *146*, 2605–2612.

Coppo, A., Colombo, M., Pazzani, C., Bruni, R., Mohamud, K.A., Omar, K.H., Mastrandrea, S., Salvia, A.M., Rotigliano, C. and Maimone, F. (1995). *Vibrio cholerae* in the horn of Africa: epidemiology, plasmids, tetracycline resistance gene amplification, and comparison between O1 and non-O1 strains. Am. J. Trop. Med. Hyg. *53*, 351–359.

Dalsgaard, A., Serichantalergs, O., Pitarangsi, C. and Escheverria, P. (1995). Molecular characterization and antibiotic susceptibility of *Vibrio cholerae* non-O1. Epidemiol. Infect. *114*, 51–63.

Dalsgaard, A., Forslund, A., Bodhidatta, L., Serichantalergs, O., Pitarangsi, C., Pang, L., Shimada, T. and Echeverria, P. (1999a). A high proportion of *Vibrio cholerae* strains isolated from children with diarrhoea in Bangkok, Thailand are multiple antibiotic resistance and belong to heterogeneous non-O1, non-O139 O-serogroups. Epidemiool. Infect. *122*, 217–226.

Dalsgaard, A., Forslund, A., Tam, N.V., Vinh, D.X. and Cam, P.D. (1999b). Cholera in Vietnam: changes in genotypes and emergence of class 1 integrons containing aminoglycoside resistance gene cassettes in *Vibrio cholerae* O1 strains isolated from 1979–1996. J. Clin. Microbiol. *37*, 734–741.

Dalsgaard, A., Forslund, A., Petersen, A., Brown, D.J., Dias, F., Monteiro, S., Molbak, K., Aaby, P., Rodrigues, A. and Sandsstrom, A. (2000a). Class 1 intergron-born, multiple-antibiotic resistance encoded by a 150-kilobase conjugative plasmid in epidemic *Vibrio cholerae* O1 strains isolated in Guinea-Bissau. J. Clin. Microbiol. *38*, 3774–3779.

Dalsgaard, A., Forslund, A., Serichantalergs, O. and Sandvang, D. (2000b). Distribution and content of class 1 integrons in different *Vibrio cholerae* O-serotype strains isolated in Thailand. Antimicrob. Agents Chemother. *44*, 1315–1321.

Dalsgaard, A., Forslund A., Sandvang, D., Arntzen, L. and Keddy, K. (2001). *Vibrio cholerae* O1 outbreak isolates in Mozambique and South Africa in 1998 are multiple-drug resistance, contain SXT element and the *aadA2* gene located on class 1 integrons. J. Anitmicrob. Chemother. *48*, 827–838.

Das, S. and Gupta, S. (2005). Diversity of *Vibrio cholerae* strains isolated in Delhi, India during 1992–2000. J. Health Popul. Nutr. *23*, 44–51.

Davey, R.B. and Pittard, J. (1975). Potential for *in vivo* acquisition of R plasmids by one strain of *Vibrio cholerae* biotype El Tor. Antimicrob. Agents Chemother. *8*, 111–116.

Dhar, U., Bennish, M.L., Khan, W.A., Seas, C., Huq Khanm, E., Albert, M.J. and Salam, M. (1996). Clinical features of, antimicrobial susceptibilities and toxin production in *Vibrio cholerae* O139 infection: comparison with *V. cholerae* O1 infection. Trans. R. Soc. Trop. Med. Hyg. *90*, 402–405.

Doganci, L., Gun, H., Baysallar, M., Albay, A., Cinar, E. and Haznedaroglu, T. (1995). Short-term quinolones for successful eradication of multiply resistant *Vibrio cholerae*. Scand. J. Infect. Dis. *27*, 425–426.

Dromigny, J.A., Rakpto-Alson, O., Rajaonatahina, D., Migliani, R., Ranjalahy, J. and Mauclere, P. (2002). Emergence and raid spread of tetracycline-resistance *Vibrio cholerae* strains, Madagascar. Emerg. Infect. Dis. *8*, 336–338.

Dupont, M.J., Jouvenot, M., Couetdic, G. and Michel-Briand, Y. (1985). Development of plasmid-mediated resistance in *Vibrio cholerae* during treatment with trimethoprim-sulfamethoxazole. Antimicrob. Agents Chemother. *27*, 280–281.

Dutta, D., Bhattacharya, S.K., Bhattacharya, M.K., Deb, A., Deb, M., Manna, B., Moitra, A., Mukhopadhyay, A.K. and Nair, G.B. (1996). Efficacy of norfloxacin and doxycycline for treatment of *Vibrio cholerae* O139 infection. J. Antimicrob. Chemother. *37*, 575–581.

Ehara, M., Nguyen, B.M., Nguyen, D.T., Toma, C., Higa, N. and Iwanaga, M. (2004). Drug susceptibility and its genetic basis in epidemic *Vibrio cholerae* O1 in Vietnam. Epidemiol. Infect. *132*, 595–600.

Falbo, V., Carattoli, A., Tosini, F., Pezzella, C., Dionishi, A.M. and Luzzi, I. (1999). Antibiotic resistance conferred by a conjugative plasmid and a class I integron in *Vibrio cholerae* O1 El Tor strains isolated in Albania and Italy. Antimicrobiol. Agent Chemother. *43*, 693–696.

Falklind-Jerkerus, S., Albert, M.J. and Weintraub, A. (2003). Lack of polymorphism in a *Vibrio cholerae* O139-specific DNA region encoding the somatic antigen in strains isolated during 1993–1998. It. J. Med. Microbiol. *292*, 505–511.

Faruque, S.M., Saha, M.N., Asadulgani, Bag, P.K., Bhadra, R.K., Bhattacharya, S.K., Sack, R.B., Takeda, Y. and Nair, G.B. (2000). Genomic diversity of among *Vibrio cholerae* O139 strains isolated in Bangladesh and India between 1992 and 1998. FEMS Microbiol. Lett. *184*, 279–284.

Faruque, S.M., Islam, M.J., Ahmad, Q.S., Biswas, K., Faruque, A.S., Nair, G.B., Sack, R.B., Sack, D.A. and Mekalanos, J.J. (2006). An improved technique for the isolation of environmental *Vibrio cholerae* with epidemic potential: monitoring the emergence of a multiple-antibiotic resistant epidemic strain in Bangladesh. J. Infect. Dis. *193*, 1029–1036.

Ferreira, E., Costa, M. and Pato, M.V. (1992). Resistance to antibiotics of *Vibrio cholerae* strains isolated in Angola. Pathol Biol. (Paris). *40*, 561–565. [French].

Finch, M.J., Morris, Jr, J.G., Kaviti, J., Kagwanja, W. and Levine, M.M. (1988). Epidemiology of antimicrobial resistance cholera in Kenya and East Africa. Am. J. Trop. Med. Hyg. *39*, 484–490.

Folgosa, E., Mastrandrea, S., Cappuccinelli, P., Uzzau, S., Rappelli, P., Brain, M.J. and Colombo, M.M. (2001). Molecular identification of pathogenicity genes and ERIC types in *Vibrio cholerae* O1 epidemic strains from Mozambique. Epidemiol. Infect. *127*, 17–25.

Gabastou, J.M., Pesantes, C., Escalante, S., Narvaez, Y., Vela, E., Garcia, L., Zabala, D. and Yadon, Z.E. (2002). Characteristics of cholera epidemic of 1998 in Ecuador during El Nino. Rev. Panam Salud. Publica. *12*, 157–164. [Spanish].

Garg, P., Chakraborty, S., Basu, I., Datta, S., Rajendran, K., Bhattacharya, T., Yamasaki, S., Bhattacharya, S.K., Takeda, Y., Nair, G.B. and Ramamurthy, T. (2000). Expanding multiple antibiotic resistance among clinical strains of *Vibrio cholerae* isolated from 1992–7 in Calcutta, India. Epidemiol. Infect. *124*, 393–399.

Garg, P., Sinha, S., Chakraborty, R., Bhattacharya, S.K., Nair, G.B., Ramamurthy, T. and Takeda, Y. (2001). Emergence of fluoroquinolone-resistant strains of *Vibrio cholerae* O1 biotype El Tor among hospitalized patients with cholera in Calcutta, India. Antimicrob. Agents Chemother. *45*, 1605–1606.

Glass, R.I., Huq, M.I., Lee, J.V., Threlfall, E.J., Khan, M.R., Aim, A.R., Rowe, B. and Glass, R.J. (1983). Plasmid-born multiple drug resistance in *Vibrio cholerae* serogroup O1, biotype El Tor: evidence for a point-source outbreak in Bangladesh. J. Infect. Dis. *147*, 204–209.

Goldstein, F., Gerbaud, G. and Courvalin, P. (1986). Transposable resistance to trimethoprim and O/129 in *Vibrio cholerae*. J. Antimicrob. Chemother. *17*, 559–569.

Gotuzzo, E., Seas, C., Echevarria, J., Carrillo, C., Mostorino, R. and Ruiz, R. (1995). Ciprofloxacin for the treatment of cholera: a randomized, double-blind, controlled clinical trial of a single daily dose in Peruvian adults. Clin. Infect. Dis. *20*, 1485–1490.

Grados, P., Bravo, N. and Battilana, C. (1996). Comparative effectiveness of co-trimoxazole, and tetracycline in the treatment of cholera. Bull. Pan. Am. Health Organ. *30*, 36–42.

Hassan, H. and Teh, A. (1993). Tetracycline-resistant *Vibrio cholerae* El Tor. Med. J. Malaysia *48*, 95–96.

Houchhut, B. and Waldor, M.K. (1999). Site-specific integration of the conjugal *Vibrio cholerae* SXT element into *prfC*. Mol. Microbiol. *32*, 99–110.

Hochhut, B., Lotfi, Y., Mazel, D., Faruque, S.M., Woodgate, R. and Waldor, M.K. (2001). Molecular analysis of antibiotic resistance gene clusters in *Vibrio cholerae* O139 and SXT element. Antimicrob. Agents and Chemother. *45*, 2991–3000.

Hofer, E., Quintaes, B.R., dos Reis, E.M., Ddos Rodrigues, P., Seki, L.M., Feitosa, I.S., Ribeiro, L.H. and Ferreira, M.R. (1999). The emergence of multiple antimicrobial resistance in *Vibrio cholerae* isolated from gastroenteritis patients in Ceara, Brazil. Rev. Soc. Bras. Med. Trop. *32*, 151–156. [Portuguese].

Hung, D.T., Shakhnovich, E.A., Pierson, E. and Mekalanos, J.J. (2005). Small molecule inhibitor of *Vibrio cholerae* virulence and intestinal colonization. Science. *310*, 670–674.

Ichinose, Y., Ehara, M., Watanabe, S., Shimodori, S., Waiyaki, P.G., Kibue, A.M., Sang, F.C., Ngugi, J. and Kaviti, J.N. (1986). Characterization of *Vibrio cholerae* isolated in Kenya in 1983. J. Trop. Med. Hyg. *89*, 269–276.

Imagawa, Y., Fukuyama, M., Kawakami, K. and Suda, O. (1994). *In vitro* antimicrobial activity of fleroxacin (FLRX) against clinical isolates of from bacterial enteritis. Kassenshogaku Zasshi. *68*, 1409–1416. [Japanese].

Israil, A., Balotescu, C., Bucurenci, N., Näfcescu, N., Cedru, C., Popa, C. and Ciufecu, C. (2003). Factors associated with virulence and survival of environmental and clinical isolates of *Vibrio cholerae* O1 and non-O1 in Romania. Roum. Arch. Microbiol. Immuno. *62*, 155–177.

Iwanaga, M., Toma, C., Miyazato, T., Insisiengmay, S., Nakasone, N. and Ehara, M. (2004). Antibiotic resistance conferred by a class 1 integron and SXT constin in *Vibrio cholerae* O1 strains isolated in Laos. Antimicrob. Agents Chemother. *48*, 2364–2369.

Jabeen, K. and Hasan, R. (2003). Re-emergence of *Vibrio cholerae* O139 in Pakistan: from a tertiary care hospital. J. Pak. Med. Assoc. *53*, 335–338.

Jain, R.C., Bhat, B.D. and Basutkar, S.H. (1993). Tetracycline resistant El Tor vibrios in Loni area. J. Asso. Physicians India. *40*, 636.

Jesudason, M.V. (2006). Change in serotypes and appearance of tetracycline resistance in *V. cholerae* O1 in Vellore, South India, Indian J. Med. Res. *24*, 152–153.

Karbyshev, G.L., Lomov, Lum., Terent'ev, A.N., Fatsailova, O.P., Golubev, B.P., Bespalov, A.I., Mazrukho, A.B. and Ivanova, N.G. (1995). The characteristics of the antibacterial therapy for cholera in Dagestan. Zh. Mikrobiol. Epidemiol. Immunobiol. *2*, 71–74. [Russian].

Khan, M.U., Shahidullah, M., Barua, D.K. and Begum, T. (1986). Epidemiological differences between cholera due to multiple antibiotic resistant and multiple antibiotic sensitive *Vibrio cholerae* infection. Trans. R. Soc. Trop. Med. Hyg. *80*, 60–63.

Khan W.A., Begum, M., Salam, M.A., Bardhan, P.K., Islam, M.R. and Mahalanabis, D. (1995). Comparative trial of five antimicrobial compounds in the treatment of cholera in adults. Trans. R. Soc. Trop. Med. Hyg. *89*, 103–106.

Khan, W.A., Bennish, M.L., Seas, C., Khan, E.H., Dhar, U., Busch, W. and Salam, M.A. (1996). Randomised controlled comparison of single-dose ciprofloxacin and doxycycline for cholera caused by *Vibrio cholerae* O1 and O139. Lancet *348*, 296–300.

Khanna, K.K., Dhanvijay, A., Riley, L.W., Sehgal, S. and Kumari, S. (1990). Cholera outbreak in Delhi: 1988. J. Commun. Dis. *22*, 35–38.

KhaÄtovich, A.B., Ved'mina, E.A., Shenderovich, V.A. and Gorelov, A.L. (1996). Dynamics of changes in antibiotic sensitivity of *Vibrio cholerae* O1, isolated from environmental objects. Antibiot. Khimioter. *41*, 29–33.

Kondo, S., Kongmuang, U., Kalnauwakul, S., Matsumoto, C., Cheng, C.H. and Nishibuch, M. (2001). Molecular epidemiologic analysis of *Vibrio cholerae* O1 isolated during the 1997–98 cholera epidemic in southern Thailand. Epdeimiol. Infect. *127*, 7–16.

Korichi, M.N., Belhocine, S. and Rahal, K. (1997). Inc J plasmids identified for the first time in *Vibrio cholerae* El Tor. Med. Trop (Mars). *57*, 249–252. [French].

Krishna, B.V., Patil, A.B. and Chandrasekhar, M.R. (2005). Fluoroquinolone-resistant *Vibrio cholerae* isolated during a cholera outbreak in India. Trans R. Soc. Trop. Med. Hyg. *100*, 224–226.

Kruse, H., Sorum, H., Tenover, F.C. and Olsvik, O. (1995). A transferable multiple drug resistance plasmid from *Vibrio cholerae* O1. Microb. Drug Resist. *1*, 203–210.

Kuwahara, S. (1978). Antibiotic resistance of *Vibrio cholerae*: special consideration of R-plasmids. Zhonghua. Min. Guo. Wei. Sheng. Wu. Xue. Za. Zhi. *11*, 99–103.

Maggi, P., Carbonara, S., Fico, C., Santantonio, T., Romanelli, C., Sforza, E. and Pastore, G. (1997). Epidemiology, clinical and therapeutic evaluation of the Italian cholera epidemic in 1994. Eur. J. Epidemiol. *13*, 95–97.

Mahon, B.E., Mintz, E.D., Greene, K.D., Wells, J.G. and Tauxe, R.V. (1996). Reported cholera in the United States, 1992–1994: a reflection of global changes in cholera epidemiology. JAMA *276*, 307–312.

Mantri, C.K. Mohapatre, S.S., Ramamurthy, T., Ghosh, R., Colwell, R.R. and Singh, D.V. (2006). Septaplex PCR assay for rapid identification of *Vibrio cholerae* including detection of virulence and int SXT genes. FEMS Microbial Lett. *265*, 208–214.

Materu, S.F., Lema, O.E., Mukunza, H.M., Adhiambo, C.G. and Carter, J.Y. (1997). Antibiotic resistance pattern of *Vibrio cholerae* and *Shigella* causing diarrhoea outbreaks in the eastern Africa: 1994–1996. East. Afr. Med. J. *74*, 193–197.

Mazel, D., Dychinco, B., Webb, V.A. and Davies, J. (1998). A distinctive class of integron in the *Vibrio cholerae* genome. Science. *280*, 605–608.

Mazel, D., Dychinco, B., Webb, V.A. and Davies, J. (2000). Antibiotic resistance in the ECOR collection: integrons and identification of a novel gene. Antimicrob. Agents Chemother. *44*, 1568–1574.

McLeod, S.M., Burrus, V. and Waldor, M.K. (2006). Requirement for *Vibrio cholerae* integration host factor in conjugative DNA transfer. J. Bacteriol. *188*, 5704–5711.

Memon, I.A. and Murtaza, G. (2002). Drug sensitivity pattern of cholera in children. J. Pak. Med. Assoc. *52*, 347–348.

Mhalu, F.S., Mmari, P.W. and Ijumba, J. (1979). Rapid emergence of El Tor *Vibrio cholerae* resistant to antimicrobial agents during first six months of forth cholera epidemic in Tanzania. Lancet *17*, 345–347.

Morita, Y., Kodama, K., Shiota, S., Mine, T., Kataoka, A., Mizushima, T. and Tsuchiya, T. (1998). NorM, a putative multidrug efflux protein of *Vibrio parahaemolyticus*, and its homolog in *Escherichia coli*. Antimicrob. Agents Chemother. *42*, 1778–1782.

Morris, J.G. Jr., Tenney, J.H. and Drusano, G.L. (1985). *In vitro* susceptibility of pathogenic *Vibrio* species to norfloxacin and six other antimicrobials. Antimicrob. Agents Chemother. *28*, 442–445.

Mwansa, J.C., Mwaba, J., Lukwesa, C., Bhuiyan, N.A., Ansaruzamman, M., Ramamurthy, T., Alam, M. and Nair, G.B. (2006). Multiply antibiotic-resistance *Vibrio cholerae* O1 biotype El Tor strains emerge during cholera outbreaks in Zambia. Epidemiol. Infect. *23*, 1–7.

Nazmul Huda, M., Morita, Y., Kuroda, T., Mizushima, T. and Tsuchiya, T. (2001). Na-driven multidrug efflux pump VcmM from *Vibrio cholerae* non-O1 a non-halophilic bacterium. FEMS Microbiol. Lett. *203*, 235–239.

Nazmul Huda, M., Chen, J., Morita, Y., Kuroda, T., Mizushima, T. and Tsuchiya, T. (2003). Gene cloning and characterization of VcrM, a Na^+ -coupled multidrug efflux pump, from *Vibrio cholerae* non-O1. Microbiol. Immunol. *47*, 419–427.

Newland, J.W., Voll, M.J. and McNicol, L.A. (1984). Serology and plasmid carriage in *Vibrio cholerae*. Can. J. Microbiol. *30*, 1149–1156.

Ng, P.P. and Taha, M. (1994). Tetracycline resistant *Vibrio cholerae* in pilgrims returning from Mecca. Med. J. Malaysia. *49*, 195.

Niyogi, S.K., Mondal, S.K., Sarkar, B.L., Garg, S., Banerjee, D. and Dey, G.N. (1994). Outbreak of cholera due to *Vibrio cholerae* O1 in Orissa State. Indian J. Med. Res. *100*, 217–218.

Niyogi, S.K., Sengupta, P.G., Bhattacharya, S.K., Garg, S., Mukhapadhayay, A.K. and Nair, G.B. (1995). Emergence of furazolidone and cotrimoxazole resistant *Vibrio cholerae* O1 in eastern India. J. Infect. *30*, 265–266.

Nizami, S.Q. and Farooqui, B.J. (1998). Cholera in children in Karachi from 1990 through 1995. A study of cases admitted to a tertiary care hospital. J. Pak. Med. Assoc. *48*, 171–173.

Nunes-Duby, S.E., Kwon, H.J., Tirumalai, R.S., Ellenberger, T. and Landy, A. (1998). Similarities and differences among 105 members of the Int family of site-specific recombinases. Nucleic Acids Res. *26*, 391–406.

Noeske, J., GuÃvart, E., Kuaban, C., Solle, J., Fonkoua, M.C., Mouangue, A. and Fouda, A.B. (2006). Routine use of antimicrobial drugs during the 2004 cholera epidemic in Douala, Cameroon. East. Afr. Med J. *83*, 596–601.

Okuda, J., Ramamurthy, T. and Yamasaki, S. (2007). The potent antibacterial activity of sitafloxacin against fluoroquinolone-resistant clinical isolates of *Vibrio cholerae* O1. Microbiol. Immunol. *51*, 467–469.

Olukoya, D.K., Ogunjimi, A.A. and Abaelu, A.M. (1995). Plasmid profiles and antimicrobial susceptibility patterns of *Vibrio cholerae* O1 strains isolated during a recent outbreak in Nigeria. J. Diarrhoeal Dis. Res. *13*, 118–121.

Oyofo, B.A., Subekti, D., Tjaniadi, P., Machpud, N., Komalatini, S., Setiawan, B., Simanjuntak, C., Punjabi, N., Corvin, A.L., Wasfy, M., Campbell, J.R. and Lesmana, M. (2002). Enteropathogens associated with acute diarrhea in community and hospital patients in Jakarta, Indonesia. FEMS Immunol. Med. Microbial. *34*, 139–146.

Petroni., A., Corso, A., Melano, R., Cacace, M.L., Bru, A.M., Rossi, A. and Galas, M. (2002). Plasmidic extended-spectrum beta-lactamases in *Vibrio cholerae* O1 El Tor isolates in Argentina. Antimicrob. Agents Chemother. *46*, 1462–1468.

Phillips, K.F. and Hailey, F.J. (1986). The use of furoxone: a perspective. J. Int. Med. Res. *14*, 19–29.

Pourshafie, M., Grimont, F., Kohestani, S. and Grimont, P.A. (2002). A molecular and phenotypic study of *Vibrio cholerae* in Iran. J. Med. Microbiol. *51*, 392–398.

Rabbani, G.H., Butler, T., Shahrier, M., Mazumdar, R. and Islam, M.R. (1991). Efficacy of a single dose of furazolidone for treatment of cholera in children. Antimicrob. Agents Chemother. 35, 1864–1867.

Radu, S., Ho, Y.K., Lihan, S., Yuherman, Rusul, G., Yasin, R.M., Khair, J. and Elhadi, N. (1999). Molecular characterization of *Vibrio cholerae* O1 and non-O1 from human and environmental sources in Malaysia. Epidemiol. Infec. *123*, 225–232.

Rahal, K. (1982). Compatibility groups of resistance plasmids form Gram negative bacteria isolated in Algeria. Pathol. Biol (Paris). *30*, 539–542. [French].

Rahal, K., Gerbaud, G. and Bouanchaud, D.H. (1978). Stability of R plasmids belonging to different incompatibility groups in *Vibrio cholerae* 'El Tor'. Ann. Mircobiol (Paris). *129*, 409–414.

Rakoto Alson, A.O., Dromigny, J.A., Pfister, P. and Mauclere, P. (2001). *Vibrio cholerae* in Madagascar: study of a multiresistant strain. Arch. Inst. Pasteur Madagascar. *67*, 6–13. [French].

Ramachandran, D., Bhanumathi, R. and Singh, D.V. (2007). Multiplex PCR for detection of antibiotic resistance genes and the SXT element: application in the characterization of *Vibrio cholerae*. J. Med. Microbiol. *56*, 346–351.

Ramamurthy, T., Pal, A., Bhattacharya, M.K., Bhattacharya, S.K., Chowdhury, A.S., Takeda, Y., Takeda, T., Pal, S.C. and Nair, G.B. (1992a). Serovar, biotype, phage type, toxigenicity, and antibiotic susceptibility patterns of *Vibrio cholerae* isolated from two consecutive seasons (1989–90) in Calcutta. Indian J. Med. Res. *95*, 125–129.

Ramamurthy, T., Pal, A., Pal, S.C. and Nair, G.B. (1992b). Taxonomical implication of the emergence of high frequency of occurrence of 2, 4-diamino-6, 7-diisopropylpteridine-resistance strains of *Vibrio cholerae* from clinical cases of cholera in Calcutta, India. J. Clin. Microbiol. *30*, 742–743.

Ramamurthy, T., Rajendran, K., Garg, P., Shimada, T., Basu, A., Chowdhury, N.R., Nandy, R.K., Yamasaki, S., Bhattacharya, S.K., Takeda, Y. and Nair, G.B. (2000). Cluster-analysis and patterns of dissemination of multidrug resistance among clinical strains of *Vibrio cholerae* in Calcutta, India. Ind. J. Med. Res. *112*, 78–85.

Ranjit, K. and Nurahan, M. (2000). Tetracycline resistant cholera in Kelantan. Med. J. Malaysia *55*, 143–145.

Recchia, G.D. and Hall, R.M. (1995). Gene cassettes: a new class of mobile element. Microbiology *141*, 3015–3027.

Reid, A.J. and Amyes, S.G. (1986). Plasmid penicillin resistance in *Vibrio cholerae*: identification of new beta-lactamase SAR-1. Antimicrob. Agents Chemother. *30*, 245–247.

Rowe-Magus, D.A. and Mazel, D. (2002). The role of integrons in antibiotic resistance gene capture. Int. J. Med. Microbiol. *292*, 115–125.

Rowe-Magnus, D.A., Guerout, A.M. and Mazel, D. (1999). Super-integrons. Res. Microbiol. *150*, 641–651.

Rowe-Magnus, D.A., Guerount, A.M. and Mazel, D. (2002). Bacterial resistance evolution by recruitment of super-integron gene cassettes. Mol. Microbiol. *43*, 1657–1669.

Rudra, S., Mathur, M., Talwar, V. and Kathuria, K. (1997). Changing pattern of *Vibrio cholerae* isolation over three consecutive cholera seasons in east Delhi. J. Commun. Dis. *29*, 15–22.

Ryzhko, I.V., Tsuraeva, R.I., Lomov, IuM., Shut'ko, A.G., Mishan'kin, B.N., Koroleva, N.S., Samokhodkina, E.D. and Shcherbaniuk, A.I. (2000). Experimental resistance of *Vibrio cholerae* El Tor to nalidixic acid and fluoroquinolones. Antibiot. Khimioter. *45*, 7–12. [Russian].

Ryzhko, I.V., Dudina, N.A., Lomov, IuM, Shut'ko, A.G., Tsuraeva R.I. and Anisimov, B.I. (2005). Activity of 22 antibacterials against O1 and O139 serogroup *Vibrio cholerae* strains isolated from humans within 1927–2005 in various regions of the world. Antibiot. Khimioter. *50*, 38–42. [Russian].

Saraswathi, K. and. Deodhar, L.P. (1990). A study of *Vibrio cholerae* strains isolated in Bombay. J. Postgrad. Med. *36*, 128–130.

Sathiyamurthy, K., Purushothaman, A. and Ramaiyan, V. (1997). Antibiotic-resistant *Vibrio cholerae* in Parangipettai coastal environs, south east India. Microb. Drug Resist. *3*, 267–270.

Sciortino, C.V., Johnson, J.A. and Hamad, A. (1996). Vitek system antimicrobial susceptibility testing of O1, O139, and non-O1, non-O139 *Vibrio cholerae*. J Clin. Microbiol. *34*, 897–900.

Scrascia, M., Maimone, F., Mohamud, K.A., Materu, S.F., Grimont, F., Grimont, P.A. and Pazzani, C. (2006). Clonal relationship among Vibrio cholera O1 El Tor strains causing the largest cholera epidemic in Kenya in the late 1990s. J. Clin. Microbiol. *44*, 3401–3404.

Scrascia, M., Forcillo, M., Maimone, F. and Pazzani, C. (2003). Susceptibility to rifaximin of *Vibrio cholerae* strains from different geographical aras. J. Antimicrob. Chemother. *52*, 303–305.

Shapiro, R.L., Kumar, L., Phillips-Howard, P., Walls, J.G., Adcock, P., Brooks, J., Ackers, M.L., Ochieng, J.B., Mintz, E., Wahlquist, S., Waiyaki, P. and Slutsker, L. (2001). Antimicrobial-resistance bacterial diarrhea in rural western Kenya. J. Infect. Dis. *183*, 1701–1704.

Shi, L., Fujihara, K., Sato, T., Ito, H., Garg, P., Chakrabarty, R., Ramamurthy, T., Nair, G.B., Takeda, Y. and Yamasaki, S. (2006). Distribution and characterization of integrons in various serogroups of *Vibrio cholerae* strains form diarrhoeal patients between 1992 and 2000 in Kolkata, India. J. Med. Microbiol. *55*, 575–583.

Singh, J., Sachdeva, V., Bhatia, R., Bora, D., Jain, D.C. and Sokhey, J. (1998). Endemic cholera in Delhi, 1995: analysis of data from sentinel centre. J. Diarrhoeal Dis. Res. *16*, 66–73.

Singh, A.K., Haldar, R., Mandal, D. and Kundu, M. (2006). Analysis of topology of *Vibrio cholerae* NorM and identification of amino acid residues involved in norfloxacin resistance. Antimicrob. Agent Chemother. *50*, 3717–3723.

Sinha, S., Chowdhury, P., Chowdhury, N.R., Kamruzzama, M., Faruque, S.M., Ramamurthy, T., Bhattacharya, S.K., Yamasaki, S., Takeda, Y. and Nair, G.B. (2001). Molecular comparison of toxigenic clinical and non-toxigenic environmental strains of *Vibrio cholerae* O1 Ogwa isolated during an outbreak of cholera in south India. Indian J. Med. Res. *114*, 83–89.

Sow, A.I., Cisse, M.F., Gaye, M., Kebe, A., Sy, O.K., Dia, N.M. and Komo, A. (1997). Bacterial diversity during the cholera epidemic in Dakar, Senegal (1995–1996). Bull. Soc. Pathol. Exot. *90*, 160–161. [French].

Steinberg E.B., Greene, K.D., Bopp, C.A., Cameron, D.N., Wells, J.G. and Mintz, E.D. (2001). Cholera in the United States, 1995–2000: trends at the end of the twentieth century. J. Infect. Dis. *184*, 799–802.

Su., B.A., Tang, H.J., Wang, Y.Y., Liu, Y.C., Ko, W.C., Liu, C.Y. and Chuang, Y.C. (2005). *In vitro* antimicrobial effect of cefazolin and cefotaxime combined with minocycline against *Vibrio cholera* non-O1, non-O139. J. Microbiol. Immunol. Infect. *38*, 425–429.

Sundaram, S. and Murthy, K.V. (1984). Transferable plasmid-mediated drug resistance among non-O1 *Vibrio cholerae* and rough strains of *Vibrio cholerae* from Tamil Nadu, India. J. Hyg (London). *92*, 59–65.

Sundaram, S.P., Revathi, J., Sarkar, B.L. and Bhattacharya, S.K. (2002). Bacteriological profile of cholera in Tamil Nadu (1980–2001). Indian J. Med. Res. *116*, 258–263.

Sur, D., Sengupta, P.G., Mondal, S.K., Dutta, P., Gupta, D.N., Ghosh, S., Pandey, A., Ramamurthy, T. and Bhattacharya, S.K. (2002). A localized outbreak of *Vibrio cholerae* O139 in Kolkata, West Bengal. India J. Med. Res. *115*, 149–152.

Taneja, N., Biswal, M., Tarai, B. and Sharma, M. (2005). Emergence of *Vibrio cholerae* O1 Biotype El Tor serotype in northern India. Jpn. J. Infect. Dis. *58*, 238–240.

Thakurta, P., Bhowmik, P., Mukherjee, S., Hajra, T.K., Patra, A. and Bag, P.K. (2007). Antibacterial, antisecretory and antihemorrhagic activity of Azadirachta indica used to treat cholera and diarrhea in India. J. Ethnopharmacol. *111*, 607–612.

Thomson, C.J., Jesudason, M.V., Balaji, V., Malathi, B., Mukundan, U. and Amyes, S.G. (1998). The prevalence of *Vibrio* spp. in dirking water and environmental samples in Vellore South India. Epidemiol. Infect. *121*, 67–76.

Thungapathra, M., Amita, Shinha, K.K., Chaudhuri, S.R., Garg, P., Ramamurthy, T., Nair, G.B. and Ghosh, A. (2002). Occurrence of antibiotic resistance gene cassettes *aac(6')-Ib*, *dfrA5*, *dfrA12*, and *ereA2*, in class I integrons in non-O1 and non-O139 *Vibrio cholerae* strainsin India. Antimicrob. Agents Chemother. *46*, 2948–2955.

Tjaniadi, P., Lesmana, M., Subekti, D., Machpud, N., Komalarini, S., Santoso, W., Siamanjuntak, C.H., Panjabi, N., Campbell, J.R., Alexander, W.K., Beecham, III H.J., Corwin, A.L. and Oyofo, B.A. (2003). Antimicrobial resistance of bacterial pathogens associated with diarrheal patients in Indonesia. Am. J. Trop. Med. Hyg. *68*, 666–670.

Urassa, W.K., Mhando, Y.B., Mhalu, F.S. and Mjonga, S.J. (2000). Antimicrobial susceptibility pattern of *Vibrio cholerae* O1 strains during two cholera outbreaks in Dar es Salaam, Tanzania. East Afr. Med. J. *77*, 350–353.

Vijayalakshmi, N., Rao, R.S. and Badrinath, S. (1997). Minimum inhibitory concentration (MIC) of some antibiotics against *Vibrio cholerae* O139 isolates from Pondicherry. Epidemiol. Infect. *119*, 25–28.

Vora, G.J., Meador, C.E., Bird, M.M., Bopp, C.A., Andreadis, J.D. and Stenger, D.A. (2005). Microarray-based detection of genetic heterogeneity, antimicrobial resistance, and the viable but nonculturable state in human pathogenic *Vibrio* sp. Proc. Natl. Acad. Sci. *102*, 19109–19114.

Wang, Y., Leung, P.C., Qian, P.Y. and Gu, J.D. (2006). Antibiotic resistance and plasmid profile of environmental isolates of *Vibrio* species from Mai Po nature reserve, Hong Kong. Ecotoxicol. *15*, 371–378.

Weber, J.T., Mitz, E.D., Conizares, R., Semiglia, A., Gomez, I., Sempertegui, R., Davila, A., Greene, K.D., Puhr, N.D. and Comeron, D.N. (1994). Epidemic cholera in Ecuador: multidrug-resistance and transmission by water and seafood. Epidemiol. Infect. *112*, 1–11.

Yamamoto, T., Nair, G.B. and Takeda, Y. (1995). Emergence of tetracycline resistance due to a multiple drug resistance plasmid in *Vibrio cholerae* O139. FEMS Immunol. Med. Microbiol. *11*, 131–136.

Young, H.K. and Amyes, S.G. (1986). Plasmid trimethoprim resistance in *Vibrio cholerae*: migration of the type I dihydrofolate reductase gene out of the *Enterobacteriaceae*. J. Antimicrob. Chemother. *17*, 697–703.

Conclusion and Future Prospects 13

Shah M. Faruque and John J. Mekalanos

Abstract

Application of molecular approaches to studying the biology of *Vibrio cholerae*, has led to remarkable progress in our general understanding of the genetics, ecology and epidemiology of the pathogen. The bulk of information generated in different areas of *V. cholerae* research has been summarized to provide an over all impression of the progress made. Future directions for research into *V. cholerae* have also been discussed with a view to understanding general themes of enteric bacterial pathogens, that cause water-borne epidemic diseases.

Distribution and function of virulence genes

The past 15 years have seen extraordinary progress in our understanding of the genetics and evolution of *Vibrio cholerae*, due mainly to the application of molecular approaches to study *V. cholerae* both under natural and laboratory settings. The species, *V. cholerae* is not homogeneous with regard to pathogenic potential. Of more than 200 known serogroups of *V. cholerae*, two serogroups, O1 and O139, account for the vast majority of cholera cases both in endemic and epidemic regions (Faruque *et al.*, 1998; Kaper *et al.*, 1995). Other serogroups of *V. cholerae* occasionally cause gastrointestinal or extra-intestinal illness utilizing similar or distinctly different virulence mechanisms (Morris, 1994). Even within the O1 and O139 serogroups, there are genetic variants that show different antibiotic resistance profile, phage resistance, and other phenotypes that may be linked to environmental fitness (Guidolin and Manning, 1987; Takeya, 1974; Watnick *et al.*, 1999).

The ability of *V. cholerae* strains to cause severe enteric infection in humans depends primarily on their virulence gene content. For example, strains capable of causing cholera invariably carry genes encoding cholera toxin (CT) and toxin coregulated pili (TCP). In addition, all pathogenic *V. cholerae* O1 and O139 strains also carry genes for a number of putative accessory virulence factors, e.g. the mannose-sensitive haemagglutinin (MSHA) pilus, the RTX toxin, and a new type IV pilus (Faruque and Mekalanos, 2003). However, the roles of the accessory virulence associated factors in cholera pathogenesis are not well established, and recent studies are beginning to reveal that at least some of these factors also play a role in environmental fitness of the pathogen and horizontal transfer of DNA. The major virulence factors themselves might also be involved in environmental persistence of the pathogen given that it was recently shown that TCP pili are involved in biofilm formation on chitin surfaces (Reguera and Kolter, 2005).

In addition to the O1 and O139 serogroups of *V. cholerae* which normally carry virulence genes, certain environmental isolates belonging to diverse serogroups have also been found to possess virulence genes or their homologues (Chakraborty *et al.*, 2000; Mukhopadhyay *et al.*, 2001). This raises questions whether the factors which have been described as virulence factors in the context of human infection with *V. cholerae* have other important roles when the bacterium resides in its environmental habitat. The role of extracellular enzymes including CT in the ecology of toxigenic *V. cholerae* is unknown, although it has been suggested that in fresh water systems, local ionic micro-environment can be controlled by toxigenic *V. cholerae* by use of toxin acting on other living cells. Further studies are required to understand more definitive roles of different virulence-associated factors and environmental selection pressures for toxigenic *V. cholerae*. The probable functions of virulence genes or their homologues in the aquatic environment, the role of putative accessory virulence factors both in environmental fitness and human disease, and above all the ecology of toxigenic *V. cholerae* which supports the emergence of new epidemic clones and maintains the seasonal pattern of cholera epidemics, remain to be more adequately explained.

Genomic diversity and evolution

Bacterial diversification is driven by a high rate of horizontal gene transfer, which introduces novel genes, confers beneficial phenotypic capabilities, and permits the rapid exploitation of competitive environment. The major virulence genes in *V. cholerae*, which are clustered in several chromosomal regions, appear to have been recently acquired by its nonpathogenic environmental progenitors through horizontal gene transfer events (Faruque and Mekalanos, 2003). Understanding of the mechanism of horizontal transfer of genes encoding the virulence factors e.g, CT and TCP can primarily explain how new pathogenic strains emerge. The transfer of other accessory genes may explain how a pathogenic strain acquire pandemic potential. The mechanism by which horizontal transfer of CT genes occur has been reasonably defined with the discovery of CTXϕ (Waldor and Mekalanos, 1996), but the mechanism involved in the transfer of the TCP pathogenicity island is still not clearly known. A previous report concluded that the TCP island also corresponds to the genome of another filamentous phage, designated VPIϕ. However, this claim was disputed since repeated attempts failed to show evidence for VPIϕ (Faruque and Mekalanos, 2003). Hence further studies are required to confirm the origin of the TCP pathogenicity island and more completely define its possible role in the aquatic ecosystems. The rationale behind our efforts to examine evolutionary events in cholera is to gain new knowledge, and eventually devise means to control cholera epidemics. Knowing how new pathogenic strains emerge is of immense scientific interest, and in this respect *V. cholerae* represents a paradigm in the co-evolution of bacteria and the virulence associated genetic elements. Future ecological studies of *V. cholerae* should be aimed at explaining more about the emergence of pathogenic strains, and factors involved in the natural selection of the species to ensure its continued existence.

All the defined pathogenicity islands or virulence gene clusters show uniquely different %GC content than the rest of the *V. cholerae* genome, suggesting that these clusters have been acquired by *V. cholerae* from exogenous sources (Faruque and Mekalanos, 2003). However, the donors of these gene clusters still remain unidentified. The emergence of the O139 serogroup is attributed to the acquisition of O139 LPS antigen genes by an El Tor strain, but there has also been instances of the transfer of O139 specific genes to non-O1 environmental strains (Faruque *et al.*, 2000). How LPS genes, pathogenicity islands, and integron genes are acquired naturally by *V. cholerae* is not clearly known. While conjugative plasmids and a single generalized transducing phage exists for *V. cholerae* (Ogg *et al.*, 1981), further efforts should be made to establish whether these or other potentially mobile genetic elements present in *V. cholerae* can be efficiently mobilized. Besides, recognition of possible common themes among bacteriophage-encoded virulence factors as well as among other mobile elements should be sought to expand our understanding of the evolution of bacterial pathogens.

Ecology and epidemiology

In spite of numerous studies over more than a century, it remains a challenging task to fully explain the ecology, epidemiology, and transmissibility of cholera. Although transmission of cholera typically occurs through water contaminated with the pathogenic organisms, the natural mechanisms that facilitate the epidemic spread of cholera and supports the survival of the pathogen in water between seasonal epidemics are not clear. The transmissibility of cholera is reportedly enhanced due to *in vivo* formation of

V. cholerae biofilms which are excreted by cholera patients (Faruque *et al.*, 2006). The epidemiological cycle of *V. cholerae* may also be driven by additional, incompletely defined ecological and genetic factors. Stools from cholera patients have been shown to contain biofilm-like aggregates of *V. cholerae* and the enhanced infectivity of *V. cholerae* shed in human stools is largely due to presence of these clumps of cells which disperse *in vivo* providing a high dose of the pathogen. It is, however, not clear which genes are involved and how are they regulated in the formation of clumps of *V. cholerae* cells *in vivo* and whether the gene expression profile of these cell clumps matches that of conventional biofilm-associated cells.

Recent studies also suggested that 'conditionally viable environmental cells (CVEC)' of pathogenic *V. cholerae* that exist in surface water but resist cultivation by conventional techniques are aggregates of cells which are possibly derived from cell clumps excreted in human cholera stools. Such CVEC can be recovered as fully virulent bacteria by inoculation of the water into rabbit intestines. A recently published method also allows conversion of CVEC from apparently nonculturable cells to normal cells under in vitro conditions, and these cells can be isolated exploiting their antibiotic resistance properties (Faruque *et al.*, 2006). In preliminary studies, this conversion process appeared to involve both dispersion of clumps and activation of partially dormant cells. However, the mechanisms supporting the formation of CVEC in the environment, and their dispersion to infectious bacteria once ingested by a victim is also not clear. Thus, it is important to conduct further studies to understand more clearly the infectious form of *V. cholerae* both in cholera stools, and the aquatic environment, as well as possible genetic regulations involved in the metabolic transformation of *V. cholerae* to support enhanced transmissibility, environmental survival and the epidemic cycle of cholera.

The acquisition of virulence gene clusters by *V. cholerae*, which is normally a marine or brakish water species, has presumably allowed the bacterium to become adapted to the intestinal environment (Faruque *et al.*, 1998). However, being primarily an environmental bacterium, toxigenic *V. cholerae* has also retained the ability to survive in the aquatic environment, and thus transit between an environmental existence and a human pathogenic lifestyle. Several studies have indicated that *V. cholerae* can persist in the environment during both adverse and favorable conditions assuming different survival forms. Further studies are required to identify and characterize the genetic determinants of these environmental survival forms, and the intricate regulations associated with the transition of the bacterium between its two lifestyles.

Multiple systems are involved in the regulation of virulence genes in *V. cholerae*. Several critical genes are coordinately regulated so that they respond in a similar fashion to specific environmental condition, and the coordinate expression results from the activity of a cascading system of regulatory factors. The complete genomic sequencing of a toxigenic *V. cholerae* strain, has facilitated the use of genomic microarrays, to study the expression profile of *V. cholerae* strains under different conditions, both *in vivo* and in vitro (Heidelberg, *et al.*, 2000; Diziejman *et al.*, 2002; Bina *et al.*, 2003; Larocque, *et al.*, 2005). These studies have great potential in revealing the role of multiple genes in the adaptation of the bacterium to different environmental conditions.

Bacterial viruses infecting *Vibrio* species (vibriophages) are known to play a critical role in the transmission of the virulence genes. Recent studies of the pathogenicity islands, and the discovery of CTXϕ have remarkably influenced our understanding of the molecular basis for the emergence of pathogenesis and natural phenomena controlling the conservation of particular genetic traits (Waldor and Mekalanos, 1996; Faruque *et al.*, 1998; Faruque and Mekalanos, 2003). However, the role of vibriophages in *V. cholerae* evolution is not limited to transfer of virulence genes alone, but the evolution and epidemiology of *V. cholerae* are driven by other interactions with vibriophages that are yet to be fully defined. The recognition that phage predation may play a role in controlling cholera epidemics (Faruque *et al.*, 2005) also suggest that changes in this organism may be driven by factors outside of the human host. These studies also indicate that the emergence and prevalence of particular epidemic serogroups of *V. cholerae* may be driven by serotype specific vibriophages. The emergence of new serogroups (e.g. O139) and the displacement of old biotypes (e.g. the classical by the El Tor) may be driven by environmental forces that include phages, propensity to form biofilms on aquatic matter and even interaction with marine animals (Colwell and Spira, 1992; Kirn *et al.*, 2005; Faruque *et al.*, 2006). It seems inevitable that the natural history of cholera will provide inspiration for research questions for years to come.

In addition to facilitating the horizontal transfer of known virulence genes, phages are likely to be critical players in the transfer of chromosomal segments that encode unique antigenic and pathobiological properties or hypothetical functions such as those conferring increased evolutionary fitness to a pandemic clone. Besides discovering novel vibriophages which are either involved in horizontal gene transfer or in modulating the epidemiology of cholera, recent studies have also identified new chromosomal islands and genes that are likely to have functions either in virulence or in environmental fitness of *V. cholerae*. Such studies also offer the opportunity to expand our understanding of the phylogenic relationships between pathogenic and nonpathogenic strains of *V. cholerae*, and the discovery of new genes that may be involved directly or indirectly in the evolution of pathogenic clones.

Novel virulence genes and chromosomal islands

V. cholerae comparative genomics has also begun to provide insights into the evolution of this organism in the past and perhaps also hints of changes that may become more common in the future. For example, the seventh pandemic El Tor O1 clone carries two chromosomal islands that are lacking in strains of the classical biotype as well as most non-O1, non-O139 isolates (Dziejman *et al.*, 2002, 2005). The exact functions of genes located in these islands are not clear at this time, but these islands are apparently moving into other *Vibrio* species and presumably may influence their virulence and epidemiology as well (O'Shea *et al.*, 2004).

Recently, some non-O1, non-O139 strains have been shown to carry a chromosomal island that encodes a type III secretion system (T3SS) (Dziejman *et al.*, 2005). This T3SS system is similar to one that has been described in a pandemic clone of *V. parahaemolyticus* and known to be involved in virulence (Park *et al.*, 2004). It has been shown that *V. cholerae* T3SS is required for intestinal colonization (Tam and Mekalanos, unpublished) but its role in pathogenesis remains to be determined. This T3SS is present in all strains of the O141 serogroup and this is quite ominous because these strains are clonal and distributed throughout the world as a cause of sporadic diarrheal disease (Dziejman *et al.*, 2005). More recently, another secretion system termed 'type VI secretion' (T6SS) has been discovered in a non-O1, non-O139 strain that caused epidemic cholera in the Sudan in 1963 (Pukatzki *et al.*, 2006). This system is responsible for macrophage cytotoxicity but for now has no known role in cholera pathogenesis per se. However, the genes for T6SS are widely distributed in pathogenic Gram negative species (Das and Chaudhuri, 2003), and most recently have been shown to be involved in the virulence of some of these (Mougous, 2006). It seems likely that these T3SS and T6SS systems will be integrated with other virulence mechanisms and result in the emergence of new pathogenic clones of *V. cholerae*. Thus, surveillance for new strains of *V. cholerae* seems a prudent undertaking and will likely help us stay one step ahead of evolutionary adaptations that may change the pattern of cholera epidemiology in the future.

References

Bina, J., Zhu, J., Dziejman, M., Faruque, S., Calderwood, S.B. and Mekalanos, J.J. (2003). ToxR regulon of *Vibrio cholerae* and its expression in vibrios shed by cholera patients. Proc. Natl. Acad. Sci. USA *100*, 2801–06.

Chakraborty, S., Mukhopadhyay, A.K, Bhadra, R.K, Ghosh, A.N, Mitra, R. Shimada, T., Yamasaki, S., Faruque, S.M., Takeda, Y., Colwell, R.R. and Nair, G.B. (2000). Virulence genes in environmental strains of *Vibrio cholerae*. App. Env. Microbiol. *66*, 4022–4028.

Colwell, R.R. and Spira, W.M. (1992). The ecology of *Vibrio cholerae*. In Cholera, D. Barua, and W.B. Greenough III eds. (New York, USA; Plenum Medical Book Co.), pp. 107–127.

Das, S. and Chaudhuri, K. (2003). Identification of a unique IAHP (IcmF associated homologous proteins) cluster in *Vibrio cholerae* and other proteobacteria through in silico analysis. In Silico Biol. *3*, 287–300.

Dziejman, M., Balon, E., Boyd, D., Fraser, C.M., Heidelberg, J.F. and Mekalanos, J.J. (2002). 'Comparative genomic analysis of *Vibrio cholerae*: genes that correlate with cholera endemic and pandemic disease.' Proc. Natl. Acad. Sci. USA *99*; 1556–61.

Dziejman, M., Serruto, D., Tam, V.C., Sturtevant, D., Diraphat, P., Faruque, S.M., Rahman, M.H., Heidelberg, J.F., Decker, J., Li, L., Montgomery, K.T., Grills, G., Kucherlapati, R., and Mekalanos, J.J. (2005). Genomic characterization of non-O1, non-O139 *Vibrio cholerae* reveals genes for a type III secretion system. Proc. Natl. Acad. Sci. USA *102*, 3465–70.

Faruque, S.M., Albert, M.J. and Mekalanos, J.J. (1998). Epidemiology, genetics and ecology of toxigenic *Vibrio cholerae*. Microbiol. Mol. Biol. Rev. *62*, 1301–1314.

Faruque, S.M., Biswas, K., Udden, S.M., Ahmad, Q.S., Sack, D.A., Nair, G.B. and Mekalanos, J.J. (2006b). Transmissibility of cholera: *in vivo*-formed biofilms and their relationship to infectivity and persistence in the environment. Proc. Natl. Acad. Sci. USA *103*, 6350–6355.

Faruque, S.M., Islam, M.J., Ahmad, Q.S., Biswas, K., Faruque, A.S., Nair, G.B., Sack, R.B., Sack, D.A. and Mekalanos, J.J. (2006a). An improved technique for isolation of environmental *Vibrio cholerae* with epidemic potential: monitoring the emergence of a multiple-antibiotic-resistant epidemic strain in Bangladesh. J. Infect. Dis. *193*, 1029–1036.

Faruque, S.M., Saha, M.N., Asadulghani, Sack, D.A., Sack, R.B., Takeda, Y. and Nair, G.B. (2000). The O139 serogroup of *Vibrio cholerae* comprises diverse clones of epidemic and nonepidemic strains derived from multiple *V. cholerae* O1 and non-O1 progenitors. J. Infect. Dis. *182*, 1161–1168.

Faruque, S.M. and Mekalanos, J.J. (2003). Pathogenicity islands and phages in *Vibrio cholerae* evolution. Trends Microbiol. *11*, 505–510.

Faruque, S.M., Naser, I.B., Islam, M.J., Faruque, A.S., Ghosh, A.N., Nair, G.B., Sack, D.A. and Mekalanos, J.J. (2005). Seasonal epidemics of cholera inversely correlate with the prevalence of environmental cholera phages. Proc. Natl. Acad. Sci. USA *102*, 1702–7.

Guidolin, A. and Manning, P.A. (1987). Genetics of *Vibrio cholerae* and its bacteriophages. Micriobiol. Rev. *51*, 285–298.

Heidelberg, J.F., Eisen, J.A., Nelson, W.C., Clayton, R.A., Gwinn, M.L., Dodson, R.J., Haft, D.H., Hickey, E.K., Peterson, J.D., Umayam, L., Gill, S.R., Nelson, K.E., Read, T.D., Tettelin, H., Richardson, D., Ermolaeva, M.D., Vamathevan, J., Bass, S., Qin, H., Dragoi, I., Sellers, P., McDonald, L., Utterback, T., Fleishman, R.D., Nierman, W.C., White, O., Salzberg, S.L., Smith, H.O., Colwell, R.R., Mekalanos, J.J., Ventor, J.C. and Frasier, C.M. (2000). DNA sequence of both chromosomes of the cholera pathogen *Vibrio cholerae*. Nature, *406*, 477–483.

Kaper, J.B., Morris, J.G., Jr and Levine, M.M. (1995). Cholera. Clin. Microbiol. Rev. *8*, 48–86.

Kirn, T.J., Jude, B.A. and Taylor, R.K. (2005). A colonization factor links *Vibrio cholerae* environmental survival and human infection. Nature *438*, 863–6.

Larocque, R.C., Harris, J.B., Dziejman, M., Li, X., Khan, A.I., Faruque, A.S., Faruque, S.M., Nair, G.B., Qadri, F., Mekalanos, J.J. and Calderwood, S.B. (2005). Transcriptional profiling of *Vibrio cholerae* recovered directly from patient specimens during early and late stages of human infection. Infect. Immun. *73*, 4488–93.

Morris, J.G. (1994). Non-O1 group 1 *Vibrio cholerae* strains not associated with epidemic disease, In *Vibrio cholerae* and cholera: molecular to global perspectives. I.K. Wachsmuth, P.A. Blake, and O. Olsvik eds. (Washington DC, USA, ASM Press), pp. 103–116.

Mougous, J., Cuff, M.E., Raunser, S., Zhou, M., Gifford, C.A., Goodman, A.L., Joachimiak, G., Ordonez, C.L., Lory, S., Walz, T., Joachimiak, A. and Mekalanos, J.J (2006). A virulence locus of *Pseudomonas aeruginosa* encodes a secretion apparatus. Science *312*, 1526–1530.

O'Shea, Y.A., Finnan, S., Reen, F.J., Morrissey, J.P., O'Gara, F. and Boyd, E.F. (2004). The *Vibrio* seventh pandemic island-II is a 26.9 kb genomic island present in *Vibrio cholerae* El Tor and O139 serogroup isolates that shows homology to a 43.4 kb genomic island in *V. vulnificus*. Microbiology *150*, 4053–63.

Ogg, J.E., Timme, T.L. and Alemohammad, M.M. (1981).General transduction in *Vibrio cholerae*. Infect. Immun. *31*, 737–741.

Park, K.S., Ono, T., Rokuda, M., Jang, M.H., Okada, K., Iida, T. and Honda, T. (2004). Functional characterization of two type III secretion systems of *Vibrio parahaemolyticus*. Infect. Immun. *72*, 6659–65.

Pukatzki, S., Ma, A.T., Sturtevant, D., Krastins, B., Sarracino, D., Nelson, W.C., Heidelberg, J.F. and Mekalanos, J.J. (2006). Identification of a conserved bacterial protein secretion system in *Vibrio cholerae* using the *Dictyostelium* host model system. Proc. Natl. Acad. Sci. USA *103*, 1528–33.

Reguera, G. and Kolter, R. (2005). Virulence and the environment: a novel role for *Vibnrio cholerae* toxin-coregulated pili in biofilm formation on chitin. J. Bacteriol. *187*, 3551–5.

Takeya, K. (1974). Lysogeny in El Tor vibrios and prophage typing, In Cholera. D. Barua, and W. Burrows eds. (Philadelphia, The W. B. Saunders Co.), pp. 74–83.

Watnick, P.I., Fullner, K.J. and Kolter, R. (1999). A role for the mannose-sensitive hemagglutinin in biofilm formation by *Vibrio cholerae* El Tor. J. Bacteriol. *181*, 3606–3609.

Waldor M.K. and Mekalanos, J.J. (1996). Lysogenic conversion by a filamentous bacteriophage encoding cholera toxin. Science *272*, 1910–1914.

Index

K

L

M

N

O

P

Q

R

www.ingramcontent.com/pod-product-compliance
Lightning Source LLC
LaVergne TN
LVHW062123200926
848484LV00007B/1750